Ion channels are crucial components of living cells. They are situated in the membranes of the cell, and allow particular ions to pass from one side of the membrane to the other. In recent years the patch clamp technique has allowed the activity of individual channels to be measured, and recombinant DNA technology has revealed fascinating detail on channel structure. Together, these technical advances have produced a great flowering of knowledge and understanding about the subject, itself leading to further breakthroughs in science and medicine. *Ion Channels* provides an introduction to this scientific endeavour. It emphasizes the molecular structure of channels as determined by gene cloning technology. This molecular approach illuminates discussions of the permeability and selectivity of channels, their gating and modulation, their responses to drugs and toxins and the human diseases caused when they do not function properly.

Ion Channels

Ion Channels

Molecules in Action

David J. Aidley

School of Biological Sciences
University of East Anglia, Norwich

Peter R. Stanfield

Department of Cell Physiology and Pharmacology
University of Leicester

CAMBRIDGE
UNIVERSITY PRESS

Published by the Press Syndicate of the University of Cambridge
The Pitt Building, Trumpington Street, Cambridge CB2 1RP
40 West 20th Street, New York, NY 10011—4211. USA
10 Stamford Road, Oakleigh, Melbourne 3166, Australia

First published 1996

Printed in Great Britain at the University Press, Cambridge

A catalogue record for this book is available from the British Library

Library of Congress cataloguing in publication data available

ISBN 0 521 49531 8 hardback
ISBN 0 521 49882 1 paperback

Contents

Preface   xi

1   Introduction   1
    The discovery of ion channels   3
    Different types of channel   7

2   Ions on the move   9
    Electricity   9
    Chemical bonds   13
    Ions in crystals   17
    Ions in solution   19
    Ions moving through membrane channels   23

3   Investigating channel activity   33
    Intracellular microelectrodes   33
    The voltage clamp technique   35
    Fluctuation analysis   39
    Artificial phospholipid bilayers   43
    The patch clamp technique   46
    Analysis of single channel current records   52

4   Molecular structures   59
    Determining channel protein sequences   59
    The nicotinic acetylcholine receptor channel   68
    Other neurotransmitter-gated channels   84
    Voltage-gated channels and their relatives   89

Channels with two membrane-crossing segments
   per subunit   99
Calcium release channels   103
Background chloride channels   106
Gap junction channels   110
Some other channels and channel-like proteins   112

5   Permeability and selectivity   121
Permeability: theoretical approaches   121
The independent electrodiffusion model   122
Interdependence of ion movements   129
Binding site models   134
The size of the selectivity filter   144
Water in the channel pore   148
The molecular basis of selectivity   152

6   Gating and modulation   161
Single channel kinetics   161
Ligand–receptor interactions   168
Gating of the nicotinic acetylcholine receptor
   channel   170
Voltage-gated channel gating   178
Gating of some other channels   199
Modulation by phosphorylation   210
Modulation by neurotransmitters and G proteins   214
Some other modulators   219
External calcium ions and surface charges   222

7   Drugs and toxins   225
Simple models for block   226
Potassium channels   229
Voltage-gated sodium channels   236
Voltage-gated calcium channels   243
Pharmacology of some neurotransmitter-gated
   channels   244

8   Dysfunctional channels in human disease   251
Voltage-gated cation channels   251
Muscle chloride channels   256
Cystic fibrosis and the CFTR   257

The glycine receptor channel   260
Gap junctions   261
The calcium-release channel   261

9   Not the last word   263

         References   267

         Index   301

Preface

Ion channels are protein molecules containing aqueous pores that can open and shut to permit ion flow through cell membranes. The concept emerged in the 1950s, but the evidence for their existence was at first limited and indirect in nature. It was 1976 before the behaviour of individual channels could be observed, and 1982 before the primary structure of the first channel protein was determined. Since then work on ion channels has burgeoned and blossomed in a most remarkable way. New discoveries about them are now reported in several thousand scientific papers each year.

This book is intended to provide an introduction to this scientific endeavour. It is too short to be comprehensive, so it does not attempt to be. We have tried to emphasize particularly the molecular aspects of the subject, since one of the really exciting aspects of the field is the way in which explanations in terms of molecular structures are beginning to provide some understanding of channel function. We have written it primarily for students and graduate students doing courses in such subjects as pharmacology, physiology, medicine, cell biology, biophysics, neuroscience and molecular biology, but it may also be useful to those just starting research in the area or to those scientists who simply wish to find out what is happening in a field different from their own.

It is important for science students to know not only where we are now but also how we got here. This does not mean that they have to study the history of the subject for its own sake, but they do need to grasp the logic of the accepted views and be aware of some of the evidence behind them. So we have often given details of how particular experiments were done, and many of our illustrations show the results of experiments rather than simply giving their conclusions. It is partly for this reason that we have given literature references for many of our statements. The other reason is that we felt that in a rapidly moving field (60% of the papers in our reference list are from 1990 or later) it was necessary to let the reader know where our information comes from.

Some biologists like equations, others find that their eyes flick rapidly downward when they meet one on the printed page. We have included a number of equations, but only where they are necessary for the argument. Much of science is concerned with quantitative testing of hypotheses, and in order to make quantitative predictions it is frequently necessary to use a mathematical approach. The precise symbolism of mathematics can also be a considerable aid to clear thinking. But the mathematics in this book, readers will find, is actually pretty easy.

We have benefited greatly from input from our colleagues in Leicester, Norwich and elsewhere, and so it is a pleasure to thank Bill Brammar, Alan Coddington, Peter Croghan, Noel Davies, Alan Dawson, Richard Keynes, Edward Lea, Philip Shelton, Michael Sutcliffe and John Thain for all the help and good advice they have given us in commenting on draft material and discussing particular points with us. It seems appropriate also to acknowledge our long-term indebtedness to Professor Sir Alan Hodgkin, who introduced us to the subject matter of this book and gave each of us considerable encouragement at crucial times. Like most who study channels, our understanding has been greatly helped by Bertil Hille's classic *Ionic Channels of Excitable Membranes*. We are grateful to the many authors and publishers who have given us permission to reproduce diagrams from their works, details of which are included in the reference list at the end of the book. Finally we thank Jessica and Pippa for their tolerance and good humoured support during the writing.

It has been a stimulating and rather enjoyable business writing a book on such a rapidly developing subject, and we have educated ourselves considerably in the process. We hope that our readers will find the result useful.

David Aidley
Peter Stanfield

1 Introduction

Ion channels are crucial components in the activity of living cells. They sit in the membranes of the cell and allow particular ions to pass through them from one side of the membrane to the other. Since they may be either open or closed, they can exert control over this ion movement by switching it on or off. Their study is one of the major endeavours of modern cell biology.

Most animal cells are in contact with moderately salty solutions such as sea water or the body fluids of the blood or the intercellular spaces. These extracellular solutions usually contain a relatively high concentration of sodium and chloride ions, and much lower concentrations of potassium, calcium and other ions. The plasma membrane acts as a barrier separating the cell contents from the outside, so that the ionic concentrations inside the cell can be maintained at levels appreciably different from those in the extracellular fluids. There is also an electrical potential difference between the cytoplasm and the external medium. This combines with the ionic concentration gradients to make an electrochemical gradient across the plasma membrane for each ion species. Cells make considerable use of these electrochemical gradients in their signalling and control systems: when the appropriate ion channels are open, the ions will flow down the gradients into or out of the cell.

The plasma membrane is composed of two layers of tightly packed lipid molecules (fig. 1.1). This structure is not readily permeable to polar molecules such as sugars or amino acids or to charged particles such as sodium or chloride ions. Such substances can pass through the membrane only via special protein molecules embedded in it. Ion channels form one group of these proteins; they permit rapid flow of ions across the membrane. Other membrane transport proteins may act as carriers for ions or other substances, transporting them at much lower rates than do channels, and sometimes up an electrochemical or concentration gradient.

An ion channel is usually composed of merely a few protein molecules,

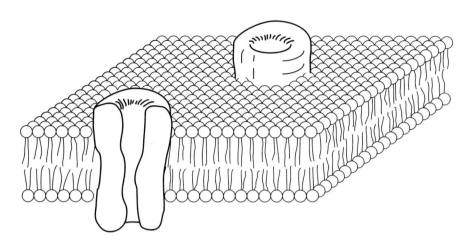

Fig. 1.1. Ion channels in the plasma membrane. The membrane phospholipids are arranged in a bimolecular layer with their polar heads on the outside and their hydrophobic tails inside. The bilayer is about 30 Å thick. Sitting in it are various intrinsic proteins, including the channels shown here.

sometimes of only one. It contains a central aqueous pore that can be opened by conformational change to allow ions to flow from one side of the membrane to the other. This arrangement allows ions to flow through the channel at rates up to 100 million ions per second when it is open. This ion movement forms an electric current that is sufficient to be measured by suitable techniques; hence we can observe the activity of an individual channel, and so of an individual molecule or molecular complex, just as it happens.

Ion channels vary considerably in their *gating*, by which we mean the factors that make them open or close. Some channels are opened by combination with particular chemicals outside or inside the cell, such as neurotransmitters or cytoplasmic messenger molecules. Others are opened by changes in the voltage across the membrane, and yet others by sensory stimuli of various kinds.

Channels show *selectivity* in the ions to which they are permeable. Some of them will permit only particular ions to pass through, such as sodium, potassium, calcium or chloride ions; others are selective for broader groups of ions, such as monovalent cations, or cations in general. Channel scientists have been much concerned with measuring this selectivity and with looking for explanations as to why it exists.

These two aspects of channel functioning – gating and selectivity – are conceptually distinct from one another. Molecular structure studies confirm this view: parts of the channel molecule concerned with gating seem to be separate from those concerned with selectivity.

We can illustrate what channels do with the example of the pancreatic β cell, as is shown in fig. 1.2. The β cells are concerned with the production and secretion of insulin, the hormone that controls blood sugar levels. Two

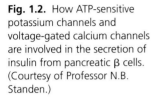

Fig. 1.2. How ATP-sensitive potassium channels and voltage-gated calcium channels are involved in the secretion of insulin from pancreatic β cells. (Courtesy of Professor N.B. Standen.)

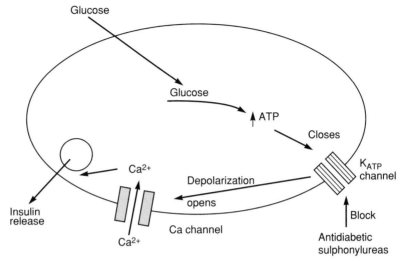

types of ion channel in the plasma membrane are involved in this. The ATP-sensitive potassium channels close when the internal concentration of ATP (adenosine triphosphate) rises to an appropriate level. The voltage gated calcium channels open when the membrane potential of the cell becomes less negative by a sufficient amount.

When blood glucose levels are low, the ATP-sensitive potassium channels in the β cell are open so that potassium ions can flow out of the cell, and this ensures that the membrane potential is kept at a negative value. When blood glucose rises after a meal, uptake of glucose into the cell leads to a rise in its ATP levels, and this makes the potassium channels close. This leads to a change in the membrane potential to less negative values, and this in turn promotes the opening of voltage-gated calcium channels. The resulting inflow of calcium ions raises the internal calcium ion concentration, which then acts as a trigger for the release of insulin from the cell.

Knowledge of how the β cell and its channels work not only is of interest in itself and as part of our understanding of how the body works, it also helps greatly in the search for cures for diabetes, the disease in which the insulin control system does not work properly. One approach for some patients is to use drugs that will block the ATP-sensitive potassium channels and so promote opening of the calcium channels.

The discovery of ion channels

The concept of ion channels was around for many years before it was possible to examine their properties as individual entities. In their classic study of the nature of the nerve impulse in 1952, Alan Hodgkin and Andrew Huxley provided a mathematical description of the flow of sodium and potassium ions through the nerve axon membrane. They found that the relations between the membrane potential and the ionic currents were very steep. It seemed likely that the changes in ionic permeability were associated with the movement of some electrically charged particles within the membrane. They could not detect this charge movement at that time whereas the currents produced by ion movement were readily observed, so they deduced there must be many ions moving across the membrane for each movable membrane charge. They therefore proposed that the ionic currents were localized at particular sites ('active patches', as they called them) in the membrane. It is these sites that later became known as voltage-gated sodium and potassium channels.

Hodgkin and Richard Keynes investigated the potassium permeability of nerve axons in 1955. They found that the movements of radioactive potassium ions across the axon membrane could best be explained if the ions passed through narrow pores in single file, and they used the word 'channel' to describe these ideas.

At the neuromuscular junction, the electrical changes following the action of the neurotransmitter substance acetylcholine on the muscle cell were investigated by Paul Fatt and Bernard Katz in 1951. It was reasonable to suppose that the ionic currents involved passed through particular sites in the muscle cell membrane, channels that were activated by acetylcholine. Both here and with the nerve impulse one could at that time measure only the currents produced by flow through some hundreds or thousands of channels at once.

The 1960s saw the discovery of a number of specific channel-blocking agents. Tetrodotoxin, for example, from the fugu puffer fish, specifically blocks voltage-gated sodium channels. This provided very convincing confirmation that the sodium and potassium channels of nerve axons really are separate from each other. It also allowed nerve potassium channels to be studied on their own, allowed estimates of channel densities in the membrane to be made, and ultimately proved crucial in the biochemical isolation of sodium channels.

The 1960s and 1970s also saw increasingly sophisticated approaches to the investigation of ion channels in nerve and muscle. Clay Armstrong used quaternary ammonium ions as blocking agents to probe the nature of potassium channels. Bertil Hille measured the permeability of channels to ions of different sizes, and so was able to estimate the minimum dimensions of the channel pore. These indirect methods gave some idea of channel properties and were highly influential in the conceptual models they produced, but it was still not possible to examine the activity of the individual channels directly.

Around 1970 some clues as to individual channel action were emerging.

Katz and Ricardo Miledi discovered that the end-plate membrane potential becomes markedly 'noisy' in the presence of acetylcholine, and they interpreted this as a series of 'elementary events' produced by the opening and closing of individual channels as they bound and released acetylcholine molecules. This led to a series of studies by Charles Stevens and others using the techniques of fluctuation analysis to gain information about the size and duration of these events.

A parallel development came from studies on artificial lipid bilayer membranes. Stephen Hladky and Denis Haydon found that when very small amounts of the antibiotic gramicidin were introduced into such a membrane, its conductance to electrical current flow fluctuated in a stepwise fashion. It looked as though each gramicidin molecule contained an aqueous pore that would permit the flow of monovalent cations through it. Could the ion channels of natural cell membranes act in a similar way? To answer this question it was first necessary to solve the difficult technical problem of how to record the tiny currents that must pass through single channels.

The breakthrough came in 1976 with the development of the patch clamp technique by Erwin Neher and Bert Sakmann. They used a glass microelectrode with a polished tip that could be applied to the surface of a cell so as to isolate a small patch of membrane. The voltage across this patch was held steady ('clamped') by a feedback amplifier so that they could measure the currents flowing through the individual ion channels in it. This revolutionary technique proved to be increasingly productive, especially after further technical improvements, so that now in each year thousands of scientific papers report results acquired by using it. In 1991 Neher and Sakmann were awarded the Nobel Prize for Physiology and Medicine for their work (see Neher, 1992; Sakmann, 1992).

Figure 1.3 shows the sort of record that is obtained by the patch clamp technique. The channel is closed for much of the time, so no current flows across the patch of membrane that contains it. But at irregular intervals the channel opens for a short time, producing a pulse of current. Successive current pulses are always of much the same size in any one experiment, suggesting that the channel is either open or closed, and not half open (although we shall see later that there are exceptions to this rule). The durations of the pulses, however, and the intervals between them, vary in an apparently random fashion from one pulse to the next. Hence the openings and closings of channels are *stochastic* events. This means that, as with many other molecular processes, we can predict when they will occur only in terms of statistical probabilities. But one of the great features of the patch clamp method is that it allows us to observe these stochastic changes in single ion channels as they actually happen: we can watch individual protein molecules in action.

The other great technical advance in the study of ion channels has been the

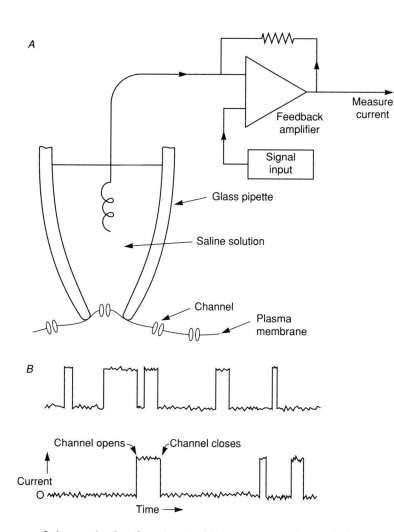

Fig. 1.3. Patch clamp records of channel opening and closing. The patch clamp technique (*A*, not to scale) allows the current flowing through an individual ion channel to be measured. The records (*B*) show short bursts of outward current while the channel is open. Currents are typically in the range of 1 to 5 pA and might last for a few microseconds or milliseconds. Notice that the current pulses are constant in amplitude but variable in duration. It is conventional to show current passing inward through the plasma membrane (i.e. from the extracellular to the cytoplasmic side) as downward deflections.

use of the methods of molecular biology to investigate their structure. The amino acid sequences of the subunits of the nicotinic acetylcholine receptor channel were determined in 1982 by teams led by Shosaku Numa, Jean-Pierre Changeux and others, and that of the voltage-gated sodium channel followed two years later. More and more sequences have been published every year since then. Sequences give us strong clues about structures, and knowledge of the structures of channels gives us an increasing understanding of the way they work.

Fig. 1.4. Gating and selectivity in various types of channels. *A* shows a neurotransmitter-gated channel that is selective to anions. *B* shows a channel selective for potassium ions that is closed by the binding of an internal ligand such as ATP. *C* shows a calcium-selective voltage-gated channel; part of the internal structure of the channel is charged and moves when the membrane potential becomes more positive inside, and this acts as the trigger for opening the channel. *A* is based loosely on the glycine receptor channel, which mediates synaptic inhibition in the spinal cord, *B* on the ATP-sensitive potassium channel of pancreatic β cells, and *C* on the voltage-gated calcium channels of neurons and secretory cells. The diagrams are much simplified: channels of the *A* and *B* types often require more than one ligand molecule for gating, and channels like *C* usually have four internal gating sections.

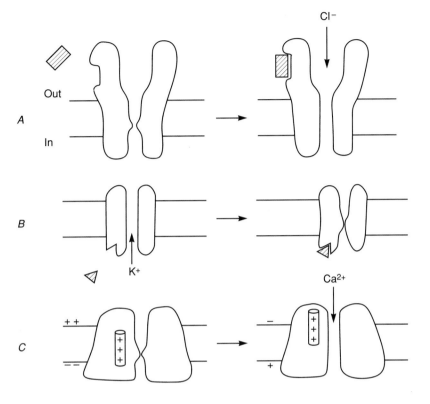

Different types of channel

We now know that there are many more different types of ion channel than was first supposed. New varieties are constantly being reported. This diversity leads to problems in their classification and nomenclature, both of which may need to be revised from time to time as the relations between different channels become clearer.

Channels are commonly described in terms of their ionic selectivity and their gating properties. Figure 1.4 gives an impression of the range of different types in these respects. The nerve axon, for example, possesses voltage-gated channels, i.e. channels that are opened by a change in the membrane potential, and these are of two main types – sodium channels and potassium channels. Voltage-gated calcium channels also occur, particularly at the nerve terminals. We can refer to all these voltage-gated channels as a group distinct from others.

Another major group of channels consists of those that are directly activated by neurotransmitters. These include the nicotinic acetylcholine receptor channel, the γ-aminobutyric acid (GABA) receptor channel, and the glycine receptor channel. A third group includes all those gated by combination with internal ligands, such as calcium ions, ATP, cyclic nucleotides and so on. Sometimes these two groups are themselves combined together as ligand-gated or chemically gated channels, to distinguish them from the voltage-gated channels.

There are other channels that fall outside these definitions. Gap junctions form channels between the cytoplasmic compartments of adjacent cells by crossing two plasma membranes. Gramicidin channels are apparently formed spontaneously when two monomers come together in the membrane into which they have been introduced.

An alternative approach is to use the molecular structure of the channels as the basis for their classification and nomenclature. Exciting discoveries have been made in this respect. The different voltage-gated channels of neurons and other cells have markedly similar structures and show appreciable similarity in their amino acid sequences. We can therefore describe them as a family of related proteins, with the implication that they have a common evolutionary origin. Further sequence determination has shown that some cyclic nucleotide-gated channels and some plant potassium channels should also be included with them to form a superfamily of related sequences.

Not all channel amino acid sequences, however, are related. Transmitter-gated channels are quite separate from the voltage-gated channels and their relatives, showing little or no sequence similarity with them and having distinct structures. Within this group, the receptor channels activated by acetylcholine, glycine, GABA and 5-hydroxytryptamine are clearly related to each other, whereas the glutamate receptors are rather different. The ABC (ATP binding cassette) superfamily forms another distinct group of proteins, only some of which are ion channels.

Knowledge of molecular structures leads to finer distinctions between channels as well as broader groupings of them. The acetylcholine receptor channels of the mammalian nervous system, for example, are of several different types, all rather different from those found at the neuromuscular junction. There are many different subvarieties of voltage-gated potassium channels, and often a number of these may be found in the same cell. Different organisms may possess similar channels with just minor differences in their amino acid sequences.

All this diversity provides a challenge for our comprehension. We need to be precise in our descriptions: it is no use talking about '*the* potassium channel' unless the context is clear and circumscribed. But the very diversity of channels demonstrates their importance to living organisms and makes them a fascinating subject for study.

2 Ions on the move

The existence of electric charge is one of the fundamental features of the universe. Single charges are associated with single subatomic particles: they are unitary and either positive or negative. Charges of the same sign repel each other and those of opposite sign attract each other. The attractive force between positive and negative charges means that it requires an appreciable amount of energy to separate them, and thus energy is released when they are allowed to come together again. Consequently we find that the normal state of matter is to be electrically neutral, with equal numbers of positive and negative charges.

The reasons for this state of affairs can be fruitfully discussed only by theoretical physicists, but the consequences are evident to all of us. When we boil water in an electric kettle, the charges that were separated some time ago at considerable energy cost in some distant power station are finally allowed to come together again, and their flow through the resistance of the kettle's heating element releases energy that we can use to make a cup of coffee. Living cells can also utilize the energy available from the attractive force between electric charges of opposite sign, and this may be evident ultimately in the flow of ions through ion channels.

We begin this chapter with a short refresher course on some aspects of the physics and chemistry of ions. Readers whose physical chemistry is in good shape may wish to skip the first few sections.

Electricity

Static electricity is the accumulation of excess positive or negative *electric charge* in some region, produced ultimately by the separation of electrons from their atoms. Quantities of charge Q are measured in coulombs, C. The positive

Table 2.1. *Some electrical quantities and their units*

Quantity	Symbol for quantity	Unit	Symbol for SI unit	Equivalent form of SI unit
Charge	Q	coulomb	C	A s
Current	I	ampere	A	C s^{-1}
Potential difference	V, E	volt	V	J C^{-1}
Energy (work)		joule	J	C V
Power		watt	W	J s^{-1}, A V
Resistance	R	ohm	Ω	V A^{-1}
Conductance	G	siemens	S	Ω^{-1}, A V^{-1}
Capacitance	C	farad	F	C V^{-1}

It is conventional to write the symbols for quantities in italics and the symbols for units in roman type. Notice particularly the difference between C, the symbol for the quantity capacitance, and C, the symbol for the coulomb, the unit of charge. The unit of time is the second, s.

Table 2.2. *Some prefixes for multiples of scientific units*

Multiple	Prefix	Symbol
10^{-2}	centi	c
10^{-3}	milli	m
10^{-6}	micro	μ
10^{-9}	nano	n
10^{-12}	pico	p
10^{-15}	femto	f
10^{3}	kilo	k
10^{6}	mega	M
10^{9}	giga	G

charge on a single sodium or potassium ion (the elementary charge, e_0) is 1.602×10^{-19} C, so one coulomb corresponds to the charge on 6.24×10^{18} univalent ions. The charge on one mole of univalent ions is given by the Faraday constant F, which is equal to Avogadro's number N_A (the number of ions, atoms or molecules in a mole, 6.022×10^{23}) multiplied by the charge on each of them, i.e.

$$F = N_A e_0$$
$$= 6.022 \times 10^{23} \times 1.602 \times 10^{-19}$$
$$= 96\,500 \text{ coulombs mol}^{-1}$$

Tables 2.1 to 2.3 collect together some of the physical quantities, units and constants used in ion channel work.

Table 2.3. *Some physical constants*

Constant	Symbol	Value	Units
Avogadro's number	N_A	6.022×10^{23}	mol^{-1}
Elementary charge (of proton)	e_0	1.602×10^{-19}	C
Faraday constant ($N_A\,e_0$)	F	9.648×10^4	$C\,mol^{-1}$
Gas constant	R	8.314	$J\,K^{-1}\,mol^{-1}$
Boltzmann constant (R/N_A)	k	1.381×10^{-23}	$J\,K^{-1}$

Values are given to four significant figures. The conventional symbol for the elementary charge is *e*, but we use the symbol e_0 in this book to avoid confusion with the root of natural logarithms, e.

Current electricity is concerned with the flow of charge from one place to another. The current *I* is equal to dQ/dt, the rate of change of charge with time; it is measured in amperes or amps (A). A current of one coulomb per second is one ampere. So if the current through a sodium channel is 10^{-12} A, or one picoamp (1 pA – see table 2.2), then the number of sodium ions flowing through it in one second will be $6.24 \times 10^{18} \times 10^{-12}$, i.e. $6.24 + 10^6$.

Current will only flow from one point to another if there is a *potential differ-ence* (V or E) between the two points and a conducting path between them. In a small torch or flashlight that is switched off, for example, no current flows through the filament of the lamp bulb because there is no potential difference across its ends. And no current flows from the positive pole of the battery to its negative pole, even though there is a potential difference between them, because there is no conducting pathway between them in the external circuit. When the torch is switched on, however, current flows from the positive pole through the filament in the bulb to the negative pole. Potential differences are measured in volts (V); when one coulomb is moved through a potential dif-ference of one volt, one joule (J) of energy is released.

The *energy* released (or work done) during current flow from one point to another is equal to the potential difference between the points multiplied by the charge transferred:

$$\text{work} = VQ \qquad (2.1)$$

The power output (i.e. the rate of energy release) is thus equal to the current multiplied by the potential difference:

$$\text{power} = VI \qquad (2.2)$$

A hand-torch with a 4 V battery, for example, might have a current of 500 mA flowing through the bulb when it is switched on; the power output, given by equation 2.2, will then be 2 watts (W). If the torch is switched on for just

10 s then, applying equation 2.1, we find that 20 J of energy are released as 5 C of electric charge flow through the light bulb filament.

The current flowing through a conductor with a particular potential difference across it is determined by its *resistance*, as was established by Ohm in 1827. According to Ohm's law, the potential V (in volts) is equal to the current I (in amps) multiplied by the resistance R. Thus,

$$V = IR$$

or
$$I = V/R \tag{2.3}$$

The units of resistance are ohms, signified by the Greek letter omega, Ω. The reciprocal of resistance is *conductance* (G), measured in siemens, S. (Older measures of conductance used the delightful name mho, or reciprocal ohm, for the siemens.) We can rewrite equation 2.3 in terms of conductance as

$$I = VG \tag{2.4}$$

In our torch, for example, with a potential difference across the bulb of 4 V and a current of 500 mA flowing through it, Ohm's law shows that the resistance of the bulb filament must be 8 Ω (from equation 2.3) and its conductance is the reciprocal of this, 125 mS (from equation 2.4).

The *resistivity* of a substance is the resistance measured between opposite faces of a 1 cm cube of it. The *conductivity* of a substance is the reciprocal of its resistivity. Resistivity is commonly measured in Ω cm, and conductivity in S cm^{-1}.

In straightforward cases the current flowing through a conductor is linearly proportional to the potential across it; here the conductance really is a constant, independent of voltage. But sometimes this is not the case, and the conductance varies with the potential difference. If the conductance changes its value when the direction of the current passing through it is changed, so that it conducts more readily in one direction than in the other, then we have an example of *rectification*.

Two plates of conducting material separated by an insulator form a capacitor. If a potential difference V is applied across the capacitor, a quantity of charge Q builds up on the plates. The charge is proportional to the potential difference, and the constant of proportionality is called the *capacitance*, denoted by C. Thus,

$$Q = VC \tag{2.5}$$

When the voltage across the capacitor is changing, charge builds up on one plate and flows away from the other, so we can speak of a current I through the capacitor, given by

$$I = C \, dV/dt \tag{2.6}$$

Fig. 2.1. Part of the periodic table of the elements, showing the main ion-forming elements of importance in ion channel studies. Group I and group II elements form monovalent and divalent cations respectively, and group VII elements form monovalent anions. The main natural permeant ions are shown in bold type, and some others that are useful experimentally are shown in italics.

H								He
Li	Be		B	C	N	O	*F*	Ne
Na	**Mg**		Al	Si	P	S	**Cl**	Ar
K	**Ca**	......	Ga	Ge	As	Se	*Br*	Kr
Rb	*Sr*	......					*I*	
Cs	*Ba*	......						
I	II		III	IV	V	VI	VII	VIII

where dV/dt is the rate of change of voltage with time. Cell membranes are thin insulators between conducting solutions, so they have a capacitance, and we shall therefore need to use equation 2.6 in describing some membrane currents.

The capacitance of a capacitor is proportional to the area of the plates of which it is made. So the capacitance of a patch of cell membrane is proportional to its area. Thus if the capacitance per unit area is C_m and the area of the membrane is a, then the capacitance C is given by

$$C = aC_m \qquad (2.7)$$

Chemical bonds

One of the triumphs of twentieth-century chemistry has been to explain the properties of chemical elements in terms of their atomic structures. Protons have positive charges and electrons have negative charges. So the nucleus of an atom, composed of protons and neutrons, possesses a positive charge (equal to its atomic number) that is balanced by the negative charges on the cloud of electrons surrounding it. A carbon atom, for example, which has an atomic number of six, contains a nucleus made up of six protons and usually six neutrons, with six electrons surrounding it.

The chemical elements can be arranged in order of their increasing atomic number to form the periodic table (fig. 2.1). As we proceed along the periodic table in sequence, electrons are added progressively into different energy levels or shells. Thus the first shell contains one electron for hydrogen, but two for helium and all larger atoms. The second shell starts with one electron for lithium, two for beryllium, three for boron, and so on until it becomes filled with eight electrons (forming neon), after which the filling of a third shell begins.

Chemical bonding between separate atoms involves only the electrons in their outermost shells; these are therefore known as the valence electrons. A *covalent bond* is formed when a pair of valence electrons is shared between two

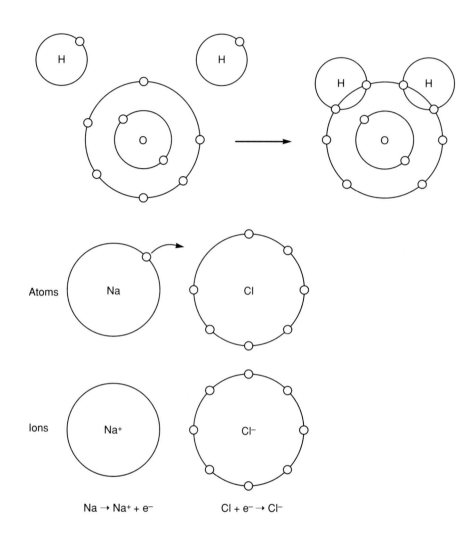

Fig. 2.2. Valence electron shells and the formation of water. The two covalent O−H bonds are each formed from a pair of shared electrons.

Fig. 2.3. Valence electron shells in sodium and chloride ions. The loss of an electron from the sodium atom and the gain of an electron by the chloride atom gives each a complete valence electron shell.

atoms. Bond formation is promoted if the sharing of electrons leads to completion of a full outer shell. In the formation of a water molecule, for example (see fig. 2.2), the oxygen atom starts with six valence electrons, two short of the eight that would make its outer shell complete. With each O−H bond, however, it receives an extra electron from a hydrogen atom, and so finishes with eight electrons in its outer shell. The two hydrogen atoms also reach an energetically more stable state: they each now have the complete number of two electrons in their shells.

A different type of link is the *ionic bond*. In sodium chloride, for example, each sodium atom has lost the single electron in its outer shell and each chloride atom has gained one, so that it now has eight instead of seven electrons in its outer shell (fig. 2.3). Both atoms are now *ions*: the sodium ion has a

Fig. 2.4. Crystal structure of sodium chloride; the sodium (black balls) and chloride (white balls) ions are arranged on a cubic lattice. When an ion enters solution it becomes surrounded by a hydration shell of oriented water molecules. (From Koryta, 1982. *Ions, Electrodes, and Membranes*, Copyright 1982. Reprinted by permission of John Wiley & Sons, Ltd.)

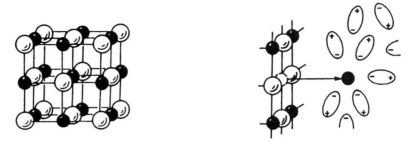

positive charge, since it has lost an electron, and the chloride ion has a negative charge since it has gained one. There is now also a strong electrostatic attraction between the two ions, and this serves to make their state of ionization energetically favourable.

In the ionic bond we have a complete transfer of charge from one atom to another. In many covalent bonds there is some approach towards this state in that the shared electrons in the bond are drawn more to one atom than to the other. In hydrogen chloride, for example, the shared electrons are drawn more to the chlorine atom than to the hydrogen atom. We can therefore regard the chlorine atom as having a partial negative charge on it, and the hydrogen atom as having a partial positive charge. To put it in another way, the centre of gravity of the negative charges in the molecule is separated in space from that of the positive charges. The molecule thus has an *electric dipole* in it, and is described as being *polar*.

Water molecules are also dipoles. The two O−H bonds are at an angle of 105° to each other. This means that the two lone pairs of electrons are on the same side of the molecule, and so there is more negative charge on that side of the molecule than on the other. The O−H bonds are also polar in that the shared electrons in them are drawn towards the more electronegative oxygen atom. Each water molecule is thus polar, with a permanent dipole built into it.

Intermolecular links can arise as a result of molecular dipoles. *Ion–dipole interactions* occur when a positive ion and the negative region of a dipole are attracted to one another, or when a negative ion is attracted to the positive end of a dipole. Thus positive ions are attracted to the negative regions (the oxygen atoms) of water molecules. *Dipole–dipole interactions* are attractions between dipoles of adjacent molecules. The polar nature of water molecules helps to make water such a good solvent for polar and ionic substances (fig. 2.4).

An intermolecular force called the *hydrogen bond* is immensely important in the chemistry of living systems. Oxygen, nitrogen and fluorine atoms are strongly attractive to electrons; they are said to be highly electronegative. When a hydrogen atom is covalently bonded to one of these atoms, the two electrons that make up the bond are pulled some way towards the other atom. This leaves the hydrogen atom with a partial positive charge, which is then able to attract a lone pair of valence electrons from another adjacent oxygen, nitrogen or fluorine atom. We can represent hydrogen bonds between oxygen and nitrogen atoms as follows:

$$O-H\cdots O \qquad O-H\cdots N \qquad N-H\cdots O \qquad N-H\cdots N$$

Hydrogen bonds are important in determining the three-dimensional structure of protein molecules, as, for example, in the α-helix structure thought to be present in so many ion channel molecules (fig. 4.5). An important feature of hydrogen bonds is that they may draw the atoms involved in them closer together than would otherwise be the case.

Water molecules can form hydrogen bonds with each other. This has considerable consequences for the supramolecular structure of liquid water, but the details of this are still not fully agreed upon. One persuasive model assumes that most of the water molecules are in clusters held together by hydrogen bonds, with the remaining non-bonded molecules (about a fifth of the total) filling the gaps between them. Such clusters would contain up to ten molecules, and individual water molecules would interchange very rapidly (within 10^{-10} s) between the cluster and non-cluster environments. The hydrogen bonding between water molecules accounts for the fact that the boiling point of water is much higher than that of similar substances such as hydrogen sulphide.

Many organic substances are soluble in water because they can form hydrogen bonds with it. Glucose and other sugars are examples of this. Compounds that do not readily form hydrogen bonds (such as benzene) are not very soluble and are likely to cluster together in an aqueous environment. The effective attractive force between such molecules is sometimes called the *hydrophobic bond*.

There are also some much weaker intermolecular forces that we should know about, known as the *van der Waals forces*. These arise because of permanent or temporary dipoles in adjacent molecules. The positions of electrons in the outer shells of atoms is altering all the time, so temporary dipoles may form, even in molecules with no permanent dipoles present. A permanent dipole in one molecule can induce a temporary dipole in its neighbour, and the two are then attracted to each other. But temporary dipoles may have a similar effect: a temporary dipole forms in one molecule, this induces a temporary dipole in the opposite sense in its neighbour, and the two are attracted together. The larger the molecules, the larger the van der Waals forces between them.

The relative strengths of the different types of chemical bond are indicated by their bond energies (or, strictly, bond enthalpies). The enthalpy of a bond is the energy released on its formation or the energy required to break it, the

Table 2.4. *Molecular interaction energies*

Type	Interaction energy (kJ mol^{-1})
Covalent bond	200–800
Ionic bond	40–400
Hydrogen bond	10–30
Hydrophobic bond	10–30
van der Waals	3–10
Ion–dipole	3–10
Dipole–dipole	0.5–3

(Data from Freifelder, 1985)

system being maintained at constant pressure. Some values are given in table 2.4.

Ions in crystals

The concept of atoms forming ions was developed by Michael Faraday in 1834, to explain the results of his experiments on the conduction of electricity by electrolyte solutions. He introduced the term *cation* for a positively charged ion, because it would move towards the cathode (the negatively charged electrode) when a solution of a salt was electrolysed, and the term *anion* for a negatively charged ion, because it would move towards the anode.

The ions in a sodium chloride crystal are arranged on a cubic lattice so that each positive sodium ion is surrounded by six negative chloride ions, and each chloride ion is itself surrounded by six sodium ions (fig. 2.4). This structure was determined by X-ray diffraction, which can provide quantitative information about the arrangement of atoms or molecules in crystals or other regularly repeating structures.

How large are ions? Precise answers to this question are not easy to obtain. One approach is to look at the distances between the centres of ions in crystals. These can be determined precisely from X-ray diffraction measurements. If we regard the ions as spheres that are just in contact with each other; then the radius of the sphere is the *crystal radius* of the ion. The distance between the sodium and chloride ions in a crystal of sodium chloride, for example, will then be the sum of the crystal radii of the two ions. Unfortunately it is clear that the crystal radii are often not precisely identical for the same ion in different compounds. Thus there is no single set of values for the Na^+, K^+, F^- and Cl^- ion radii that will fit all the distances between the ion centres in the crystals of NaF, KF, NaCl and KCl.

Table 2.5. *Crystal radii of various ions,*
as calculated by Pauling (1927, 1960)

Ion	Crystal radius (Å)
Li^+	0.60
Na^+	0.95
K^+	1.33
Rb^+	1.48
Cs^+	1.69
Be^{2+}	0.31
Mg^{2+}	0.65
Ca^{2+}	0.99
Sr^{2+}	1.13
Ba^{2+}	1.35
F^-	1.36
Cl^-	1.81
Br^-	1.95
I^-	2.16

Nevertheless it is common for biophysicists and others to utilize estimates of crystal radii in discussing how ions move through channels. Linus Pauling in 1927 was one of the first to produce such estimates, calculated on the basis of the expected forces between ions with different electronic configurations. Some of his values are shown in table 2.5. The radii in the table are given in ångströms, a unit of length often used in describing atomic and molecular distances; 1 Å is equal to 0.1 nm.

Pauling's estimates show reasonable but not exact agreement with measured interionic distances in crystals. For example he calculated that the average crystal radius of the sodium ion is 0.95 Å and that of the chloride ion is 1.81 Å. So if these figures were exact and always applied, the distance between the centres of the sodium and chloride ions in a crystal of common salt would be 2.76 Å. It is actually 2.81 Å; the extra 0.05 Å is probably attributable to mutual repulsion between nearby chloride ions, which are bigger than the sodium ions and so come closer to each other in the crystal lattice.

More recently, X-ray diffraction techniques have allowed high resolution electron-density maps to be determined for some crystals. These suggest that the earlier calculations tended to produce figures that were too low for cations and too high for anions. Nevertheless Pauling's figures are still used for many purposes and they may well be the appropriate ones for use in ion channel studies. The problems in determining ionic radii have been usefully discussed by Adams (1974) and Shannon (1976).

It may not be very realistic to think of atoms or ions as having fixed definite radii. One should remember that each consists of a central nucleus surrounded by a cloud of electrons whose positions can be described only in terms of probabilities. However, since our knowledge of the structures of ion channels is very much less precise than that of inorganic crystals, the uncertainties in the figures used for crystal radii may not in fact matter very much.

In spite of these difficulties, it is possible for us to know some useful facts about ionic sizes. The crystal radius of a cation is smaller than the radius of the atom from which it is derived (as determined, for example, by X-ray analysis of pure metals) because it has lost its outer shell of electrons. Thus atomic sodium has a metallic radius of 1.80 Å whereas ionic sodium has a crystal radius of 0.95 Å. The crystal radius of an elemental anion such as chloride is larger than the radius of the corresponding atom because of the increased mutual repulsion of the electrons it contains. We shall see later that the size of an ion may well be of crucial importance in determining whether or not it can pass through a particular ion channel.

Ions in solution

When a solute substance A dissolves in a solvent B, we must assume that the cohesive links between solute molecules and between solvent molecules are not much greater than those between solvent and solute molecules, otherwise the substance would not dissolve. In an *ideal solution*, the cohesive forces between A and A, between B and B, and between A and B would all be equal, so that the spatial distribution of all the molecules would be completely random and the solution would be totally homogeneous. Real solutions are not ideal, especially at higher solute concentrations.

Solutions of ionic substances depart from the ideal condition at much lower concentrations than do those of non-electrolytes. The reason for this is that ions of the same sign repel one another, whereas those of opposite sign attract one another. Their distribution is thus not random, except at very low concentrations when the distance between adjacent ions is very great. A consequence is that thermodynamic equations that include concentration terms are not precisely accurate. Instead, it is desirable to use the *activity* of the ion, rather than its concentration; the ratio between the two is the *activity coefficient*. Thus, for an ion X, at a concentration [X] at which the activity is a and the activity coefficient is y,

$$a = y[X]$$

Activity coefficients for the common ions fall from 1 at infinite dilution to lower values at physiological concentrations. In pure solutions of sodium chloride, for example, the sodium ion activity coefficient is 0.965 when the concentration is 1 mM, falling to 0.774 at 100 mM. However, in mixed solutions the activity coefficient is dependent on the total ionic strength of the solution, not on the concentration of the individual ion species. In a solution containing 1 mM sodium chloride and 100 mM potassium chloride, for example, the activity coefficient of the sodium ion will be close to 0.77. Since most cells are in osmotic equilibrium with their surrounding media, the ionic strengths on the two sides of their plasma membranes are usually very similar. This means that for any particular ion species the activity coefficients on the two sides of the membrane are also very similar even though there may be markedly different concentrations. Hence it is common for biologists to use concentrations rather than activities in equations such as 2.9 and 5.4.

A theoretical treatment of activity in dilute ionic solutions was produced by Debye and Hückel in 1923. They assumed that positive ions are surrounded by an 'atmosphere' of negative ions and negative ions are surrounded by an 'atmosphere' of positive ions as a result of the attractive forces between them, and worked out the consequences of this. The mean radius of this 'atmosphere' is called the Debye length. In a pure sodium chloride solution it falls from 96 Å at 1 mM to 9.6 Å at 100 mM, and in most physiological solutions it is likely to be in the range 4 to 9 Å.

Ion hydration

We have seen that the ions in a sodium chloride crystal are arranged so that each sodium ion is surrounded by six chloride ions, and each chloride ion is itself surrounded by six sodium ions. This is a very stable array: the melting point of sodium chloride is 801 °C and the boiling point as high as 1413 °C, which shows how hard it is to break the bonds between the sodium and chloride ions. How, then, is it so easy for these bonds to be broken when the salt crystals dissolve in water?

The polar nature of water provides the answer to this question. Water molecules can take the place of the ions in the crystal lattice. Instead of forming ionic bonds with ions of opposite charge, the ions form ion–dipole links with water molecules. When common salt dissolves in water, for example, the sodium ions become surrounded by the negative sides of water molecules instead of by negative chloride ions, and the chloride ions become surrounded by the positive sides of water molecules instead of by positive sodium ions (fig. 2.4).

Just how the water molecules are arranged in a hydrated ion is not firmly established. Most models postulate an inner primary hydration shell where the water molecules are relatively tightly bound by ion–dipole links and precisely oriented, surrounded by an outer secondary hydration shell where the binding and orientation is much looser (see, for example, Edsall & Mackenzie, 1978).

These hydration shells must not be thought of as rigid structures; the individual water molecules involved are changing all the time.

Estimates of the number of water molecules associated with each ion (the hydration number) are not easy to obtain, and the results vary with the technique used to do so. Burgess (1978), for example, quotes values between 2 and 22 for the lithium ion and between 1 and 7 for the potassium ion. Nevertheless most methods agree on the order in which the hydration numbers for different ions are arranged. Thus, hydration numbers fall with increasing ionic size, as in the sequence $Li^+ > Na^+ > K^+ > Rb^+ > Cs^+$. Small ions have a greater charge density at their surfaces than do larger ones; they therefore form more links with water molecules and so their hydration shells are larger. Hydration numbers also increase with increasing charge, e.g. $Na^+ < Mg^{2+} < Al^{3+}$, for similar reasons.

The differences in the size of hydrated ions are accompanied by differences in the velocity at which they will move through a solution in an electric field. We can set up an electric field in the solution by placing two electrodes in it and applying a potential difference across them. The cations will then move towards the cathode and the anions towards the anode. For any particular ion species in dilute solution the velocity of movement between two electrodes is proportional to the potential gradient. The velocity in a gradient of 1 V cm^{-1} is called the *mobility* of the ion. The greater the hydration of an ion species, the less is its mobility: ions that have to drag larger numbers of water molecules with them will move more slowly. Thus the mobilities of the lithium, sodium and potassium ions at 25 °C are 4.01, 5.19 and $7.61 \times 10^{-4} \text{ cm s}^{-1}$, respectively.

Electrodiffusion

A quantity of gas introduced into a closed volume rapidly expands to fill the space available to it. If we add a drop of concentrated salt solution to a volume of distilled water then the ions in the drop will move into the surrounding water so that eventually the ion concentrations are equal throughout the whole solution. These phenomena are examples of *diffusion*.

Molecules in liquids are in constant motion due to thermal agitation. At room temperature typical molecular velocities are of the order of 100 m s^{-1}. The average centre-to-centre distance between water molecules is about 2.85 Å. So a water molecule will not travel very far (ångströms) or for very long (picoseconds) before it hits another one and changes direction. Similarly, ions or other particles in solution or in suspension in water will be buffeted from all directions by the impact of water molecules. For particles that can be seen through the microscope, the ensuing random movement is called Brownian motion.

L   R

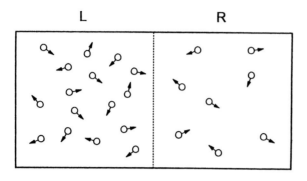

Fig. 2.5. Diagram to illustrate diffusion after removal of a barrier between two compartments. Solute particles move around at random, but the sum of all their random walks leads to net movement from regions of high concentration into regions of low concentration. There are more molecules moving to the right in L than there are moving to the left in R, so when the barrier between them is removed or perforated there will be a net movement from L to R.

In 1909 Einstein showed that diffusion results from the sum of single particles moving about at random under thermal agitation (see Einstein, 1926). If we look at a single particle in a solution, we cannot say whether it will move next to the left or to the right. But, with a large number of particles in a uniform solution or suspension, the number moving to the left will be approximately equal to those moving to the right.

Now imagine we have a container with two compartments of equal size separated by a removable barrier, as is shown in fig. 2.5. In the right hand compartment R is a solution of molecules of a substance at a particular concentration, and in the left hand compartment L is one with twice the concentration. So there will be twice as many molecules moving to the right in L as there are molecules moving to the left in R. If we now remove the barrier between the two compartments, then twice as many molecules will move from L into R as will move from R into L. There will be net movement, therefore, from the region of higher concentration to that of lower concentration. This is what is meant by diffusion.

The rate of the net movement from L to R is proportional to the area of the interface through which they are moving and to the concentration gradient. This statement is called Fick's first law of diffusion. It can be stated mathematically as follows:

$$-\frac{\mathrm{d}n}{\mathrm{d}t} = DA\frac{\mathrm{d}c}{\mathrm{d}x} \qquad (2.8)$$

Here n is the number of molecules so $\mathrm{d}n/\mathrm{d}t$ is the rate of net transfer across the interface; A is the area of the interface; c is the concentration and x is distance so $\mathrm{d}c/\mathrm{d}x$ is the concentration gradient. The constant of proportionality D is called the diffusion coefficient; it is usually measured in $\mathrm{cm}^2\,\mathrm{s}^{-1}$.

We can introduce another component into our model by making the molecules ions and adding an electrical field to the system, making compartment L positive to compartment R. If the ions are positively charged they will move from left to right when the barrier is removed, if they are negatively charged they will move in the opposite direction. These electrophoretic movements will add to or subtract from the diffusive movements. The total gradient is called the *electrochemical gradient*, and the total movement is called *electrodiffusion*.

Ions moving through membrane channels

Let us modify further the model described in the previous paragraph. We introduce pores in the barrier that will allow the solvent and solute ions to pass through when they are opened. Then instead of removing the barrier we simply open the pores in it. When this happens the solute ions will still cross from one side to the other under the influence of the electrochemical gradient, but the rate at which they do so will be greatly affected by the characteristics of the pores.

We can apply these ideas to living cells. The plasma membrane is the barrier to ion movement, and ion channels are the aqueous pores through which the ions can move. The *direction* in which they move is determined by the electrochemical gradient. However the *rate* at which ions move across the membrane is determined by a number of factors:

(1)   the magnitude of the electrochemical gradient,
(2)   the nature of the ion and the characteristics of the ion channels (properties which can be described in terms of permeability and selectivity),
(3)   the number of ion channels present per unit area of membrane, and
(4)   the proportion of them that are open.

The electrochemical gradient

How can we describe the electrochemical gradient in quantitative terms? Consider a model system with two compartments that contain different concentrations of a potassium salt KA; the concentration in compartment 1 is higher than that in compartment 2 (fig. 2.6). They are separated by a membrane that contains a number of ion channels embedded in it. When the aqueous pores through the middle of these are open they will let the potassium ions through but not the anions. The channels, in other words, are selectively permeable to potassium ions, and any current flowing through them will be carried by potassium ions only. A voltmeter allows us to measure the potential difference between the two compartments.

If the channels are closed then there can be no movement of ions across the membrane, so there is no excess of positive charges over negative charges in either of the compartments, and hence there is no potential difference between them. A new situation arises if the channels suddenly open. Potassium ions flow down their concentration gradient through the channels from compartment 1

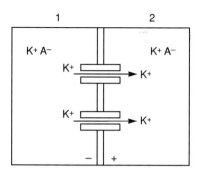

Fig. 2.6. A concentration cell produced by electrodiffusion through channels in a membrane. The solute is a potassium salt KA, at higher concentration in compartment 1 than in 2. The channels are permeable to potassium ions but not to A⁻. Hence potassium ions move from 1 to 2 until the potential produced by their excess positive charges in 2 is sufficient to stop further net movement.

into compartment 2. In so doing they carry their positive charge with them, so there is some depletion of positive charge in compartment 1 and some excess in compartment 2. This cannot be compensated for by a corresponding movement of negative charge since the ion channels do not allow the A^- anions to pass through. Consequently a potential difference arises between the two compartments, with compartment 2 being positive to compartment 1. But this potential difference itself now affects the movements of the potassium (K^+) ions: the excess positive charge in compartment 2 tends to drive them back from 2 to 1. The system soon reaches an equilibrium position where the concentration gradient tending to push the potassium ions from 1 to 2 is balanced by the electrical gradient tending to push then in the opposite direction. The potential difference between the compartments at this point is called the *equilibrium potential* for potassium ions, and is represented by E_K. A system of this type is known as a concentration cell.

What is the value of the equilibrium potential? The answer was supplied by Nernst in 1888, and the equation he produced is commonly known as the *Nernst equation*. For an ion X it is written as follows:

$$E_X = \frac{RT}{zF} \ln \frac{[X]_1}{[X]_2} \tag{2.9}$$

Here R is the gas constant (8.314 J K^{-1} mol^{-1}), T is the absolute temperature (K), z is the charge number of the ion ($+2$ for calcium ions, -1 for chloride ions, for example) and F is the Faraday constant (96 500 C mol^{-1}). $[X]_1$ and $[X]_2$ are the concentrations (or, strictly, activities) of X in the two compartments. E_X is the potential in volts of compartment 2 measured with respect to compartment 1. Sometimes the equilibrium potential E_X is called the Nernst potential of the system.

If we divide the Faraday constant F by Avogadro's number, N_A, we get the unit of elementary charge e_0, 1.6022×10^{-19} C. If we divide the gas constant R by N_A we get the Boltzmann constant k, 1.3807×10^{-23} J K^{-1}. Thus e_0 and k are the molecular equivalents of the mole-related constants F and R. Clearly kT/e_0 is equal to RT/F, so an alternative form of equation 2.9 is

$$E_X = \frac{kT}{z e_0} \ln \frac{[X]_1}{[X]_2}$$

Equation 2.9 can be simplified by enumerating the constants to become, at 20 °C,

$$E_X = \frac{25}{z} \ln \frac{[X]_1}{[X]_2}$$

or, converting from natural to base ten logarithms,

$$E_X = \frac{58}{z} \log_{10} \frac{[X]_1}{[X]_2} \qquad (2.10)$$

where E is now given in millivolts. For example, suppose $[X]_1$ is ten times greater than $[X]_2$, then E will be $+58$ mV if X is K^+ and $+29$ mV if X is Ca^{2+}. If X is an anion, z becomes negative, so that E will be -58 mV if X is Cl^- and -29 mV if X is SO_4^{2-}. If the temperature is different, the factor of 58 mV changes; it is 56 mV at 10 °C and 61.5 mV at 37 °C, for example.

How many X ions have to cross the membrane to set up the potential? The answer depends on the capacitance of the membrane. The larger the membrane capacitance, the greater the number of ions that have to move to produce a particular potential. Suppose our membrane has a capacitance C_m of 1 μF cm^{-2} and a potential V of 100 mV across it. Then the charge Q on 1 cm^2 is given by

$$Q = C_m V$$

where Q is measured in coulombs, C_m in farads and V in volts. The number of moles of X moved will be $C_m V / zF$. In this case, if X is monovalent

$$\frac{C_m V}{zF} = \frac{10^{-6} \times 0.1}{96500}$$

$$= 10^{-12} \text{ mol cm}^{-2}$$

This is a very small quantity in chemical terms; how much difference would it make to the ionic concentrations in the system? Suppose we are dealing with a spherical cell 20 μm in diameter. The area of the plasma membrane will be 12.6×10^{-6} cm^2, so there will be a net inflow of 12.6×10^{-18} moles into a volume of 4.2×10^{-9} cm^3, giving a concentration change of 3 μM. For sodium or potassium ions, with internal concentrations perhaps 5 mM or 145 mM, respectively, this would have a negligible effect on the internal concentration of the ion. However, for calcium ions (for which the inflow would be half since $z = +2$), with an internal concentration less than 0.1 μM, such an increase would be highly significant. For this reason calcium channels are often important components of cellular signalling systems.

Table 2.6. *Ionic concentrations and Nernst potentials in a squid axon and a typical mammalian cell*

	Squid axon			Mammalian cell		
Ion	Blood (mM)	Axoplasm (mM)	E_X (mV)	Plasma (mM)	Cytoplasm (mM)	E_X (mV)
Na^+	440	50	+55	145	12	+67
K^+	20	400	−75	4	140	−95
Ca^{2+}	10	10^{-4}	+145	1.8	10^{-4}	+131
Mg^{2+}	54	10	+21	1.5	0.8	+8
Cl^-	560	40	−66	115	4	−90

E_X is calculated for 20 °C for squid axon and 37 °C for the mammalian cell. The cytoplasmic figures for calcium are for free calcium ions; total calcium is about 0.4 mM in squid and up to 5 mM in mammals. Internal organic anions are not listed. Data simplified after Hodgkin (1958) and West (1991).

So far we have been looking at equilibrium conditions, as described in equations 2.9 and 2.10. But what if the potential difference between the two compartments is not equal to the equilibrium potential? In that case the potassium ions will be driven through the open channels down the electrochemical gradient. Suppose, for example that the concentration of potassium ions in compartment 1 is 10 mM whereas that in 2 is 100 mM, and the potential of 2 is held at −10 mV with respect to 1. The value of E_K is −58 mV, so the potential across the membrane is 48 mV away from its equilibrium value. Potassium ions will be driven down the electrochemical gradient from 2 to 1, i.e. from the negative to the positive side of the membrane. If the potential of 2 is −100 mV with respect to 1, however, then potassium ions will move in the reverse direction, from 1 to 2.

Ionic gradients in cells

Each living cell is bounded by a plasma membrane, which separates the contents of the cell from the external medium in which it lives. The concentrations of particular ions inside the cell are almost always different from those in the external medium. Table 2.6 demonstrates this for a squid giant axon and a typical mammalian cell.

The cell as a whole must be electrically neutral, so that the total numbers of positive and negative ionic charges must be effectively equal. Consequently the concentrations of anions and cations must balance in this respect. It is usual for the intracellular negative charges to be associated largely with a variety of organic anions to which the plasma membrane is not permeable.

For some ions the concentration in free solution is much less than the total amount present; the total calcium bound in most cells is much greater than the amount of free ionic calcium, for example.

In animal cells the osmotic concentration of the intracellular contents is usually similar to that of the extracellular fluid. This may not be so for plant cells and others with inextensible cell walls.

How do the ionic gradients illustrated in table 2.6 arise? In the case of sodium ions, the answer is clear: there is an active transport system that continually extrudes sodium ions from the cell. This extrusion process, or *sodium pump* as it is commonly called, is one of the most important processes in the animal body. It is carried out by a membrane-bound sodium–potassium ATPase (see Skou, 1989). Three sodium ions are extruded and two potassium ions are drawn into the cell for each ATP molecule that is broken down.

The potassium ion concentration gradient arises in part from the uptake associated with the activity of the sodium pump, and partly by a passive movement of potassium ions through open potassium-selective channels. In many cells chloride ions are largely passively distributed (i.e. they distribute themselves in accordance with equation 2.9); in other words the negative membrane potential ensures that the internal chloride concentration is much less than that outside the cell. Calcium ions are actively extruded from most cells and they are also sequestered in intracellular compartments such as the endoplasmic reticulum and the mitochondria. This leads to very low concentrations of free calcium ions, typically in the region of 0.1 μM or less.

Membrane potentials

Most cells possess a *membrane potential*, a voltage across the plasma membrane such that the inside is (usually) some tens of millivolts negative to the outside. In the squid axon it is about -60 mV in the resting condition, when the axon is not conducting nerve impulses. Is it possible to explain this potential in terms of ionic gradients?

We can calculate the Nernst potentials for the various different ions involved in the squid axon system, using equation 2.9. The results for the more important univalent ions, using the concentrations given in table 2.6, are -75 mV for E_K, $+55$ mV for E_{Na} and -66 mV for E_{Cl}. This shows that sodium ions are very far from being in equilibrium, and thus that the membrane is not likely to be very permeable to sodium ions in its resting state. The Nernst potential for potassium ions is not too far from the actual membrane potential, suggesting that there could be a number of open potassium-selective channels in the resting membrane and that these are important in determining the resting membrane potential.

In excitable cells (a group that includes nerve, sensory and muscle cells) the

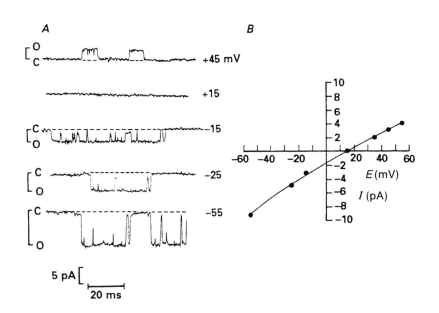

Fig. 2.7. Reversal potential and current–voltage relation for single channel currents. *A* shows patch clamp records from sodium-activated potassium channels in cultured chick neurons at different membrane potentials. C is the current baseline with the channel closed, O the open channel current. *B* shows the current–voltage relation, derived from *A* and similar records. The reversal potential (when the current *I* is zero) is +15 mV, very close to the calculated Nernst potential for potassium ions, +17 mV. (From Dryer *et al.*, 1989.)

membrane potential may change quite rapidly as a result of the opening of other channels. The action potentials of nerve axons, for example, arise as a result of the opening of sodium-selective channels triggered by an electrical stimulus. Sodium ions flow through them into the cell so that the membrane potential becomes inside-positive and approaches the value for the sodium equilibrium potential, +55 mV. Synaptic potentials involve the opening of a different set of channels as a result of the action of neurotransmitter substances.

The voltage gradient or electric field across the membrane is enormous. If there is a potential difference of 60 mV across a membrane 3 nm thick, for example, then the voltage gradient is $200\,000$ V cm^{-1}. Clearly this must have appreciable consequences for the movement of ions through the channels and the distributions of charges in the channel proteins.

Current flow through open channels

We can use Ohm's law to determine the conductances of ion channels. We need to know the single channel current and the driving voltage. But just what is this driving voltage? We can answer this question by considering the experiment shown in fig. 2.7 (Dryer *et al.*, 1989). The patch clamp records show currents flowing through a single potassium-selective channel at different membrane potentials. At -55 mV the currents are inward and over 9 pA in size. At -25 mV they are smaller (just under 5 pA), at -15 mV they are smaller still and at $+15$ mV they cannot be detected. At $+45$ mV they can be seen again but they are now outward currents, just over 3 pA. In other words the single channel currents reduce in size as the membrane potential approaches +15 mV, and reverse in direction as it passes beyond this level. Hence +15 mV is known as the *reversal potential* of the single channel current, symbolized by E_{rev}. The driving voltage for use in applying Ohm's law to the system is then the difference between the actual membrane potential E and the reversal potential E_{rev}.

It is usual to use the symbols i to denote single channel current and γ to denote single channel conductance. Then, applying Ohm's law, as in equation 2.4, we get

$$i = \gamma(E - E_{rev}) \qquad (2.11)$$

In fig. 2.7, for example, we see single channel currents averaging 4.8 pA at a membrane potential of −25 mV, and the reversal potential is +15 mV, giving a value of 40 mV for $(E - E_{rev})$. We can rearrange equation 2.11 to give the single channel conductance:

$$\gamma = \frac{i}{(E - E_{rev})} \qquad (2.12)$$

i.e.

$$\text{conductance (siemens)} = \frac{\text{current (amps)}}{\text{driving voltage (volts)}}$$

so

$$\gamma = \frac{4.8 \times 10^{-12}}{40 \times 10^{-3}}$$

$$= 120 \text{ pS}$$

Why should the reversal potential be at +15 mV? If the channel is permeable only to potassium ions, then the reversal potential should be equal to the potassium equilibrium potential E_K. In the experiment shown in fig. 2.7, the potassium ion concentration $[K]_i$ on the inside surface of the patch was 75 mM and the external concentration $[K]_o$ was 150 mM. We use equation 2.9 to calculate E_K, and the answer is +17 mV, effectively equal to the observed reversal potential, and incidentally providing good evidence that the channel really is highly selective for potassium ions. For a different potassium ion concentration gradient, we would calculate a different value of E_K and see a different reversal potential. With $[K]_i$ at 120 mM and $[K]_o$ at 5 mM (a ratio much nearer to the natural situation in the body), for example, E_K and the reversal potential would be −80 mV.

In other channels that are highly selective and so permeable to only one ion, the reversal potential is similarly equal to the Nernst potential for that ion. In an idealized sodium channel, for example, E_{rev} is equal to E_{Na}. This is not the case for channels permeable to more than one ion. The nicotinic acetylcholine

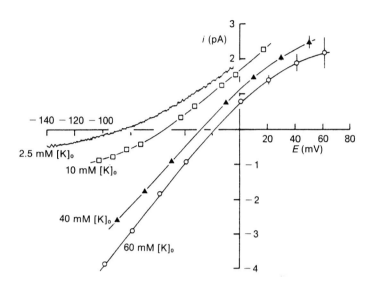

Fig. 2.8. Current–voltage curves for individual ATP-sensitive potassium channels from frog muscle at different external potassium ion concentrations. The curves were obtained from inside-out patches with different potassium ion concentrations in the patch pipette, by measuring the single channel current i at different membrane potentials E or, in the case of the 2.5 mM $[K]_o$ curve, by measuring the current continuously while slowly changing the membrane potential. $[K]_i$ was at 120 mM throughout. (From Spruce *et al.*, 1987a.)

receptor channel, for example, is permeable to both sodium and potassium ions. Here the reversal potential is commonly near -15 mV in frog muscle, in between E_{Na} at about $+60$ mV and E_K near to -100 mV. At the reversal potential the net outflow of potassium ions precisely balances the net inflow of sodium ions.

By plotting a graph of i at different values of E, as in fig. 2.7B, we obtain the *current–voltage relation*. The slope of this line is the single channel conductance γ,

i.e. $$\gamma = di/dE \qquad (2.13)$$

so now we have two ways of calculating γ. Equation 2.13 gives us the *slope conductance* (the gradient of the tangent to the current–voltage curve at a particular value E), whereas equation 2.12 gives us the *chord conductance* (the gradient of the chord connecting i at E to $i = 0$ at E_{rev}). The two are always equal if the current–voltage relation is a straight line, but this is not so if it is not. The current–voltage curve shown in fig. 2.7 shows a slight curvature, indicating that γ is not quite independent of the potential across the membrane. For some channels γ may be more markedly affected by membrane potential, as in inward rectifier potassium channels.

Figure 2.8 shows current–voltage relations for some ATP-sensitive potassium channels at different external potassium ion concentrations (Spruce *et al.*, 1987a). Notice how the reversal potential becomes less negative as $[K]_o$ is increased, roughly (but not precisely) in accordance with the Nernst equation for potassium ions. The discrepancy is evident at the lower values of $[K]_o$, as is illustrated in fig. 5.1, and is probably due to some permeability of the channel to sodium ions. If this is so we would expect a small inward flow of sodium ions when the membrane potential is at E_K (where there will be no net flow of potassium ions); this will be balanced by a small outward flow of potassium ions at E_{rev}, which is a few millivolts less negative than E_K.

Another feature that affects the single channel conductance is the concentration of permeant ions in the solutions adjacent to the channels. Dilute solutions are usually associated with lower conductances, since there are fewer ions to carry the current. Thus in fig. 2.8 the slopes of the curves increase as the external potassium ion concentration is increased, so that γ (calculated at the reversal potential using equation 2.13) is 14.8 pS with $[K]_o$ at 2.5 mM and 42.3 pS with $[K]_o$ at 60 mM. In comparing conductances of different channels, therefore, we need to know what the concentrations of the permeant ions were.

How many ions pass through an open channel? Consider a channel selective for sodium ions, with a conductance of 16 pS and a driving force $(E - E_{Na})$ of 100 mV; the single channel current i will be given by

$$i = \gamma(E - E_{Na})$$

so here $i = 1.6$ pA. The number of ions flowing per second will be given by

$$\frac{\text{current in amps} \times \text{Avogadro's number}}{\text{Faraday's constant}}$$

For our single channel current of 1.6 pA, then, this number is

$$\frac{1.6 \times 10^{-12} \times 6 \times 10^{23}}{96\,500}$$

$$= 10^7 \text{ ions s}^{-1}$$

So if the channel is open for just 1 ms, 10 000 ions will pass through it in that time.

Conductances in parallel add together. So (to continue our example) a patch of membrane containing five open channels each with a conductance of 16 pS will have a conductance of 80 pS with a total current (at a clamped membrane potential 100 mV away from the reversal potential) of 8 pA passing through it. A whole cell with 2000 such channels will show a membrane current of 3.2 nA when they are all open at the same time. If for some period of time the probability of any individual channel being open is only 0.2, then the average whole cell current during that time will be 640 pA.

3 Investigating channel activity

Much of the progress in science depends upon the development of new techniques, both in physical apparatus and in methods of analysis. So in this chapter we look briefly at how methods for recording and analysing the electrical activity of cells and their ion channels have evolved. Its object is not to show you how to use the various techniques in the laboratory: excellent books are available for that (e.g. Sakmann & Neher, 1983, 1995; Rudy & Iverson, 1992; Ogden, 1994a). But it is necessary to have some idea of how the various methods work in order to understand the experimental results they produce.

The first measurements of electrical activity in animal tissues were made in the nineteenth century, but the limitations of moving coil galvanometers made them seriously lacking in time resolution. The development of thermionic valves and the cathode ray oscilloscope allowed rapid and reproducible records to be obtained from the 1920s onwards. At first these measurements were made with extracellular electrodes such as silver wires. However, such electrodes can only measure events in the electric field outside the active cell. Since this electric field arises from current flow through ion channels in the plasma membrane, we really need to put an electrode inside the cell, so that the potential across the plasma membrane can be measured directly.

Intracellular microelectrodes

The first intracellular measurements on animal cells were made in 1939 by Hodgkin and Huxley at Plymouth and by Curtis and Cole at Woods Hole, Massachusetts. They used the giant nerve fibres of squids, which had then just recently been discovered, and which have the great advantage of being up to 1 mm in diameter. They pushed fine glass capillaries filled with sea water or with an isotonic potassium chloride solution into the cut end of the axon, so that their recording system measured the potential between the tip of the intracellular electrode and an external electrode in the sea water outside the axon.

Ten years later the glass intracellular microelectrode was invented by Ling & Gerard (1949). This consists of a hard glass tube pulled out to a very fine tip and filled with a strong electrolyte solution such as 3 M potassium chloride. The tip is less than 1 μm in diameter and can be pushed through the plasma membrane so as to record the potential inside the cell; the membrane apparently reseals around the tip of the electrode as it enters the cell.

To make the microelectrode, a length of hard glass tube is heated by a wire element so that it softens in the middle, and it is then pulled out strongly as it cools. Machines for doing this are available. The softened portion gets harder and thinner as it is pulled out and cools, and eventually it breaks. There are various ways of filling the electrode with the electrolyte solution. A good method is to make the electrode from tubing with a fine glass filament fused inside it; the acute angle between the filament and the inner wall of the tube leads to a strong capillary action so that the electrode can easily be filled to the tip by injecting the electrolyte solution into the shaft.

Measurements from a wide variety of medium-sized cells, such as motor neurons, muscle fibres or secretory cells, can be made with these electrodes. Figure 3.1 shows the arrangement for recording the end-plate potential (see below) from a vertebrate neuromuscular junction. The isolated muscle, with its motor nerve attached, is held in a bath of physiological saline, which may contain the paralytic alkaloid curare. The tip of the microelectrode is inserted into the end-plate region of the muscle cell, and the voltage across the membrane is measured as the potential between the tip of the microelectrode and a second electrode outside the cell.

At rest the microelectrode records a resting potential of about -90 mV from the muscle cell. When the motor axon serving it is stimulated, a depolarization (a reduction in the membrane potential to a less negative value) of several millivolts occurs. This is called the end-plate potential; it is produced by the flow of ions through the acetylcholine receptor channels, which are clustered on the muscle cell surface at this point. In the absence of curare the end-plate potential would be much larger, and would give rise to a propagated action potential involving voltage-gated sodium and potassium channels.

Intracellular microelectrodes have a high resistance, typically in the range 10 to 100 MΩ, and the input stage of the amplifier has to be of a type suitable to deal with this. Its input resistance (the effective resistance between the two input terminals) must be very high; if it is not the membrane potential measurement will be distorted. Suppose, for example, that the input resistance is just 1 MΩ, a typical value for an oscilloscope amplifier, and the micro-

Fig. 3.1. Using an intracellular microelectrode to record the end-plate potential from a frog muscle cell. The recording system in the upper diagram consists of a preamplifier with a high input resistance and an oscilloscope. The end-plate potential (lower diagram) is reduced in size because of the curare present. (Based on Fatt & Katz, 1951, from Keynes & Aidley, 1991.)

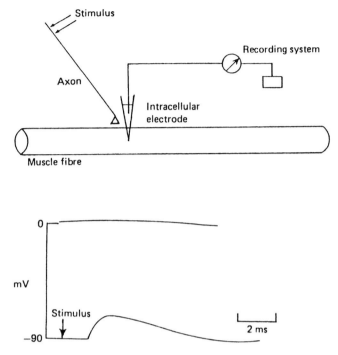

electrode resistance is 10 MΩ. Then the voltage drop across the tip of the electrode will be ten times that across the amplifier terminals, and so the recorded voltage will be only one eleventh of what it should be. With an input resistance of 1000 MΩ, however, the recorded voltage will be 99% of the potential across the membrane. High input resistances, up to 1000 GΩ for example, can be obtained by using field effect transistors or appropriate operational amplifiers as input stages.

The voltage clamp technique

Measurements of rapid membrane potential changes are electrically complex. They reflect currents through the membrane capacitance C_m (i.e. accumulation of charge on one side of the membrane and its depletion from the other) as well as movements of ions through ion channels. The total membrane current I_m is the sum of these two components:

$$I_m = I_{ionic} + C_m \frac{\mathrm{d}V}{\mathrm{d}t} \tag{3.1}$$

It would be very useful if one could get rid of the capacitance current and measure the ionic current alone. Clearly the way to do this is to hold the

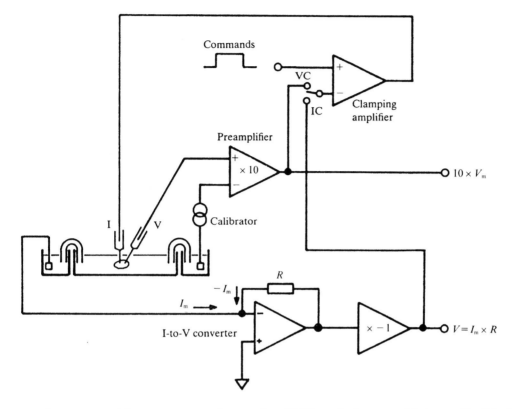

membrane potential at a constant value, so that $\mathrm{d}V/\mathrm{d}t = 0$; there would then be no capacity current and the measured current would be entirely ionic.

This can be done by means of the *voltage clamp* technique, first invented by Marmont and Cole (see Cole, 1968; Hodgkin, 1992) and greatly developed and exploited by Hodgkin and Huxley in 1952. This system uses a negative feedback loop to hold the membrane potential constant at some desired level, while the current flowing through the membrane is measured. It is usual to have two intracellular electrodes in the cell, one to pass the feedback current and the other to measure the membrane potential.

Figure 3.2 outlines an arrangement for the voltage clamp technique using two microelectrodes; it might, for example, be applied to a nerve cell in a molluscan ganglion. Microelectrode V is used to measure the membrane potential, using a suitable amplifier. The output from this amplifier is fed into one of the input terminals of another amplifier (labelled the clamping amplifier in fig. 3.2), which receives a command voltage at its other terminal. The output from this is passed through the second microelectrode I and so through the muscle cell membrane. We thus have a negative feedback loop: if the value of the membrane potential is different from that of the command voltage, current produced via the clamping amplifier output flows

Fig. 3.2. A voltage clamp system using two microelectrodes. The potential-recording electrode V and the current-passing electrode I are inserted into the cell under investigation, which is held in a bath of physiological saline. Outputs are the membrane potential of the cell V_m and the current flowing through its membrane I_m. The circuit can be switched between the voltage clamp mode (VC) and the current clamp mode (IC). (From Halliwell *et al.*, 1994. Reproduced with permission, copyright Company of Biologists Ltd.)

Fig. 3.3. The end-plate currents recorded from a frog muscle cell under voltage clamp conditions, using a two microelectrode clamp. The membrane potential (the six upper traces) was clamped at various levels from −120 mV to +38 mV. Stimulation of the motor nerve axon produced the corresponding currents shown in the lower traces, with inward currents shown as downward deflections and outward ones upward. Each current is the sum of the individual currents through some tens of thousands of nicotinic acetylcholine receptor channels. (From Magleby & Stevens, 1972.)

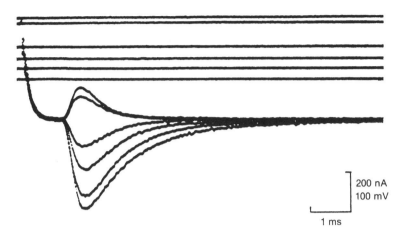

200 nA
100 mV

1 ms

through microelectrode I so as to reduce the difference to zero. The membrane potential is thus held ('clamped') at the value determined by the command voltage. The current flowing through microelectrode I is the same current as is flowing through the cell membrane, and also through the resistance R. So it can be measured from the potential drop across R, again using a suitable amplifier.

The system has been used to measure the end-plate currents in frog muscle fibres, as is shown in fig. 3.3. Shortly after the stimulus to the nerve axon, there is a brief flow of current, inward at negative membrane potentials and outward at positive ones. This current is entirely ionic and represents the summed current through all the open acetylcholine receptor channels at the muscle fibre end-plate.

The voltage clamp is particularly useful for investigating responses to changes in membrane potential, especially when these involve the activation of voltage-gated channels. If the membrane potential is driven by the voltage clamp system from one membrane potential to another at which it is then held constant, then the capacitance current $C_m\, \mathrm{d}V/\mathrm{d}t$ lasts only for a very short period of time so the remainder of the current flow is ionic (equation 3.1 above), passing through the membrane channels.

An example of this use comes from Hodgkin & Huxley's classic work on squid axons. They used axial wires as internal electrodes, allowing much larger currents than can be passed through microelectrodes, and permitting the membrane potential to be clamped over an appreciable area of membrane rather than just at a point. In fig. 3.4 the axon is depolarized under voltage clamp by about 65 mV. Coincident with the depolarization and lasting for about 60 μs is a brief outward blip of current that can be ascribed to discharge of the membrane capacitance. After this there is a transient inward current due to inflow of sodium ions through the voltage-gated sodium channels,

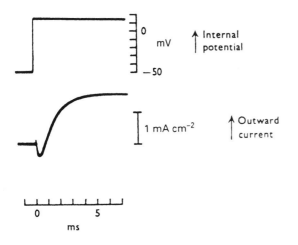

Fig. 3.4. A voltage clamp record from a squid giant axon. The lower trace shows the current flow through the membrane when it is depolarized by 65 mV (upper trace). Notice the initial brief blip of outward capacity current, followed by the transient inward ionic current and maintained outward ionic current. (From Hodgkin, 1958, after Hodgkin & Huxley, 1952b.)

followed by a maintained outward current carried by potassium ions moving through the voltage-gated potassium channels.

The evidence for this interpretation is shown in fig. 3.5. The total ionic current is the sum of the sodium and potassium currents:

$$I_{\text{ionic}} = I_{\text{Na}} + I_{\text{K}}$$

(there is also a small leakage current that has been removed from the traces in fig. 3.5 and which we need not consider further). Trace *a* shows the normal response to depolarization by 56 mV. Trace *b* shows the response when the external sodium concentration was reduced to one tenth of its normal value. This changed the sodium equilibrium potential E_{Na} (equation 2.9) so that it became equal to the depolarization at which the membrane was clamped. There would then be no net current through the sodium channels, so all the current would be the potassium current. By subtracting the potassium current (trace *b*) from the total current (trace *a*), we get the sodium current, shown in trace *c*.

By repeating this type of analysis at depolarizations of different magnitude, Hodgkin & Huxley were able to determine the relations between membrane potential and ionic conductance. From this they calculated the form of the action potential and the time course of the sodium and potassium conductance changes upon which it is based. These changes in ionic conductances reflect the activity of the ion channels in the membrane: the sodium conductance rises because many of the sodium channels open, and then it falls again when they close or become inactivated. Similarly the potassium conductance rises because potassium channels open.

The voltage clamp system in its original form measures the current flow through an area of membrane no smaller than several hundred square micrometres and usually rather larger than this. The currents recorded are the

Fig. 3.5. Hodgkin & Huxley's separation of the ionic currents in a squid axon voltage clamp record. Trace *a* shows the ionic current in response to a depolarization of 56 mV with the axon in sea water. Trace *b* is a similar response with the axon in 10% sea water, so that the membrane potential during depolarization is now at E_{Na}; because of this there is no sodium current so the trace shows the potassium current. Trace *c* is obtained by subtracting *b* from *a* and so shows the sodium current. (From Hodgkin, 1958, after Hodgkin & Huxley, 1952b.)

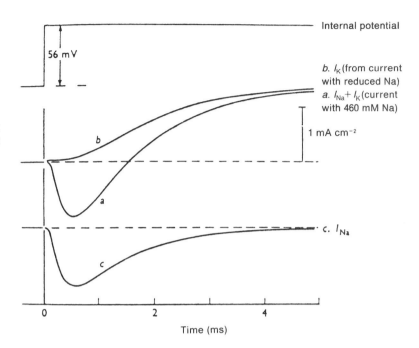

Internal potential

56 mV

b. I_K (from current with reduced Na)
a. $I_{Na} + I_K$ (current with 460 mM Na)

1 mA cm⁻²

c. I_{Na}

Time (ms)

summed currents of several hundred or more channels, so they are sometimes called 'macroscopic' currents. The electrical activity of single channels is not accessible by this means. Three further extensions of the voltage clamp method were developed for investigating single channel activity: fluctuation analysis, the use of artificial lipid bilayers, and the patch clamp technique. Let us have a look at them.

Fluctuation analysis

In an area of membrane containing a large number of channels, each of which may be open or closed at any particular instant, the total number of channels open will fluctuate in a random fashion from one moment to the next. Consequently the current flowing through the membrane will also fluctuate from one moment to the next: it will be 'noisy'. Fluctuation analysis is a method for getting information about the nature of the individual events from statistical descriptions of this noise.

Figure 3.6 shows an example of the current noise produced in a membrane as the number of open channels varies. The two-microelectrode voltage clamp system was used, with the electrodes inserted into the end-plate region of a frog muscle, and acetylcholine was applied to it. The acetylcholine molecules would combine with the acetylcholine receptors at the end-plate and

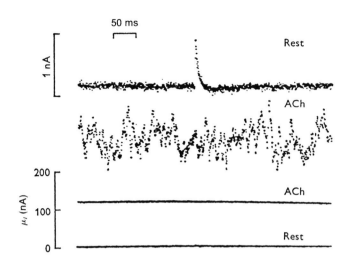

Fig. 3.6. Fluctuations in membrane current recorded from a frog muscle end plate in the presence and absence of acetylcholine (ACh). The two lower traces are at a relatively low amplification and show the mean end-plate current of 120 nA produced by acetylcholine. The two upper traces are at higher amplification and show the 'noisy' fluctuations in the current. The high noise level in the presence of acetylcholine is produced by random fluctuations in the number of open acetylcholine receptor channels. Acetylcholine was applied from a nearby micropipette by iontophoresis, a technique in which current is passed through the pipette and the positively charged acetylcholine ions migrate out of it in the electric field. The membrane potential was clamped at −100 mV throughout. (From Anderson & Stevens, 1973.)

result in the random opening of their channels; channel closures occur at random also. Statistical fluctuations in the numbers of open channels result in the noisiness of the current trace.

A full description of fluctuation analysis would be out of place in this book, but we can usefully consider some of its more elementary procedures. Thorough treatments are given by Neher & Stevens (1977) and DeFelice (1981).

We assume that all the channels whose currents contribute to the noise come from a single homogeneous population, and that each channel can be either open or closed. Suppose there are a total of N channels in this population of which r are open at any one time. Suppose the probability of any particular channel being open is p_O. The probability of any particular channel being closed is then $(1 - p_O)$ and the population statistics are described by the binomial distribution. Then,

$$\text{mean value of } r = Np_O \tag{3.2}$$

and

$$\text{variance of } r = Np_O(1-p_O) \tag{3.3}$$

The membrane conductance G will be the number of open channels (r) times the single channel conductance γ, i.e.

$$G = \gamma r$$

So

$$\text{mean value of } G = \gamma Np_O$$

and

$$\text{variance of } G = \gamma^2(\text{variance of } r)$$

$$= \gamma^2 Np_O(1 - p_O)$$

Hence

$$\frac{\text{variance of } G}{\text{mean of } G} = \gamma(1 - p_O)$$

If p_O is very low this simplifies to give

$$\frac{\text{variance of } G}{\text{mean of } G} = \gamma \tag{3.4}$$

Equation 3.4 was used by Anderson and Stevens (1973) to estimate the single channel conductance of the acetylcholine receptor channel at the neuromuscular junction. They clamped the membrane potential at the end-plate using a system similar to that in fig. 3.2, and measured the current fluctuations produced by low concentrations of acetylcholine. In one experiment the mean conductance was 2×10^{-6} S and its variance was 40×10^{-18} S^2, giving a value for γ of 20 pS.

Records of current fluctuations can also give information about the duration of channel openings. We assume that each open channel has a constant probability of closing in any time interval:

$$\text{open} \xrightarrow{\alpha} \text{closed}$$

Here α is the rate constant for closing; it is the reciprocal of the mean open time. If we take measurements at very short time intervals (i.e. at a high frequency) most of the channels open at the beginning of an interval will still be open at its end, so there will not be much change in the numbers of channels open. But at longer time intervals (longer than $1/\alpha$, for example) most of the channels open at the beginning of an interval will have closed by its end, so the fluctuations in channel numbers will be greater. Hence the variance of the noise should decrease as we make measurements at higher and higher frequencies.

Fourier analysis of the current fluctuations leads to the production of a power density spectrum, such as is shown in fig. 3.7. The individual points on this curve show the amplitude of the noise at their particular frequencies, each measured (to be technical) as the sum of the squared amplitudes of the sine and cosine components of the current fluctuations at that frequency. The form of the curve is described by the following relation:

$$S(f) = \frac{S_0}{1 + (2\pi f/\alpha)^2} \tag{3.5}$$

where f is the frequency and S_0 is a constant, the value of $S(f)$ when $f = 0$. We can extract α from this by setting a frequency f_c at which the power is half its maximum. Then,

$$S(f_c) = \frac{S_0}{2} = \frac{S_0}{1 + (2\pi f/\alpha)^2}$$

Hence $\qquad\qquad (2\pi f_c/\alpha)^2 = 1$

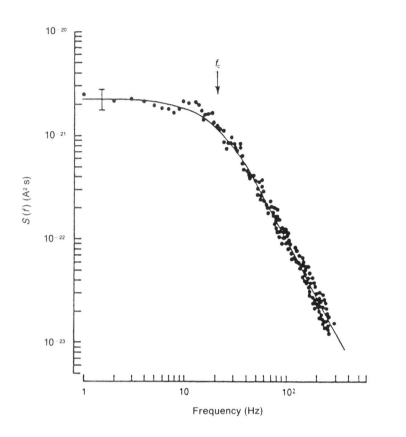

Fig. 3.7. Power density spectrum of acetylcholine noise at a frog muscle end-plate. f_c is the frequency at which the power is half its maximum value. (From Anderson & Stevens, 1973.)

and so
$$f_c = \alpha/2\pi$$

In fig. 3.7, for example, f_c is 21 Hz, giving a rate constant α of 130 s^{-1}. The mean channel opening time will then be $1/\alpha$, or 7.6 ms.

The method of analysis just outlined is called *stationary* fluctuation analysis, since it assumes that the probabilities of channels opening or closing do not change with time. Another method, called non-stationary or ensemble analysis, is used when this is not so.

In *non-stationary fluctuation analysis* we obtain a large set of similar responses to some appropriate stimulus, and compare the variance of the current with its mean at the same times after the stimulus. If i is the current through an individual channel when it is open, and $I(t)$ is the total current at time t, then, from equations 3.2 and 3.3,

$$\text{mean value of } I(t) = Nip_O(t)$$

and
$$\text{variance of } I(t) = Ni^2p_O(t)[1 - p_O(t)]$$

Hence, substituting for p_O,

$$\text{variance of } I = iI - I^2/N \qquad (3.6)$$

Fig. 3.8. Variance–mean plot from non-stationary fluctuation analysis of the sodium currents in a frog nerve fibre. The inserts show the mean sodium current I (lower trace, 1 nA per small division) and its variance var(I) (upper trace, 2×10^{-22} A^2 per small division) in response to a series of 20 ms clamped depolarizations to -5 mV. Potassium currents were eliminated with the blocking agent tetraethylammonium. (From Sigworth, 1980a.)

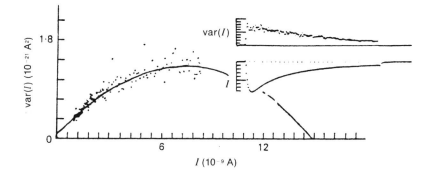

Equation 3.6 is a parabola that hits the I axis at iN and has a slope of i at the origin. So by plotting var(I) against I it is possible to estimate i and N. The variance of I is of course zero when either none of the channels is open or all of them are, and is at its maximum when half of them are open, as is also evident from equation 3.3.

This method was used by Sigworth (1980a) to determine single sodium channel currents at the node of Ranvier in frog nerve. From a large number of depolarizations of the same extent he was able to determine the time course of the total sodium current I and its variance. Figure 3.8 shows the plot of var(I) against I in one experiment, giving values of 43 000 for the number of sodium channels at the node, 0.4 pA for the magnitude of their individual currents at -5 mV, and 5.8 pS for the single channel conductance.

For many purposes, fluctuation analysis has been superseded by the patch clamp method, which gives a much more direct measure of single channel characteristics. It is still useful where channel densities are high or where the membrane may not be accessible to a patch clamp electrode. It may give misleading results if the channel population is not homogeneous in its properties, and it has difficulty in dealing with some of the complexities of channel gating that have been revealed by patch clamp records.

Artificial phospholipid bilayers

Lipid molecules of the type found in cell membranes readily form monolayers when spread on the surface of water. By putting two such monolayers back to back we arrive at an artificial model of the cell membrane (Miller, 1983, 1986; Williams, 1994). The bilayers are often called planar lipid bilayers to show that they are flat and to distinguish them from other artificial membrane systems such as liposomes. Pure lipid bilayers show inert behaviour and a high electrical resistance because they are not permeable to ions. This situation changes markedly if channel proteins are incorporated into

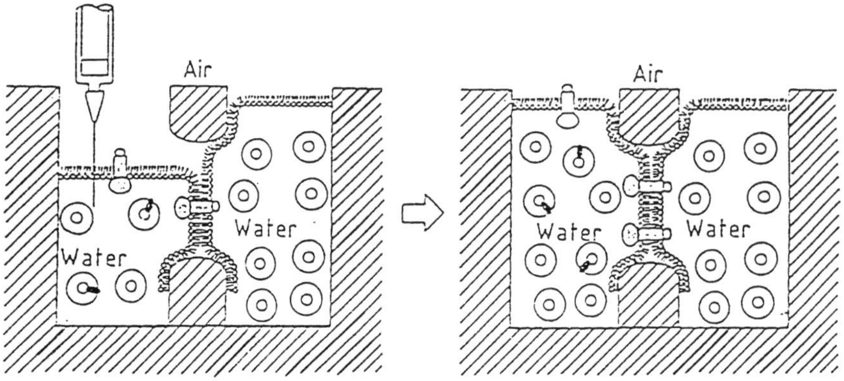

Fig. 3.9. Formation of an artificial phospholipid bilayer in Schindler's version of the monolayer folding method. The chamber contains two compartments separated by a Teflon partition with a small hole in it. Monolayers are formed on the surface of water in the two compartments by the spontaneous spreading of vesicles. First one side is raised up past the hole in the partition, then the other. Self-assembly of the two monolayers at the hole leads to bilayer formation. Channel proteins in the vesicles in the left compartment orient one way in their surface monolayer, and so are oriented one way only in the bilayer. Not to scale. (After Pattus, Lakey & Dargent, 1988. Bacterial proteins which form membrane pores: a biophysical approach. Reproduced from *Microbiological Science* with copyright permission of Blackwell Science Ltd.)

them, a process called *reconstitution,* when single channel currents can readily be measured.

These artificial membrane models can be made in various ways. In one experimental arrangement a partition made of Teflon or another suitable plastic separates two compartments containing aqueous solutions, and the partition has a small hole in it. The bilayer is made by painting the hole with a solution of lipid in a suitable solvent such as decane. The solvent disperses, leaving a bilayer consisting largely of the lipid molecules.

Alternatively the bilayer can be formed by coating the hole with two mono-layers, one from each side of the partition; this is sometimes called the 'folded monolayer' procedure. This method allows asymmetrical membranes to be made, with different lipid compositions on the two sides. In Schindler's version, the two compartments contain suspensions of artificial lipid vesicles or vesicles formed from natural membranes (Schindler, 1980; Schindler & Quast, 1980). Monolayers form on the surface of the compartments by spon-taneous breakdown of the vesicles. By raising the solution level past the hole, first on one side and then on the other, a bilayer is formed across the hole. The vesicles in one compartment contain channel proteins, and these are incor-porated into the bilayer as it is formed; since they usually take up the same ori-entation in the monolayer, their orientation in the bilayer is correspondingly uniform (fig. 3.9).

Another way of making a folded monolayer is to start from monolayers formed on the surface of the two compartments by spreading, on the water surface, a solution of phospholipid in a volatile solvent such as hexane, and then raising each of them up past the hole to form a bilayer (Montal & Mueller, 1972). Channel proteins are incorporated into the membrane by adding vesicles containing them to one of the compartments after the bilayer has been made. The vesicles then tend to associate with, or bind to, the bilayer and will fuse with it if they can be made to swell by establishing an osmotic

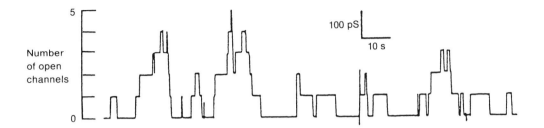

Fig. 3.10. Gramicidin channel activity in a lipid bilayer. The step changes in current across the membrane represent the opening or closing of single gramicidin channels. (From Hladky & Haydon, 1984.)

gradient across them. The vesicles may be obtained by cell fractionation techniques (as for sarcoplasmic reticulum membranes, for example) or by mixing purified detergent-extracted protein with artificial lipid vesicles (liposomes).

It is conventional to call the side of the bilayer in the compartment to which the vesicles are added the *cis* side, and the other side the *trans* side. The *trans* compartment is regarded as being at zero voltage. With vesicles from natural membranes one usually finds that the channels are all inserted into the membrane the same way round, so that one side of the bilayer corresponds to the outer side of the cell membrane and the other to the cytoplasmic side.

In a third method a patch pipette electrode is withdrawn through a monolayer at an air–water interface and then dipped into it again, so that one layer of lipid is folded onto the other (Coronado & Latorre, 1983). This is sometimes called the 'tip-dip' or 'double dip' method.

When bilayers are formed on patch clamp electrodes, then patch clamp electronic circuits are necessary for measuring the channel currents. For the two-compartment method, a similar low noise, fast response voltage clamp circuit is required, but there may be some differences in detail. The hole in the Teflon septum is likely to be 40 μm or larger in diameter, and so the capacitance (equation 2.7) and intrinsic noise level of the membrane will be much larger than for the membrane in a patch electrode. Hence the method is better adapted for use with channels with relatively high conductances.

An alternative to the use of planar bilayers is to incorporate channels into phospholipid vesicles (liposomes) and then apply the patch clamp technique to them. Normal liposomes are too small for this, but cycles of freezing and thawing or drying and rehydration may produce larger structures to which the patch electrode can be applied (Tank & Miller, 1983; Tomlins & Williams, 1986).

Lipid bilayers played an important role in the development of channel concepts, since discrete single channel currents were first observed in them, initially from a bacterial protein extract (Bean *et al.*, 1969), and then from the antibiotic gramicidin (Hladky & Haydon, 1970; see fig. 3.10). Since then bilayer methods have been particularly useful in looking at channels extracted from otherwise rather inaccessible sites, such as intracellular membranes like the sarcoplasmic reticulum. They are also important in testing the properties of purified channel proteins.

The patch clamp technique

The patch clamp technique was invented by Neher & Sakmann in Göttingen in order to measure directly the currents flowing through single channels. The amplitude of these currents is very small: 1 or 2 pA is a typical value. In conventional voltage clamp systems the extraneous noise level – the spontaneous fluctuations in the baseline current in the absence of any determinable signal – is much larger than this, so the single channel currents cannot be seen. The noise level in a voltage clamped muscle fibre, for example, using intracellular microelectrodes and in the absence of acetylcholine, is about 70 pA root-mean-square, i.e. about 200 pA peak-to-peak (Anderson & Stevens, 1973). This would have to be reduced by about 100-fold for us to see individual channel currents and by at least 1000-fold to measure them accurately.

Some of this noise arises from the thermal agitation of electrical charges; it is often known as Johnson noise after the man who first investigated it. The variance of the current noise is given by

$$\sigma_I^2 = 4kTf_c/R \tag{3.7}$$

where k is the Boltzmann constant, T is the absolute temperature, f_c the bandwidth of the measurement (i.e. the upper cut-off frequency) and R the internal resistance of the signal source. If we want to measure single channel events with currents of the order of 1 pA and durations of around 1 ms, then we need a root-mean-square value for the noise as low as 0.1 pA (10^{-13} A) and a bandwidth of at least 1 kHz. Then,

$$\sigma_I^2 = (10^{-13})^2$$

So, substituting in equation 3.7,

$$10^{-26} = 4 \times 1.381 \times 10^{-23} \times 10^3 \times 280/R$$

or

$$R = 4 \times 1.381 \times 10^{-23} \times 10^3 \times 280/10^{-26}$$

$$= 1.5 \times 10^9 \ \Omega$$

What this means is that the internal resistance of the signal source must be 2 GΩ or higher if we are to be able to record single channel currents satisfactorily. Since the resting membrane resistance of a typical cell membrane is about 1000 Ω cm² (i.e. 10^{11} Ω μm²), the membrane area we record from must be less than 50 μm² in area. This rules out conventional microelectrode recordings, which can only be made from cells whose surface area is considerably larger than this.

Fig. 3.11. The first patch clamp records of ion channel activity, from the acetylcholine receptor channels of denervated frog muscle. Channel openings are represented by pulses of inward current, seen as downward deflections from the baseline. The patch pipette contained suberyldicholine, an analogue of acetylcholine, which induces very long-lived channel openings. (From Neher & Sakmann, 1976. Reprinted with permission from *Nature* **260**, p. 800, Copyright 1976 Macmillan Magazines Limited.)

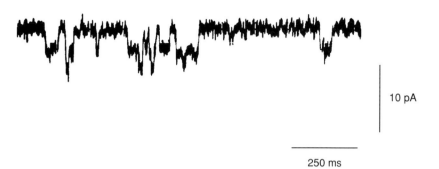

10 pA

250 ms

These considerations led Neher & Sakmann to make measurements from a small patch of cell membrane. Their material was frog muscle fibres that had been denervated some time previously; under these circumstances nicotinic acetylcholine receptors appear scattered widely over the surface of the fibre. They used a glass micropipette a few micrometres in diameter at the tip as an electrode, with its terminal rim smoothed by fire polishing. It was filled with saline solution and pushed against the cell membrane. It was important to get a good seal between the end of the pipette and the membrane. Without this there would be leakage between the interior of the pipette and the bath, so the effective source resistance of the electrode would be reduced and the current noise consequently increased. To help to make a good seal, the muscle fibres were treated with collagenase and protease to remove the extracellular material surrounding them. The interior of the electrode contained Ringer's solution and a low concentration of acetylcholine or an agonist such as suberyldicholine. A silver wire coated with silver chloride connected the saline in the electrode to the recording equipment.

The first results with the patch clamp technique were published by Neher & Sakmann in 1976 (fig. 3.11). They showed discrete square-wave pulses of current, suggesting that the acetylcholine receptor channels were either open or closed, and that they moved from one state to the other in a stochastic manner. It was a great day for channel science.

These early patch clamp records were still not fully satisfactory. The high noise level disguised small or rapid channel openings. Many channels appeared to be partly under the rim of the pipette, so that their currents were reduced, leading to a wider dispersion of recorded current amplitudes than would be expected from identical channels. This led to intensive work at the Göttingen laboratory to make the system better, with eventual triumphant success. As Neher described it in his Nobel lecture:

> By about 1980, we had almost given up on attempts to improve the seal, when we noticed by chance that the seal suddenly increased by more than two orders of magnitude when slight suction was applied to the pipette. The resulting seal

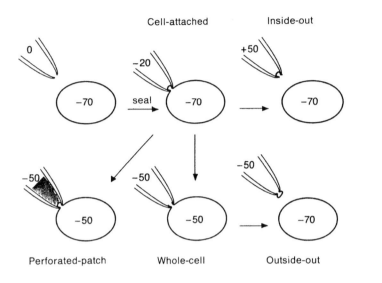

Fig. 3.12. Different patch clamp configurations. All begin with the formation of a gigaseal in the cell-attached configuration. Withdrawal of the pipette electrode produces an inside-out patch. Breakage of the patch while the pipette is attached to the cell gives whole-cell recording. Withdrawal from this produces an outside-out patch. A perforated patch (produced with nystatin or amphotericin in the pipette) also produces whole-cell records. The numbers show the electrode voltages required to produce a patch or whole-cell membrane potential of −50 mV if the cell resting potential is initially −70 mV. (From Cahalan & Neher, 1992.)

was in the gigaohm range, the so-called 'Gigaseal'. It turned out that a gigaseal could be obtained reproducibly when suction was combined with some simple measures to provide for clean surfaces, such as using a fresh pipette for each approach and using filtered solutions. (Neher, 1992)

The gigaseal is highly effective in reducing the noise level of the recording, and so it greatly increases the resolution with which the single channel currents are observed (Hamill *et al.*, 1981). The distance between the glass rim of the pipette electrode and the cell membrane probably falls to around 1 Å when the gigaseal forms. In accordance with this, it is found that small molecules do not diffuse through the seal; thus acetylcholine outside the pipette does not activate acetylcholine receptor channels in the patch. Light microscopy and electron microscopy show that the cell membrane is drawn into the pipette in an Ω shape for a distance of some few micrometres.

The gigaseal proved to be a physically robust phenomenon, permitting different configurations of the membrane in relation to the pipette electrode (fig 3.12). The original arrangement whereby the patch of membrane remains in position in the cell after the gigaseal has formed is called the *cell-attached* configuration. If the pipette electrode is now pulled away from the cell, the patch is excised from the cell but is still in place in the electrode and open to investigation. The arrangement is called the *inside-out* configuration since the extracellular face of the membrane patch is in contact with the solution in the pipette and its cytoplasmic face is bathed by the external solution. It has the great advantage that the solutions in contact with both sides of the membrane can now be controlled, as can the potential across the membrane.

Strong suction or a brief voltage pulse of up to 1 V applied in the cell-attached configuration will break the patch but retain the gigaseal. This means that the patch electrode is in contact with the interior of the cell, so it will record voltage or current changes from the whole cell membrane. This *whole cell recording* system allows voltage clamp records to be made from much smaller cells than is possible with conventional intracellular microelectrodes. Furthermore the cell contents rapidly equilibrate with the solution in the electrode, so it is possible to control the internal concentrations of ions, second messengers and other soluble compounds. Of course the area of membrane recorded from is much larger than in a small patch, so the noise level is too high for single channel currents to be discerned.

An alternative method of recording from the whole cell is provided by the *perforated patch* configuration. Here the electrolyte solution in the patch electrode contains a pore-forming antibiotic – amphotericin or nystatin. When the gigaseal is formed in the cell-attached configuration, the amphotericin acts on the membrane patch to make it very permeable to monovalent ions. This means that the patch now has a low resistance to current flow so that the whole cell can be voltage clamped. But, unlike the whole cell clamp, most of the soluble cytoplasmic constituents do not diffuse out of the cell into the patch pipette; second-messenger systems, for example, remain intact.

Single channel recording from an *outside-out* patch is also possible. Here we start from the whole cell recording configuration and pull the patch electrode away from the cell. This draws out a 'neck' of membrane, which then breaks and re-forms as a patch with the extracellular surface bathed by the external solution and the cytoplasmic surface in contact with the solution in the pipette. Outside-out patches have to be larger than inside-out ones since the initial cell-attached patch cannot be broken if the pipette tip diameter is too small.

Patch pipettes are made by drawing out glass tubing, usually in two stages and by an electrode puller designed for the job, so as to produce a tip diameter of 1 to 2 μm or sometimes a little larger. Microelectrode tubing with a glass filament is useful since it makes filling with saline easier. They are coated near their tips with Sylgard resin so as to reduce the capacitance of the pipette wall and thus avoid extra noise and unwanted capacitative transients. Their tips are polished by bringing them close to a red-hot heater filament for a few seconds, using a microscope and micromanipulators to control the operation. They are filled with a suitable saline solution, usually by backfilling the bulk of the electrode from a syringe.

The cells to be patch clamped need to be free from extracellular material such as collagen or basement membrane. Hence it may be necessary to treat them with collagenase and sometimes with other proteases (although proteases should be used with care, since they could destroy or modify the very cell membrane proteins that we want to look at). These procedures may not

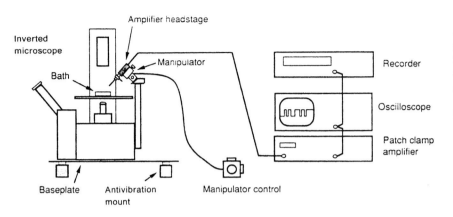

Fig. 3.13. The main components of a patch clamp set-up. (From Standen & Stanfield, in *Monitoring Neuronal Activity: A Practical Approach*, ed. J.A. Stamford, 1992. Reproduced by permission of Oxford University Press.)

be needed if the cells are from embryos, cultured cell lines or primary tissue culture. In skeletal muscles, one can produce blebs of plasma membrane by loading the muscle with potassium chloride and protease treatment (Spruce *et al.*, 1987a). Nerve cells in brain slices can be exposed by blowing away the overlying material in a gentle stream of extracellular solution (Sakmann *et al.*, 1989).

Figure 3.13 shows the arrangement for a patch clamp experiment (see Standen & Stanfield, 1992; Ogden & Stanfield, 1994). A good microscope is needed to see the cells, with phase-contrast or Nomarski optics. Firm micromanipulators with remote controls are needed to carry the electrodes and the recording amplifier headstage. The whole has to be carried on a massive base with antivibration mounting.

In order to make the most of the low noise properties of a patch pipette with a gigaseal, it is necessary to have a correspondingly low noise recording system. The crucial first component of this is the headstage, the current-to-voltage converter that receives its input directly from the patch electrode. This involves one or more operational amplifiers with field effect transistors as their input stages.

The more commonly used type of headstage involves an operational amplifier with a high resistance feedback, as is shown in fig. 3.14. The operational amplifier has two input terminals, non-inverting ($+$) and inverting ($-$). A voltage V_{ref} is applied to the non-inverting input, and the amplifier passes current through its feedback resistor to make the voltage of the inverting input (V_p) the same as V_{ref}. V_{ref} thus acts as the command for voltage clamping the membrane patch isolated by the tip of the pipette. The amplifier draws no current through its input, since it has an extremely high input resistance. So the current through the feedback resistor (i_f) will be equal to the current collected by the patch pipette (i_p) but opposite in sign, since the sum of i_p and i_f is zero.

Fig. 3.14. Patch clamp circuit. The high gain operational amplifier A is connected so that the current through the patch, i_p, can be measured as the voltage drop, V_p, across the feedback resistance R_f. (Based on Ogden & Stanfield, 1994. Reproduced with permission, copyright Company of Biologists Ltd.)

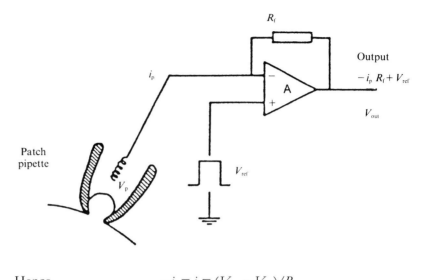

Hence

$$-i_p = i_f = (V_{out} - V_{ref})/R_f,$$

or

$$V_{out} = -i_p R_f + V_{ref},$$

so measurement of the difference between V_{out} and the reference voltage V_{ref} (this is done by a subsequent differential amplifier) gives the current flow through the patch electrode. R_f is usually fixed at a high value such as 10 GΩ so as to reduce the current noise arising from it (see equation 3.7) and to make the changes in V_{out} large enough to be measured.

There is inevitably some stray capacitance C_f associated with the feedback resistor, usually of the order of 0.1 pF. A step change in current through R_f will thus produce an exponential change in V_{out}, with a time constant τ equal to $C_f R_f$:

$$V_{out} - V_{ref} = -i_p R_f [1 - \exp(-t/\tau)]$$

Since τ will be about 1 ms, some compensation for this effect is needed. This can be done by differentiating V_{out} and adding it to the response:

$$V_{out} + \tau d V_{out}/dt = -i_p R_f$$

This procedure is often called 'high frequency boost'.

Circuit requirements for patch clamping are discussed by Sigworth (1995), Ogden (1994b) and Levis & Rae (1992), and suitable amplifiers are commercially available. Additional stages following the headstage give a final amplification up to 1 V/pA, which is easily handled by the display and analysis equipment. The final output is usually filtered to cut out high frequency changes (above 4 kHz, for example) and then digitized by sampling at some frequency above the filter cut-off rate (at 40 kHz, for example) and stored on disc ready for computer analysis. For long sequences the output may be recorded on magnetic tape prior to digitization. Computer software is available for statistical analysis of the records. It is important for the experimenter not to be too dazzled by the computer output, since one needs to know just how the raw data have been modified in their pathway through the analysis system.

Analysis of single channel current records

The first question to ask about a series of single channel currents is 'What sort of channel is producing them?' In some cases, as for example when we have introduced a particular reconstituted channel into a bilayer membrane, there should be no problem. In other cases the nature of the channel may need to be identified, and we might then seek answers to the following questions (Ogden & Stanfield, 1994):

(1)   What makes the channel open? Neurotransmitter-gated channels will open in the presence of the particular ligand, such as acetylcholine or glutamate, at the external surface of the membrane patch. Channels opened or closed by intracellular constituents, such as calcium ions or cyclic nucleotides, will respond to changes in the concentration of the particular ligand at the internal surface. Voltage-gated channels will respond to changes in membrane potential, usually to a depolarization so that the internal surface becomes less negative with respect to the external surface.

(2)   What ions flow through the open channel? This is tested by varying the concentration gradient across the membrane for different particular ions and looking for changes in the reversal potential. If E_{rev} changes by 58 mV when we change the external potassium ion concentration tenfold, for example, then we are dealing with a potassium channel (see equation 2.10). If it does not change at all under these circumstances, then the channel is probably not permeable to potassium ions. If it changes by some intermediate value, then the channel may well be a non-selective cation channel, permeable to other cations such as sodium, as well as to potassium.

(3)   What is the single channel conductance? Channels with otherwise similar gating and selectivity properties may differ in this respect. Examples include different calcium-activated potassium channels, and foetal and adult muscle acetylcholine receptor channels.

(4)   Are there specific blocking agents? Voltage-gated sodium channels are blocked by external tetrodotoxin, whereas amiloride-sensitive sodium channels are blocked by amiloride. Most, but not all, nicotinic acetylcholine receptor channels are blocked by α-bungarotoxin. Many further examples are given in chapter 7.

Fig. 3.15. Histogram showing the amplitudes of 395 single channel currents from a frog muscle endplate. The membrane potential was −91 mV, acetylcholine receptor channels were activated by suberyldicholine. The Gaussian curve (or normal distribution) fitted to the data has a mean of 2.61 pA and a standard deviation of 0.08 pA. The observed distribution is rather more sharply peaked than the Gaussian curve. (From Colquhoun & Sigworth, 1983. Reproduced with permission, copyright Plenum Publishing Corporation.)

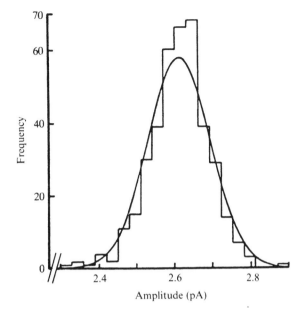

Measuring single channel currents

When a patch of membrane contains just a single channel that is either open or shut and the noise level is low, then measurement of the unitary current is relatively straightforward. We simply measure the difference between the baseline and the open channel current, perhaps by moving cursors on a computer screen display of the record. In fig. 3.15, for example, the amplitudes of 395 single channel currents are plotted as a histogram with a bin width of 0.03 pA. The data here are quite closely fitted by a Gaussian distribution with a mean of 2.61 pA and a standard deviation of 0.08 pA (Colquhoun, 1994).

It is quite common, however, for things to be more complicated than this. The noise level may be appreciable, there may be more than one channel present in the patch, and there may be occasions when the recorded open channel current is less than the normal maximum. One way of dealing with these problems is to use a *point-amplitude histogram*. We measure the amplitude of the current for each data point in the digitized record and include it in a histogram so as to build up a frequency distribution of the various current levels.

Figure 3.16 illustrates the situation when there is more than one channel in the patch, from an investigation on the ATP-sensitive potassium channels of frog muscle (Spruce *et al.*, 1987a). The current flowing through the patch fluctuates in a stepwise fashion between different levels, and the steps are equal in size. This effect reflects the fluctuations in the number of channels that are

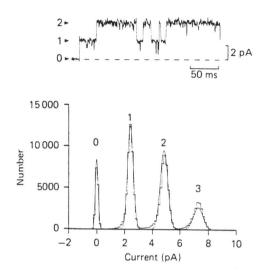

Fig. 3.16. Point amplitude histogram showing the currents produced by up to three channels open at once. The current record (a section is shown above) shows the activity of ATP-dependent potassium channels in an inside-out patch of frog muscle plasma membrane. The individual data points from the digitized record over a continuous length of time are plotted in the histogram. Each data point is a measure of the transmembrane current at a particular time, and there were 2000 of these measurements per second for 64 seconds. The curves are drawn so that the areas under them are in accordance with the binomial distribution, as in equation 3.8, with $p_O = 0.50$ and $N = 3$. Their variances (σ^2) were predicted by fitting Gaussian curves to the baseline current, giving σ_B^2, and the current when one channel was open, giving σ_1^2. Then $\sigma_2^2 = 2\sigma_1^2 + \sigma_B^2$ and $\sigma_3^2 = 3\sigma_1^2 + \sigma_B^2$. (From Spruce *et al.*, 1987a.)

open at any instant. Two open channels produce twice as much current as one, and three open channels produce three times as much; in electrical terms, the channels are in parallel. The point-amplitude histogram shows three peaks in addition to the peak for the zero baseline. This suggests that there were three channels in the patch and that at different times none, one, two or all three of them were open. The individual channel current in this experiment was 2.4 pA, and the peaks at 4.8 and 7.2 pA represent occasions when two and three channels were open. The area under the graph about each peak (or, more generally, between any two current values) is proportional to the time spent in that range.

There is inevitably some noise associated with patch clamp records, so that the zero current baseline and the various current levels are not rock steady but show some variation from one data point to the next. This is reflected in the spread of the peaks in the point-amplitude histogram. We would expect the noise level associated with single channel opening to be additive, so that the higher peaks, corresponding to more and more channels open, should show a wider spread. This also is evident in fig. 3.16.

If we have three channels opening independently of one another in a patch, how often would we expect to see one, two or all three open? This is the sort of situation described by the binomial theorem. Suppose that there are N channels in the patch. Let the probability of any particular channel being open at any particular instant be p_O. (This means, for example, that if over an appreciable period of time the channel is open 20% of the time then p_O is 0.2.) We assume that p_O is the same for each of the N channels. Then the binomial theorem states that the probability $P(x)$ that x of the channels will be open is given by

Fig. 3.17. Subconductance states in glycine-activated channels. The records in *A* show fluctuation of single channel currents between the baseline, the main level of 1.39 pA and several sublevels. The histogram in *B* shows the amplitudes of all currents lasting longer than 0.4 ms during a sample period of time. Cell-attached patch of a cultured spinal neuron, depolarized to −3 mV. Time scales (the bar under a_1) are 20 ms for a_1, 75 ms for a_2, 40 ms for a_3 upper, and 5 ms for a_3 lower. (From Bormann *et al.*, 1987.)

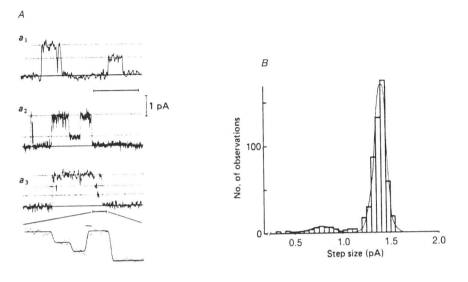

$$P(x) = \frac{N!}{x!(N-x)!}(p_O)^x(1-p_O)^{(N-x)} \tag{3.8}$$

For the record whose point-amplitude histogram is shown in fig. 3.16, equation 3.8 fits the data well with $p_O = 0.5$ and $N = 3$.

Notice that if p_O is very low the chances of more than one channel being open at a time are much reduced. If we have a patch with two channels each with $p_O = 0.01$, for example, then the probability of both being open at the same time is only 0.0001. A relatively short length of record (containing less than 100 channel openings, say) might easily show no occasions when both channels were open, leading to the erroneous conclusion that only one channel is present.

A further complication comes from the existence of *subconductance levels*. In many channels the currents sometimes adopt levels intermediate between the shut and the fully open currents. Figure 3.17 shows an example, from the glycine-activated channels of spinal neurons (Bormann *et al.*, 1987). The histogram shows the height distribution of the various substates as measured in about 600 different events. The main state peaks at 1.39 pA whereas the substate peaks at 56% of this, 0.78 pA.

The time characteristics of channel current records

A patch clamp record of the activity of a single channel will show a sequence of alternate open and shut periods of different durations. We can call the duration of any particular open or closed event its dwell time, denoted by t. A common method of analysing single channel records is to sort the individual

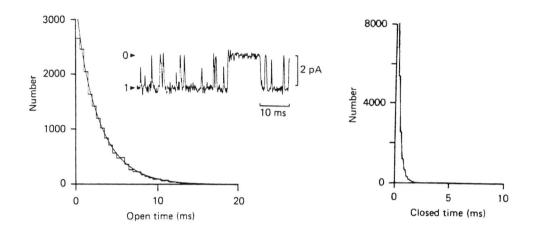

events according to their dwell times, putting them into a series of bins of constant width. After several hundred or more openings have been sorted in this way, normally by computer, we will have produced a *dwell time histogram* for the open state of the channel. A similar procedure is used to produce a closed state histogram. Figure 3.18 shows open and closed dwell time histograms for an ATP-sensitive potassium channel.

These histograms can usually be fitted by one or more exponential functions in the general form

$$f(t) = a \exp(-t/\tau_1) + b \exp(-t/\tau_2) + \ldots \tag{3.9}$$

where a and b are constants and τ_1 and τ_2 are time constants. We shall see in chapter 6 that these exponential terms may be related to the kinetics of the channels and so may be useful in modelling channel behaviour.

Sigworth & Sine (1987) have pointed out that the fitting of multiple exponentials to dwell time histograms is made much easier if the bin widths are not equal for different values of the dwell time t but are proportional to the logarithm of t. Such histograms are plotted on a logarithmic time axis and with a square-root scale for the ordinate. Individual exponential components then appear with separate peaks in the histogram, with maxima at their time constants, as shown in fig. 3.19. The square-root ordinate is used because it makes the expected random scatter in bin numbers much the same across the range of values of t.

There are inherent limitations on the degree of time resolution that can be obtained from patch clamp records of single channel activity. Such records are usually filtered to remove high frequency noise and then digitized by sampling. Some removal of high frequencies by filtering is an inevitable consequence of the recording system: all amplifiers and other electronic devices have some frequency limit above which they cannot respond. There is also appreciable

Fig. 3.18. Open and closed dwell time histograms for ATP-sensitive potassium channels. The inset shows part of the record that produced the histograms. The open-time histogram is fitted by a curve (equation 3.9) that is the sum of two exponential terms, whereas the curve for the closed times has three terms. (From Spruce *et al.*, 1987a.)

Fig. 3.19. Dwell time histograms with logarithmic time axes and square root ordinates. The channels were *Torpedo* acetylcholine receptors expressed in a mouse fibroblast cell line. Records were obtained by patch clamp in the cell-attached mode, filtered at 12 kHz; part of one is shown at the top. To make the histograms, the records were digitally sampled at 94 kHz and accumulated in bins whose width was proportional to the logarithm of the event duration. Events below 16 μs (log $t = -4.8$ s) were not included. The open time histogram shows 2790 events, fitted (continuous line) with a single exponential ($\tau = 144$ μs). The closed time histogram shows 3190 events, fitted with two exponentials, $\tau_1 = 11.9$ μs for the brief closings and $\tau_2 = 119$ ms for the long ones (see equation 3.9). The patch pipette contained acetylcholine at a low concentration. (From Sine *et al.*, 1990. Reproduced from *The Journal of General Physiology* 1990, **96**, pp. 395–437, by copyright permission of The Rockefeller University Press.)

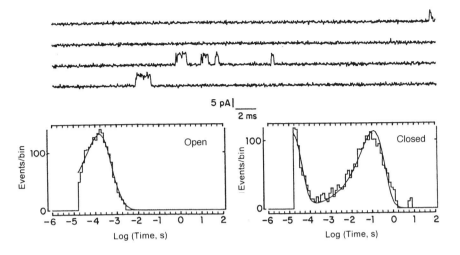

background noise in the recording system and this has to be removed in order to observe single channel events.

The characteristics of a filter can be described in terms of its cut-off frequency, f_c, which is the frequency at which the output falls by 3 dB, i.e. the square of the output current is one half of the maximum. Values of 1 to 4 kHz are commonly used in patch clamp experiments. The sampling rate of the digitizer should be considerably higher than the filter cut-off frequency, otherwise it will introduce its own distortion into the system. A sampling rate of 40 kHz is commonly used.

Limits of time resolution can sometimes be allowed for. It is common to eliminate from the analysis all events below a particular duration near to the limit of resolution. The number of openings that we would expect to fall in these empty bins can then be estimated once exponentials have been fitted to the dwell time histogram, and this estimate can then be used to correct the mean value of the dwell times. The difficulties increase when substates have to be taken into account. Is a brief blip simply a full opening or closing that is too brief to be fully recorded, or does it represent a rather longer time at some intermediate level? Colquhoun (1994) and Heinemann (1995) discuss some of the ways of dealing with problems of this type.

Molecular structures

One of the fascinating features of biology is the way structures interact with functions. Whether we are talking about the circulation of the blood or the activity of a protein molecule, we cannot understand how the system works without knowing something about its structure. At present we do not know as much about the structure of ion channels as we would like to. But what we do know, arising largely from the use of the powerful techniques of molecular biology, has proved to be very illuminating. In this chapter we consider the overall structures of a number of different ion channels. Particular aspects of the molecular bases of permeability and of gating we defer to chapters 5 and 6.

Ion channels are composed of one or more protein molecules. A common pattern, we shall see, is to have an oligomeric arrangement whereby the channel consists of a rosette of four, five or six similar subunits with an aqueous pore at its centre. Each subunit, like all protein molecules, is a chain of amino acids linked together in a linear sequence by peptide bonds. There are 20 different amino acids and there may be some hundreds of residues arranged in a unique sequence in the molecules of any particular protein. The amino acid sequence determines the nature of the protein, so if we want to understand how the protein works we first need to know what this sequence is.

Determining channel protein sequences

A few proteins have had their amino acid sequences determined directly by chemical analysis, but the procedure is lengthy and laborious. Much more rapid sequencing methods have been perfected in recent years for nucleic acids. Hence protein sequences are now usually determined indirectly from the base sequences of the nucleic acids which code for them.

Table 4.1. *Some landmarks in the development of molecular biological techniques*

1951	Sequencing of the insulin B chain by Sanger
1953	Discovery of the structure of DNA by Watson & Crick
1958	Demonstration that DNA replication is semiconservative
1960	Discovery of DNA-dependent RNA synthesis
1961	Discovery of messenger RNA and its role in protein synthesis
1966	Determination of the genetic code
1970	Discovery of reverse transcriptase in RNA tumour viruses
1970	Discovery of first restriction enzymes
1972	First recombinant DNA molecules (simian virus 40 with inserted phage and bacterial DNA) made by Jackson, Symons & Berg
1975	Screening of DNA libraries with specific hybridization probes
1976	Synthesis and cloning of complementary DNA from mRNA
1977	Development of rapid sequencing methods for nucleic acids
1977	Discovery of non-coding regions (introns) in eukaryotic genes, and of RNA splicing to remove them before translation of the mRNA sequence
1978	Development of site-directed mutagenesis methods begins
1982	First sequencing of a channel protein: the nicotinic acetylcholine receptor channel from *Torpedo* electric organ, deduced from its cDNA sequence
1985	Development of the polymerase chain reaction for amplification of DNA

To get the nucleic acids for sequencing, the various techniques of recombinant DNA technology must be used. It would be out of place here for us to delve too far into the details of how this is done; many biologists will already be familiar with the methods and there are excellent accounts elsewhere (see, for example, Watson *et al.*, 1992; Alberts *et al.*, 1994; Nicholl, 1994; Old & Primrose, 1994; Lodish *et al.*, 1995). Nevertheless, a brief look at the logic of the process seems appropriate. Table 4.1 gives some landmarks in the development of our knowledge in this field.

From gene to protein

Genetic information is stored in the DNA molecules, which constitute the material basis of heredity. DNA molecules are two-stranded polymers in the form of a double helix. Each strand is a long sequence of nucleotides connected together so that their bases project from a sugar-phosphate backbone. The four bases are adenine (A), cytosine (C), guanine (G) and thymine (T). The two strands are held together by hydrogen bonds between the opposing bases: adenine is always paired with thymine and guanine is always paired with cytosine. The consequence of this base-pairing is that the two

Fig. 4.1. The 'central dogma' of molecular biology: sequence information flows from DNA to RNA to protein. DNA consists of two complementary strands. RNA is produced by complementary base-pairing from one of these (transcription). The protein sequence is produced by translation of the genetic code inherent in the base sequence of RNA.

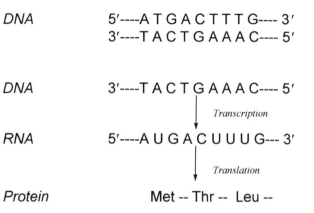

DNA   5'----A T G A C T T T G---- 3'
            3'----T A C T G A A A C---- 5'

DNA   3'----T A C T G A A A C---- 5'
               Transcription

RNA   5'----A U G A C U U U G--- 3'
               Translation

Protein   Met -- Thr -- Leu --

strands of the double helix are *complementary* to one another. Thus, if one strand has a particular sequence of bases, we know that the other strand paired with it will have a precisely complementary sequence. For example, if part of one strand has the sequence ATGACTTTG (as in fig. 4.1) then the part of the other strand that is paired with it must have the sequence TACT-GAAAC.

The phosphate groups linking the sugars to which the bases are connected in DNA and RNA are attached at one end to the fifth (5' or five-prime) carbon atom of the sugar and at the other to the third (3'). This means that the nucleic acid chains have directionality built into them. At one end of a chain the free phosphate group is attached to the 5' atom of the first sugar, at the other the 3' atom of the last sugar has no phosphate group attached. The two ends are thus known as the 5' (five-prime) and 3' (three-prime) ends, respectively. The two strands in DNA are antiparallel, i.e. they run in opposite directions so that at a free end of the double helix we will find the 5' end of one strand and the 3' end of the other.

This complementarity of the two strands of DNA provides the information whereby DNA replicates during cell division. The two strands unwind and pull apart, then each acts as the template for the production of a new complementary strand, so that two identical copies of the original DNA molecule are produced. In this way the genetic information can be passed on to daughter cells and to succeeding generations.

The information inherent in the sequence of bases in the DNA becomes meaningful when it is converted into the amino acid sequences of protein molecules. RNA acts as an intermediate here. It is synthesized as a sequence complementary to one of the strands of DNA, with uracil (U) pairing with adenine in place of thymine, but with A, C and G pairing with T, G and C. Thus the sequence of bases on the DNA is subject to *transcription* to produce the complementary sequence on the RNA. The primary transcript RNA is

Table 4.2. *The amino acids of proteins. They have the general formula R-CH(NH$_2$)COOH, where R is the side-chain or residue. Proline is actually an imino acid. Hydropathy index from Kyte & Doolittle (1982)*

Type	Amino acid	Side-chain	Abbreviations[a]		Hydropathy index
Non-polar	Isoleucine	$-CH(CH_3)CH_2CH_3$	Ile	I	4.5
	Valine	$-CH(CH_3)_2$	Val	V	4.2
	Leucine	$-CH_2CH(CH_3)_2$	Leu	L	3.8
	Phenylalanine	$-CH_2C_6H_5$	Phe	F	2.5
	Methionine	$-CH_2CH_2SCH_3$	Met	M	1.9
	Alanine	$-CH_3$	Ala	A	1.8
	Tryptophan	$-CH_2C(CHNH)C_6H_4$	Trp	W	−0.9
	Proline	$-CH_2CH_2CH_2-$	Pro	P	−1.6
Uncharged polar	Cysteine	$-CH_2SH$	Cys	C	2.5
	Glycine	$-H$	Gly	G	−0.4
	Threonine	$-CH(OH)CH_3$	Thr	T	−0.7
	Serine	$-CH_2OH$	Ser	S	−0.8
	Tyrosine	$-CH_2C_6H_4OH$	Tyr	Y	−1.3
	Histidine	$-CH_2C(NHCHNCH)$	His	H	−3.2
	Glutamine	$-CH_2CH_2CONH_2$	Gln	Q	−3.5
	Asparagine	$-CH_2CONH_2$	Asn	N	−3.5
Acidic	Aspartic acid	$-CH_2COO^-$	Asp	D	−3.5
	Glutamic acid	$-CH_2CH_2COO^-$	Glu	E	−3.5
Basic	Lysine	$-(CH_2)_4NH_3^+$	Lys	K	−3.9
	Arginine	$-(CH_2)_3NHC(NH_2)NH_3^+$	Arg	R	−4.5

[a] Triple-letter (left) and single-letter (right) systems.

modified in eukaryotes to produce messenger RNA (mRNA), which is then exported from the nucleus into the cytoplasm.

The mRNA base sequence acts as a template for *translation* to produce the amino acid sequence of the protein. Thus, the flow of information is from DNA to RNA to protein; this concept is sometimes known as the 'central dogma' of molecular biology. A linear sequence of bases in a particular section of the DNA determines a linear sequence of amino acids in a particular protein molecule (fig. 4.1).

The central dogma carries with it the idea of the 'genetic code', a set of rules that relates groups of bases on RNA to the amino acids which they determine in the protein. The code, it turns out, is quite simple: three sequential bases (forming a codon) code for each amino acid. Clearly there are 64 (i.e. 4^3) possible combinations of any three of the four bases, and these have to code for the 20 amino acids. Table 4.2 lists the amino acids and table 4.3 gives details of the code. All of the possible triplets are used, so the code is degenerate, i.e. some amino acids are coded for by more than one codon. Three

Table 4.3. *The genetic code. The nucleotide bases code in triplets for the amino acids. For example, the codons for cysteine are UGU and UGC*

First position 5'	Second position				Third position 3'
	U	C	A	G	
U	Phe	Ser	Tyr	Cys	U
	Phe	Ser	Tyr	Cys	C
	Leu	Ser	STOP	STOP	A
	Leu	Ser	STOP	Trp	G
C	Leu	Pro	His	Arg	U
	Leu	Pro	His	Arg	C
	Leu	Pro	Gln	Arg	A
	Leu	Pro	Gln	Arg	G
A	Ile	Thr	Asn	Ser	U
	Ile	Thr	Asn	Ser	C
	Ile	Thr	Lys	Arg	A
	Met	Thr	Lys	Arg	G
G	Val	Ala	Asp	Gly	U
	Val	Ala	Asp	Gly	C
	Val	Ala	Glu	Gly	A
	Val	Ala	Glu	Gly	G

triplets are known as stop codons, since they bring translation to an end. Methionine is the first amino acid laid down at the beginning of a new protein chain; its codon AUG acts as the signal to start the chain, provided it is in an appropriate sequence context.

During translation the mRNA sequence is read sequentially from the 5' end. As each successive amino acid is added to the developing protein chain, its α-amino group combines with the α-carboxyl group of the previous member to form the peptide bond linking the two. Thus the first member of the chain has a free amino group, forming the N terminus, and the last has a free carboxyl group, forming the C terminus.

We can define a gene as a segment of DNA necessary to make a functional polypeptide chain or RNA molecule. This sequence will include transcription control regions as well as the part that forms the primary RNA transcript. Most eukaryotic genes contain further non-coding regions, so the sequence making up a gene contains a number of sections called *exons*, which encode parts of the protein sequence, separated by *introns* ('intervening sequences'), which do not. The *Shaker* potassium channel gene in *Drosophila*, for example, is at least 65 kilobases long (1 kb is a thousand bases); it contains at least seven different exons, whose total lengths are less than 3 kb and which together produce a protein 616 amino acid residues long (Papazian *et al.*, 1987; Tempel *et al.*, 1987). Many of the mammalian potassium channel genes, however, do not possess introns.

The introns are initially transcribed into RNA but are removed by RNA splicing before the mRNA is exported from the nucleus prior to the translation process. Sometimes the various exons making up the coded part of a gene can be put together in different ways. This phenomenon is called *alternative splicing*; it means that somewhat different proteins can sometimes be made from the same gene.

In eukaryotes there is also much non-coding DNA between the gene sequences as well as within them. The human genome is about three billion base-pairs in size, and contains somewhere between 100 000 and 500 000 genes. Over half of these are probably specific to the brain (Chaudhari & Hahn, 1983; Changeux, 1985), and an important component of them comprises those that code for channel proteins.

The amino acid sequences of modern proteins have arisen as a result of evolution over millions of years. The sequences of two proteins derived from a common ancestral sequence will differ from each other roughly in proportion to the length of time since the divergence occurred. Two proteins are regarded as homologous if they are clearly derived from a common ancestor, and the term *homology* is commonly used by molecular biologists to indicate the degree of similarity between two genetically related sequences. Thus, if we say that two protein sequences show 80% identity, we mean that they can be aligned so that the same amino acid residues are present at 80% of the corresponding sites. If we say that they show 80% homology, we further imply that the two proteins have a common evolutionary ancestry (see Patterson, 1988; Donoghue, 1992).

Recombinant DNA technology

Recombinant DNA technology is also known as gene cloning or genetic engineering. Its essential purpose is to take a single DNA molecule or fragment and from this produce large numbers of molecules with identical sequences. Such a DNA fragment is joined with a vector DNA sequence, which can replicate when introduced into a host cell. The single recombinant DNA molecule, vector plus inserted DNA fragment, is then replicated in the host cell so that large numbers of identical DNA molecules are produced. The replication systems include vectors such as plasmids or bacteriophages, and these are introduced into host cells such as the bacterium *Escherichia coli*. Figure 4.2 summarizes the main processes in the cloning of a gene; let us have a look briefly at how it is done.

The discovery of *restriction enzymes* was an essential step in the development of gene cloning techniques. These are bacterial enzymes whose natural

Fig. 4.2. An outline of gene cloning.

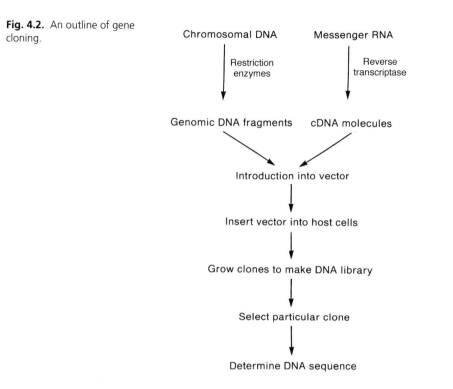

Chromosomal DNA   Messenger RNA

Restriction enzymes   Reverse transcriptase

Genomic DNA fragments   cDNA molecules

Introduction into vector

Insert vector into host cells

Grow clones to make DNA library

Select particular clone

Determine DNA sequence

function is apparently to break down any foreign DNA that may enter the cell. They do so by recognizing particular short (4 to 8) base sequences and cutting the DNA molecule there. This results in fragments of DNA that may be up to a few thousands of base-pairs long; they are reproducible and the right size for the various recombinant handling techniques. Many restriction enzymes cut DNA with a slight stagger between the two chains, so that short single-stranded 'sticky ends' are produced. The enzyme *Eco*RI, for example, recognizes the sequence GAATTC and cuts it after the first base; its complementary sequence is CTTAAG (i.e. GAATTC reading from 5′ to 3′) and this is cut before the last base, so that each half of the DNA molecule is left with a single-stranded 5′ projecting 'sticky end' of AATT.

DNA molecules can be joined together with the enzyme DNA ligase, so new DNA molecules can be made; the process is easier if the pieces to be joined have similar sticky ends. In this way segments of foreign DNA can be introduced into a suitable vector, a carrier DNA molecule that will replicate when put into a suitable host cell. Plasmids are frequently used as vectors. They are small circular DNA molecules found in bacterial cells, and they have been engineered to contain a gene for antibiotic resistance so that cells containing them can be readily selected for. DNA segments up to about 20 kb long can be inserted into them for cloning. Alternative vectors are bacteriophage λ (lambda) and (for longer DNA segments) hybrid systems called cosmids.

There are two main starting points for gene cloning: chromosomal or genomic DNA, and mRNA. To make vectors containing fragments of genomic DNA, then, we start with DNA extracted from the chromosomes in the organisms that we are interested in, cut it up with some suitable restriction enzyme, and combine it with vector DNA that has been cut with the same enzyme. Alternatively we can start with mRNA, which has the advantage that we are dealing with genes that are actually being expressed in the cells we are interested in. In certain tumour viruses there is an exception to the general rule in that information flows from RNA to DNA, and this is promoted by an enzyme called reverse transcriptase. This enzyme is used to make *complementary DNA* (cDNA) from mRNA. Complementary DNA differs from genomic DNA in that it contains only exons. It is usually introduced into a vector after attaching short linker sections to its ends.

One of the most useful components in the recombinant DNA toolbox is the *DNA library*. This is a collection of up to a million or so different DNA molecules held as separate clones in bacteriophage λ or a plasmid vector. A fly head cDNA library, for example, is made from the mRNA extracted from *Drosophila* heads, and will therefore contain cDNA coding for many brain proteins including, we might expect, various ion channels. A *Drosophila* genomic DNA library contains a large number of DNA fragments derived from *Drosophila* chromosomes.

To make a human genomic DNA library, DNA is extracted from some suitable tissue (embryonic material or sperm, for example), and partially digested by a restriction enzyme so as to produce fragments about 20 kb long. These are introduced into the λ phage DNA by replacing its middle section, so that we have large numbers of phages carrying these 20 kb pieces of the human genome. The human genome is about 3×10^9 base-pairs long so, allowing for overlaps and random selection, we need about a million separate 20 kb fragments in phages to ensure that more than 90% of it is included in the library. The phages are grown in square Petri dishes on a 'lawn' of *E. coli*. Each individual phage infects and replicates in a single bacterial cell, lyses it and releases many progeny which then infect the adjacent cells; the process is usually stopped when a visible plaque of lysed cells has formed. Since up to 50 000 plaques can be accommodated separately on a single dish, only 20 to 30 dishes are needed to hold the whole library.

How do we get the right DNA out of the library? Of all the million or so λ phage plaques in our library, how do we find the particular one that contains the particular DNA we are interested in? We *screen* the library by nucleic acid hybridization. To do this we make an oligonucleotide probe, a short stretch of DNA or RNA that has a base sequence that is complementary to part of the

DNA fragment that we are looking for, and which contains radioactive ^{32}P as a tracer. We make a replica of the pattern of plaques on a Petri dish by overlaying it for a minute or so with a filter sheet of nitrocellulose. This picks up ('blots') some of the phage particles in a replica of their arrangement on the bacterial lawn in the Petri dish. We denature the DNA in alkali, neutralize, and bake it at 80 °C *in vacuo* (all this converts it to the single-stranded form and binds it to the filter), and then we incubate it at 65 °C with the ^{32}P-labelled probe. Then we wash off the excess probe and lay the filter on a sheet of X-ray film. After a while we develop the film to see where the radioactive probe has bound. If it has done so we can go back to the original plate and pick off the plaque in the same position, and produce more DNA of the required type by purification and propagation of the recombinant clone or by the polymerase chain reaction (PCR).

Getting the right probing sequence is clearly crucial to this process. It is done in one of three ways. The most direct way is to determine part of the amino acid sequence of the channel protein, and then to make a set of short oligonucleotide sequences that will code for it (the set of related sequences is required because the genetic code is degenerate so that for many amino acids there is more than one codon). This was the method used for the first channels to be sequenced, such as the nicotinic acetylcholine receptor and the voltage-gated sodium channel, both of them extracted from fish electric organs (Noda *et al.*, 1982, 1984), and the calcium channel of rabbit skeletal muscle (Tanabe *et al.*, 1987). This method is feasible only if there is enough channel protein for it to be isolated in a biochemically pure state, and is necessary only if we have little or no genetic information about the channel or its near relatives.

A quite different method is based on genetic analysis. If we know the position on the chromosome of the gene coding for a particular channel (or, more likely, for a particular mutant defect which might well be caused by a defective channel), then it is possible to sequence the genomic DNA in this region and look for segments that might act as suitable probes for our DNA library. This method was used in the identification of the *Shaker* potassium channels in *Drosophila* (Papazian *et al.*, 1987; Tempel *et al.*, 1987) and the human cystic fibrosis conductance regulator (CFTR; Riordan *et al.*, 1989; Rommens *et al.*, 1989).

A third method follows from either of these two. If we know the sequence for a particular channel then we can look for similar channels by probing a cDNA library using a low stringency hybridization method. This detects sequences that show some similarity to the original one but are not identical with it. In this way the various mammalian voltage-gated sodium channels were found, by homology with the electric eel sodium channel (Noda *et al.*, 1986; Kayano *et al.*, 1988), and whole groups of potassium channels related to the *Shaker* channel were found, both in *Drosophila* and in mammals (Wei *et al.*, 1990; Pongs, 1992).

Having got the right DNA clone it is then necessary to determine its sequence. Standard rapid methods for determining the sequences of lengths of DNA up to several hundred base-pairs long are commercially available. The DNA may first be broken into suitable lengths at precise points by using restriction enzymes. If necessary, the DNA can be replicated *in vitro* by PCR before sequencing.

Replication of DNA chains is usually remarkably free from error. On occasion, however, something goes wrong and the new DNA molecule is not identical with its parent. Such mutations are of course the raw material of evolution, since sometimes they may produce a new viable protein with desirable properties which is then selected for. Random mutations can be induced by ionizing radiation or chemical methods and have been used for many years in genetic analysis. One of the triumphs of the more recent molecular cloning methods has been to introduce specific changes into DNA molecules, so as to produce proteins with particular amino acids changed or removed (Smith, 1994). This *site-directed mutagenesis* has proved to be a powerful method for investigating ion channel function. If we think that a particular amino acid residue is crucial to some aspect of channel functioning, then we can delete it or change it to something else and see how this affects the properties of the channel. One method of doing this is outlined in fig. 4.3.

It is worth introducing here a convenient method of describing particular mutations. A particular residue can be described by its three-letter or single-letter code (table 4.2) and its position in the protein chain. For example, we shall see later that an important residue in the CFTR channel is the phenylalanine at position 508; we can call this Phe-508 or F508. The commonest mutation in the disease cystic fibrosis is the deletion of this residue, and this mutation is called ΔF508. Experiments using site-directed mutagenesis have been used to change the lysine (K) residues at positions 95 and 335 to aspartic acid (D) and glutamic acid (E) respectively, so these mutations are called K95D and K335E.

The nicotinic acetylcholine receptor channel

The nicotinic acetylcholine receptor (nAChR) channels of the postsynaptic regions of vertebrate muscle cells are among the most well known of channels. Electrical stimulation of the nerve supplying a muscle releases enough acetylcholine to combine with large numbers of nAChRs. Their channels open and allow sodium and potassium ions to flow through, and the resulting depolarization acts as a trigger for the activation of voltage-gated channels

Fig. 4.3. Site-directed mutagenesis. The plasmid single-stranded DNA template contains a cloned gene (heavy line). A synthetic DNA oligonucleotide contains a mutant codon flanked by regions complementary to the cloned gene. After annealing to the template the rest of the plasmid is made double stranded, producing a hybrid with a mismatch in the mutated region. Replication in *E. coli* produces double-stranded copies of the wild-type (WT) and mutant (M) forms. The mutant carries the original mutant codon (white star) and its complementary base sequence (black star). (From Nicholl, 1994.)

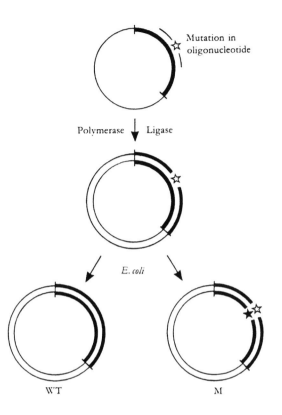

and ultimately for muscular contraction. They are called nicotinic receptors because they can also bind nicotine. (There is another type of acetylcholine receptor that will bind muscarine but not nicotine, and is hence called the muscarinic receptor; it does not have an intrinsic ion channel as part of the molecule.) Very similar nAChR channels are found in the electric organs of fishes.

The electric organs of the electric ray *Torpedo* provide a rich source of nAChRs. They can be isolated by utilizing their specific binding to the snake venom α-bungarotoxin. By 1980 protein biochemists had shown that the receptors are pentameric proteins (i.e. they consist of five subunits that are separate polypeptide chains) with a total molecular weight of about 290 000. The subunits are called, in order of their increasing molecular weights, the α (alpha), β (beta), γ (gamma) and δ (delta) chains. There are two α chains in each receptor and one of each of the others. Direct sequencing of the first 50 or so amino acid residues in each of the subunits showed that they are homologous, with similar but not identical amino acid sequences (Raftery *et al.*, 1980). The binding sites for acetylcholine are located on the α chains (see Claudio, 1989; Changeux *et al.*, 1992).

The nAChR from *Torpedo* electric organ was the first ion channel to have its amino acid sequence determined. Some of the first results were published in 1982 by a team led by Shosaku Numa at Kyoto University (Noda *et al.*, 1982). They used cDNA cloning methods to determine the base sequence of the α subunit mRNA, and from this they could determine the amino acid sequence of the subunit itself. Sequences for the other three subunits soon followed, and confirmatory sequences were published by teams at the Salk Insitute in California and the Institut Pasteur in Paris (Claudio *et al.*, 1983; Devillers-Thiéry *et al.*, 1983; Noda *et al.*, 1983a, b).

Numa's team began by isolating 2.3 μg of mRNA from the electric organ of *Torpedo californica*, and made cDNA from it with reverse transcriptase. At this stage the cDNA would be in the form of many thousands of different sequences up to a few kilobases long, and only a very small number of them would be derived from mRNA coding for nAChR subunits. These cDNA molecules were incorporated into plasmids containing a gene for ampicillin resistance, and the plasmids were introduced into *E. coli* cells. The *E. coli* culture was then plated out and grown on a medium containing ampicillin, so that only those cells into which the plasmids had been successfully incorporated would survive. The result was a library of about 200 000 transformant colonies, each one containing its own particular cloned cDNA. To find which of these was derived from mRNA coding for the acetylcholine receptor, replicas of the plates were screened by hybridization at 40 °C with two radioactive oligonucleotide probes, synthetic lengths of single-stranded DNA 15 and 18 bases long that would code for a short section of the amino acid sequence as determined by the protein chemists. Colonies identified by this method contained plasmids with a single cDNA species about 2000 bases long. The cDNA was extracted and cut with restriction enzymes to give overlapping shorter lengths for sequencing.

The amino acid sequence is readily deduced from the cDNA base sequence by application of the genetic code. Sequences of the four subunits are shown in fig. 4.4*A*. There is very considerable identity between them; in many cases we find the same amino acid at corresponding points in two or more of the chains. This suggests that the different subunits are similar in structure, and also that they have a common evolutionary origin. The mature α, β, γ and δ chains contain 437, 469, 489 and 501 amino acid residues, respectively. The calculated molecular weight for the α chain is just over 50 000 and that for the whole $\alpha_2\beta\gamma\delta$ complex is nearly 270 000; glycosylation (connection of sugars to the chain) and other modifications brings this up to a total of about 290 000.

Interpreting the sequence

What can the amino acid sequences of the protein chains tell us about their structure? Information from proteins whose structure is relatively well known

Fig. 4.4. (right) *A* shows amino acid sequences of the subunits of the nicotinic acetylcholine receptor from the electric organ of the electric ray *Torpedo californica*. Amino acids are shown in the single letter code (see table 4.2). Residues at the beginning (<) and end (>) of the four putative transmembrane segments M1 to M4 are shown. *in* and *ex* refer respectively to intracellular and extracellular portions of the protein chains. Residues that are labelled by acetylcholine binding site-directed reactions are underlined. The signal sequences, of up to 24 hydrophobic residues, that are removed from the N terminus after the protein chain has been inserted in the membrane, are not shown. The inset, *B*, shows the probable transmembrane topology of the polypeptide chain; the putative membrane-crossing segments M1 to M4 are indicated. (From Karlin, 1993. Reprinted with permission from *Current Opinion in Neurobiology*, copyright Current Biology Ltd.)

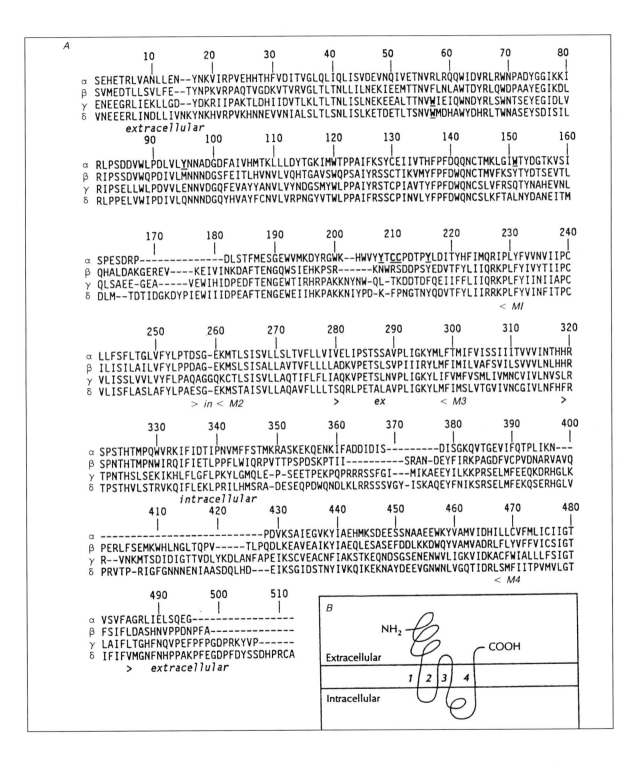

can give us some clues. Cysteine residues may pair to form disulphide bridges, which provide firm links between two different parts of the protein chain. N-glycosylation (binding of sugars via $-NH_2$ groups) may occur at asparagine residues in an Asn-X-Ser/Thr sequence exposed on the outside of the cell. Phosphorylation may occur at serine or threonine residues in a consensus sequence such as -Arg-Arg-X-Ser/Thr- or similar, if this part of the chain is exposed to the cytoplasm.

The 20 amino acid residues of protein chains have different properties. Two of them are basic, with positive charges, two are acidic, with negative charges, eight are polar but not charged, and eight are non-polar (table 4.2). These properties affect the way in which the protein chain is folded. Non-polar residues tend to occur in the middle of the molecule or in association with lipids in the cell membrane. Polar and charged residues are more likely to be found on the outside of the molecule in contact with the aqueous environment. Hydrogen bonds, electrostatic interactions and disulphide links all serve to hold the chain in its folded configuration.

One particular configuration of the protein chain that is important for our understanding of membrane protein structure is the α-helix (fig. 4.5). Here the protein chain forms a helix held in position by hydrogen bonds connecting the peptide links of its backbone; the carbonyl oxygen atom in each peptide bond is hydrogen-bonded to the hydrogen on the amide group of the third amino acid residue away. The amino acid residues emerge radially from the helical backbone. Since there are as many hydrogen bonds as there are amino acid residues, and since they connect atoms in the backbone of the protein chain rather than in its side-chains, a length of α-helix is a surprisingly stable structure. The helix is right-handed, with a rise per residue of 1.5 Å and a pitch height of 5.4 Å, corresponding to 3.6 residues per turn. Some amino acid residues are more likely to be found in α-helices than others; thus alanine, glutamate, leucine and methionine are good α-helix formers, whereas proline, glycine, tyrosine and serine are poor ones (Branden & Tooze, 1991; Creighton, 1993).

Twenty successive hydrophobic amino acid residues would produce an α-helix 30 Å long. This should be sufficient to cross the cell membrane from one side to the other. Direct evidence that α-helices really do form transmembrane segments in membrane proteins is derived from just two proteins at present. The reason for this sparsity is that high resolution studies on molecular structure require large numbers of molecules in the same orientation, such as are found in crystals, and membrane proteins do not normally form crystals. An exception is the photosynthetic reaction centre from the purple bacterium *Rhodopseudomonas*, in which X-ray diffraction studies show that there are 11 membrane-crossing α-helices in the molecule (Deisenhofer *et al.*, 1985).

The other example where we have some evidence that α-helices can form

Fig. 4.5. Structure of the α-helix. C, N and O atoms are labelled, hydrogen atoms are shown as small spheres. The positions of the amino acid side-chains are represented by the spheres R. Hydrogen bonds are dotted lines. (From Yudkin & Offord, 1980.)

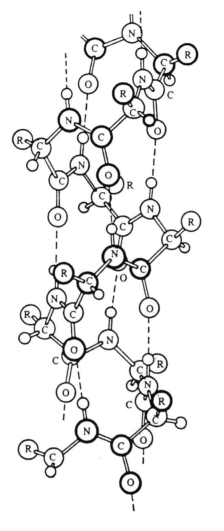

membrane-crossing structures is bacteriorhodopsin, which occurs in highly ordered arrays in the purple membranes of *Halobacterium*. Image reconstruction methods on electron micrographs of these arrays show seven membrane-crossing columns in each molecule. These agree well with the occurrence in the amino acid sequence of seven segments of 20 or so non-polar amino acid residues which would be expected to form α-helices of the requisite length (Henderson & Unwin, 1975; Dunn *et al.*, 1987; Henderson *et al.*, 1990).

A quite different membrane-crossing structure is found in bacterial porins, proteins that contain rather large pores that allow diffusion of small solutes across the outer membrane of Gram-negative bacteria. Their structure is based on the β-sheet conformation, where adjacent chains of amino acid

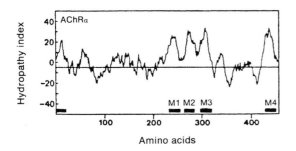

Fig. 4.6. Hydropathy profile of the nicotinic acetylcholine receptor (nAChR) α subunit. The horizontal axis shows the position of amino acid residue *i* in the chain, the vertical axis shows the sum of the hydropathy indices (see table 4.2) from *i* − 8 to *i* + 8. Black bars show the positions of the hydrophobic regions: the signal sequence at the left and the four putative membrane-crossing segments M1 to M4. The signal sequence is removed in the mature protein. (From Schofield *et al.*, 1987. Reprinted with permission from *Nature* **328**, p. 223, Copyright 1987 Macmillan Magazines Limited.)

residues are linked to each other, again via hydrogen bonds between the CO and NH groups of their peptide bonds. X-ray crystallography of porin crystals from *Rhodobacter* shows that in each subunit the protein chain forms a β-barrel, a set of 16 antiparallel β-chains hydrogen-bonded to each other so as to form a cylinder, as is shown in fig. 4.28 (Weiss *et al.*, 1991). It may well be that β structures occur in the membrane-crossing regions of channels to a greater extent than was previously thought (Unwin, 1993a,b).

In order to locate the hydrophobic membrane-crossing segments in the amino acid sequence of a protein, we need a quantitative measure of just how hydrophobic or hydrophilic the 20 different amino acids are. One such measure is provided by Kyte & Doolittle's (1982) hydropathy index, shown in table 4.2. The scale runs from an index of +4.5 for isoleucine (which is very hydrophobic) to −4.5 for arginine (which is strongly charged and very hydrophilic). A useful approach for membrane proteins is to measure the mean hydropathy of a window of say 19 adjacent amino acid residues (the expected length of a membrane-crossing α-helix), and to move this window one residue at a time along the length of the protein chain. This produces a graph (a hydropathy or hydrophobicity plot) in which the peaks are regions with a succession of hydrophobic residues and the troughs indicate hydrophilic ones.

Figure 4.6 shows a hydrophobicity plot for the α subunit of the nAChR; plots for the other subunits are very similar. The original interpretation of this was that in each subunit the protein chain crosses the membrane four times, at hydrophobic regions labelled M1 to M4, and that both N and C termini are on the extracellular side of the membrane. Different models of just how the chain is folded across the membrane have been considered at various times, but the original scheme with four putative membrane-crossing segments (fig. 4.4*B*) seems to be the most generally accepted arrangement (Claudio, 1989; Changeux *et al.*, 1992). The use of the word 'putative', meaning 'commonly supposed to be', recognizes the indirect nature of the evidence about these segments.

As in all proteins, synthesis of the amino acid chain proceeds in a linear fashion from the N terminus to the C terminus. The first part to be formed is a sequence of 16 hydrophobic amino acid residues that is absent from the mature nAChR protein. This group is called the signal sequence; it seems to be used to insert the beginning of the chain into the membrane so that it can be fed through as it is synthesized. The signal sequence is removed by enzymic action once the chain is completed, allowing the N terminus of the mature protein to take its place on the outside of the cell.

The long hydrophilic section of over 200 amino acid residues from the N terminus to the beginning of the M1 segment is probably all on the outside of the cell, facing the synaptic cleft. It contains sites for glycosylation and disulphide cross-linking, and in the α subunit it contains the acetylcholine and α-bungarotoxin binding sites. Then, proceeding along the chain, we have the M1, M2 and M3 putative membrane-crossing segments, with short connecting sections between them. There is a fairly extensive cytoplasmic section between the M3 and M4 segments, which contains sites for the attachment of phosphate groups. Finally there is a short extracellular section finishing at the C terminus.

Transmembrane topologies different from that shown in fig. 4.4*B* have been proposed from time to time. One of these suggested that an amphipathic MA segment, part of the sequence between M3 and M4, crossed the membrane and lined the channel pore; this would imply that the C terminus was cytoplasmic (Finer-Moore & Stroud, 1984). Another postulated the existence of extra membrane-crossing segments before M1 (Criado *et al.*, 1985). Evidence that seemed to confirm some of these suggestions came from experiments using antibodies to particular parts of the polypeptide chain; application of the antibody to the extracellular or cytoplasmic surface ought to indicate aspects of the transmembrane topology. However, there are problems with this immunolocalization methodology, especially regarding the specificity of the antibodies used, and the results are not always self-consistent (McCrea *et al.*, 1988).

Further evidence for the generally accepted model of the transmembrane topology of the receptor subunits comes from various sources. Glycosylation sites are necessarily on the outside of the membrane, as are the neurotransmitter and toxin binding sites. When extra potential glycosylation sites were introduced into the N terminus region by site-directed mutagenesis, they were in fact glycosylated (Chavez & Hall, 1991). The phosphorylation sites in the M3–M4 link must necessarily be cytoplasmic (Huganir, 1988). Evidence that the C terminus really is extracellular comes from the δ subunit, which in *Torpedo* possesses a cysteine residue near its end and which forms a disulphide link so as to form nAChR dimers; this disulphide link can be broken by enzymes applied to the extracellular surface only (DiPaola *et al.*, 1989).

An ingenious method of determining membrane protein topologies was

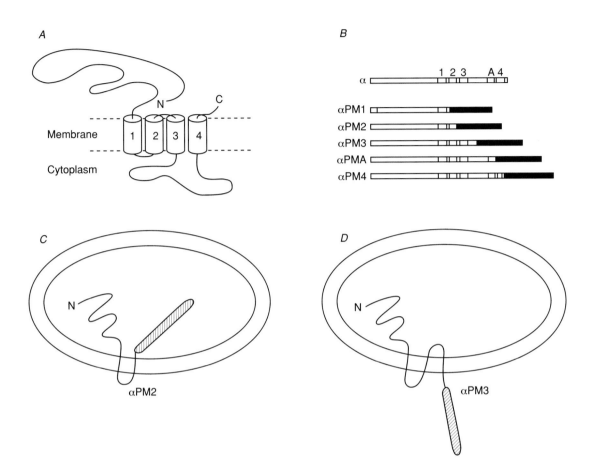

applied to the nAChR by Chavez & Hall (1992). They made fusion proteins in which a fragment of the hormone prolactin replaced various C-terminal lengths of nAChR subunits, and expressed them in pancreas microsomes using an *in vitro* translation system. The microsomes are made by homogenizing pancreatic secretory cells and are vesicles formed from endoplasmic reticulum membrane; the lumen (inside) of the microsome is topologically equivalent to the extracellular space. Thus, in some fusion proteins the prolactin section would be on the cytoplasmic face of the membrane, and accessible to a proteolytic enzyme, whereas in others it would be inside the microsome and so inaccessible to proteolysis (see fig. 4.7). Recovery of a polypeptide containing the prolactin fragment (identified by immunoprecipitation, for which specific and reliable antibodies are available) would thus show that it had not been attacked by the enzyme and therefore that the point of attachment of the prolactin fragment was on the extracellular face of the membrane. The method is sometimes called an epitope protection assay, since the prolactin fragment acts as an epitope (a recognition and binding region) for the antibody.

Fig. 4.7. (left) The epitope
protection method for
determining transmembrane
topology of the nAChR. *A*
shows the generally accepted
transmembrane topology as
predicted by the
hydrophobicity analysis. *B*
shows the fusion proteins, in
which a prolactin fragment
(black) is attached downstream
of part of the α subunit chain,
just after the four putative
membrane-crossing segments
M1 to M4 and also after the
amphipathic MA segment.
These were inserted into
pancreatic microsomes, in
which the inside of the
microsome corresponds to the
extracellular face of the
channel; *C* shows the expected
topology for αPM2 and *D*
shows that for αPM3, with the
prolactin fragment cross-
hatched. So a proteolytic
enzyme would remove the
prolactin fragment from αPM3
but not from αPM2, indicating
that the C-terminal end of M3
is cytoplasmic whereas that of
M2 is extracellular. In the
experiments, prolactin
fragments were still present
after proteinase digestion for
the αPM2 and αPM4 fusion
proteins, but not for αPM1,
αPM3 and αPMA. Similar
results were obtained with the
δ subunit, showing that the
model in *A* is correct. (*A* and *B*
from Chavez & Hall, 1992.
Reproduced from the *Journal
of Cell Biology* 1992, **116**, pp.
385–93, by copyright
permission of The Rockefeller
University Press.)

The results of this experiment were quite clear. If the prolactin domain was attached just after the M2 or M4 segments, it could be recovered, whereas if it was attached just after the M1, M3 or MA segments, it could not. So the model shown in figs. 4.4*B* and 4.7*A* really does seem to be the correct one.

There is strong evidence that the M2 α-helix lines the channel pore at its narrowest. It is possible to label the nAChR with radioactive molecules of the non-competitive blocking agent chlorpromazine, whose action appears to be due to binding to the walls of the channel pore. The label becomes attached to certain amino acids in the M2 segments, suggesting that they line the pore (Revah *et al.*, 1990; see Changeux *et al.*, 1992). Mutations introduced into the M2 sequence produce marked effects on the permeability of the ionic channel, as we shall see in the next chapter.

Quaternary structure

The amino acid sequence of a protein is sometimes known as its primary structure. The term secondary structure applies to the arrangement of short lengths of the chain, as in α-helix and β-sheet structures. Tertiary structure applies to the arrangement of the whole chain in a single subunit, and quaternary structure applies to the overall shape of the whole molecule complex, including all its subunits.

Electron microscopy of *Torpedo* electric organ postsynaptic membranes gives us some idea of the quaternary structure of the receptors. They appear to consist of a number of subunits arranged round a central pit. We would expect there to be five subunits and some approach to pentamerous symmetry. This has been confirmed by applying a quantitative image analysis technique to electron micrographs of artificial tubular crystalline arrays of receptors embedded in ice (Brisson & Unwin, 1985; Toyoshima & Unwin, 1988; Unwin, 1993a, 1995).

The results of this technique are in the form of three-dimensional contour maps of image density, as is shown in fig. 4.8 and on the cover of this book. They show that the receptor is a pentameric structure with its axis of pseudosymmetry at right angles to the plane of the cell membrane. Much of the mass of the receptor is clearly on the outside of the cell, projecting into the synaptic cleft. This outer part is hollow, with a central entrance tube 20 to 25 Å in diameter, leading into a transmembrane pore that is too narrow to be precisely resolved by the image analysis method.

Application of the imaging technique at its highest resolution has produced some surprises. α-Helices will appear in the contour maps as dense rods. In the bilayer-spanning part of the receptor, there is only one dense rod per subunit, and this is very close to the wall of the transmembrane pore. This must be the M2 segment, which we know lines the pore, confirming that it really is in the

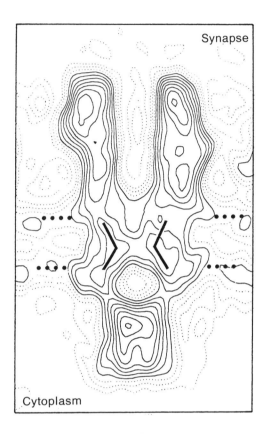

Fig. 4.8. Quaternary structure of the nAChR as determined by Fourier transform analysis from diffraction patterns of multiple electron microscope images. Solid contours show electron density of the receptor. Large dots show the limits of the membrane bilayer, 30 Å apart. The bent transmembrane rods, thought to be the M2 segment, are shown. The square structure at the bottom is a protein attached to the receptor but not part of it. (From Unwin, 1993a.)

form of an α-helix. The maps show that the M2 segment is kinked at its middle, and the kink apparently forms the narrowest part of the pore.

The other material at the level of the bilayer, as displayed by the imaging technique, does not appear to be in the form of α-helices. Unwin (1993a) suggests that the M1, M3 and M4 segments form a β-barrel structure of antiparallel strands holding the whole receptor together. These segments are hydrophobic in nature and are clear candidates for the material that surrounds the five M2 α-helices at the bilayer level; perhaps they each consist of one or three β-sheet strands. Such an arrangement is found in certain bacterial toxins, where a pentameric doughnut-shaped structure is formed in which five α-helices surround a central pore and are themselves surrounded and held together by a β-barrel of 30 antiparallel β-sheet strands (Sixma *et al.*, 1991; Stein *et al.*, 1992).

Unwin points out a design advantage for the β-barrel model: a common plan could be utilized by a wide range of channel subtypes. Subunits would associate by hydrogen bonding between the polypeptide backbone groups, rather than by bonding between the residues of adjacent α-helices. This would permit greater variation in amino acid sequence while still retaining the essential ground plan.

Evidence that is difficult to reconcile with the β-barrel model has been provided by some studies with 1-azidopyrene (1-AP), a hydrophobic fluorescent probe that will bind to proteins when activated by light. 1-AP bound to the M1, M3 and M4 segments in all subunits of the *Torpedo* nAChR, supporting the view that these three segments are in contact with the lipid part of the bilayer. Another lipid-soluble label known as TID (3-trifluoromethyl-3-(-*m*-[^{125}I]iodophenyl)diazirine) bound specifically to the Cys-412, Met-415, Cys-418, Thr-422 and Val-425 residues of the α subunit M4 segment, and to some of the corresponding residues in other subunits. The periodicity of these binding sites is just what one would expect to see if the label became attached to one side of an α-helix, whereas a β structure would produce binding sites at every second residue. Similar TID binding periodicities were found in the M3 segment, suggesting that it too is an α-helix in contact with the lipid bilayer (Blanton & Cohen, 1992, 1994). So why are α-helices not evident in the image analysis maps? Blanton & Cohen suggested that their position at the protein–lipid interface could make their structures less highly ordered than at the central axis of the receptor.

Use of Fourier transform infrared spectroscopy suggests that 40% of the membrane-crossing region is in the β-structure form and 50% is α-helical (Görne-Tschelnokow *et al.*, 1994). One interpretation of this is that the M2 and M4 segments are α-helices and the M1 and M3 segments are β-sheet strands. Clearly the long discussion about just how the nAChR is put together is not yet at an end (Popot, 1993; Hucho *et al.*, 1994).

A large part of the nAChR projects into the synaptic cleft on the outside of the cell. It is made up of the long N-terminal regions of all five subunits, and must contain the two acetylcholine binding sites in the α subunits. The high resolution image analysis revealed three dense rods in the synaptic part of each subunit, and these are presumably α-helices (Unwin, 1993a). It seems likely that in the α subunits they are involved in the acetylcholine binding site (see chapter 6).

Expression in oocytes

The oocytes of the African clawed frog *Xenopus* have provided a most useful test-bed for looking at the functioning of many channel proteins. Oocytes are large cells which are just about to become mature eggs ready for fertilization. They have all the normal translation machinery of living cells and so they will respond to the injection of mRNA by making the protein for which it codes.

Figure 4.9 shows the results of an experiment in which nAChRs are expressed in an oocyte. First, mRNA from *Torpedo* electric organ is injected into the oocyte. A few days later it responds to application of acetylcholine with an inward flow of current that is very similar to what is produced in the

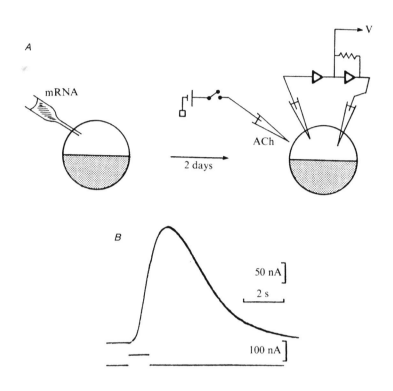

Fig. 4.9. Expression of nAChR in *Xenopus* oocyte. Messenger RNA from *Torpedo* electric organ is injected into an oocyte (*A*). Two days later the oocyte membrane is voltage clamped while acetylcholine is applied by iontophoresis. The record (*B*) shows the inward current in response to acetylcholine; the lower trace monitors the iontophoresis current. (From Barnard *et al.*, 1982.)

electrocytes of the fish by nervous stimulation (Barnard *et al.*, 1982). Patch clamp records from the oocyte surface membrane show single channel currents just like those from native cell membranes.

Injection of their individual mRNAs allows the role of the different nAChR subunits to be investigated. Only when mRNAs for all four *Torpedo* subunits were injected into it would the oocyte later respond fully to the application of acetylcholine (Mishina *et al.*, 1984). This suggests that all four subunits are needed to make a functioning receptor. It also shows that no extra components are required, and thus provides pleasing confirmation for the conclusions of the recombinant cDNA work.

The Kyoto University group used site-directed mutagenesis to look for sensitive regions in the α subunit (Mishina *et al.*, 1985). They found that deletions from any of the M1 to M4 segments removed the response to acetylcholine. The response was also removed by mutations at the asparagine residue at 141 and the cysteine residues at 128, 142, 192 and 193. Asn-141 is important in glycosylation, and the cysteine residues will form disulphide links; both features are probably crucial to the structure of the acetylcholine binding site. For deletions elsewhere in the chain the response was either normal or reduced.

Expression in mammalian cell lines

Instead of the somewhat laborious process of injecting small quantities of mRNA into individual oocytes, it may be desirable for some purposes to introduce DNA coding for the nAChR into large numbers of mammalian cells in tissue culture. This process of *transfection* involves the use of a vector such as a plasmid that is introduced into the cells by various tricks such as calcium phosphate treatment, electric shocks or encapsulation in liposomes. Incorporation of plasmids into the cytoplasm leads to transient transfection, in which a high proportion of the cells express the nAChR, but this feature is lost over succeeding generations as the plasmids are not replicated when the cells divide. A much lower proportion of cells may incorporate the DNA into their chromosomes. If the vector contains some suitable marker gene then these cells can be selected for by using an appropriate growth medium, leading to the establishment of a stable cell line in which all the cells and their descendants will continue to express the nAChR.

The transient transfection method has been used by Kreienkamp and his colleagues (1995) to introduce various chimeric and wild-type (unaltered) nAChR subunits into cells from a human embryonic kidney line, so as to investigate their assembly into the complete receptor or parts of it. The large number of cells involved allowed them to harvest the expression products. They found that α and γ subunits would together produce $\alpha\gamma$ dimers or $\alpha\gamma\alpha\gamma$ tetramers, whereas α and δ subunits together would produce only $\alpha\delta$ dimers. This suggests that the γ subunit can bind to the α subunit on two of its sides, and thus that it is probably the γ subunit that sits between the two α subunits in the complete receptor.

Stable transfection of *Torpedo* nAChR into cells has been achieved by Claudio and her colleagues (1987) using a mouse fibroblast line. Such cells can be used for a variety of pharmacological and biochemical experiments (see Claudio, 1989) and have been used for detailed investigations on nAChR channel kinetics (as in fig. 3.19), assembly (e.g. Green & Claudio, 1993) and immunology (Loutrari *et al.*, 1992).

Diversity among nicotinic acetylcholine receptors

The amino acid sequences of corresponding subunits from related animals show a high degree of similarity or homology (Numa, 1986). The α subunit chains from humans and calves, for example, show 97% identity (i.e. the same amino acid occurs in both chains at 97% of the positions), whereas those from humans and *Torpedo* show only 80% identity. This reflects the fact that the evolutionary divergence between humans and cows is much more recent than that between humans and cartilaginous fish. Corresponding figures for the γ

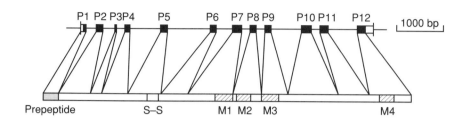

Fig. 4.10. Structural organization of the human nAChR γ subunit gene. The upper line represents a length of chromosomal DNA about 7000 base-pairs long. It has 12 exons, P1–P12, which code for various regions of the protein chain as indicated. Positions of the prepeptide signal sequence, the disulphide bridge and the four putative membrane-crossing segments (M1–M4) are shown. (From Shibahara *et al.*, 1985).

subunit are 92% and 55%, and the figures for the β and γ subunits are similar, suggesting that evolutionary changes in the α subunit have been slower than in the other subunits. The presence of the acetylcholine binding site must make the α subunit less tolerant of mutational changes than are the other subunits.

Homologies between the different subunits in one species are usually less than those between the same subunit in different species. The *Torpedo* α and β subunits, for example, show only 42% identity. This suggests that they diverged (probably as a result of gene duplication) at an early stage in the evolutionary history of animals. A rather different subunit, called the ε (epsilon) subunit, occurs in adult mammals; it takes the place of the γ subunit, which is present in foetal mammals (Takai *et al.*, 1985).

Genes for nAChR subunits contain appreciable quantities of non-coding DNA as introns. The human α subunit gene, obtained by screening a human genomic DNA library with calf nAChR cDNA probes, contains nine exons and the γ subunit gene contains 12, as is shown in fig. 4.10 (Noda *et al.*, 1983c; Shibahara *et al.*, 1985). Chicken subunit genes have similar arrangements. Their exons show strong homology with those of mammalian species, but their introns do not. This is to be expected since there is little or no evolutionary selection against the accumulation of mutations in introns, whereas most mutations in exons are likely to result in non-functional proteins. Genes for the different subunits may be on different chromosomes; in the mouse α is on chromosome 17, β is on 11, and γ and δ are on 1 (Heidmann *et al.*, 1986).

So far we have been considering nAChRs found in electric organs and muscle. There are also, however, nicotinic acetylcholine receptors in the nervous system, and these have a rather different make-up (Role, 1992; Sargent, 1993). These neuronal nAChRs seem to have just two general types of subunits: those which bind acetylcholine, called α subunits, and those which do not, called non-α or β subunits. The distinction is usually made on the basis of the presence of adjacent cysteines a few residues before the start of the M1 segment, corresponding to Cys-192 and Cys-193 in the *Torpedo* α subunit. These paired cysteines are assumed to be diagnostic of the acetylcholine binding site.

Neuronal nAChR sequences have been found by probing either cDNA

Fig. 4.11. Evolutionary tree showing the probable relationships of various nAChR subunits, based mainly on computerized comparisons of their amino acid sequences. The time scale at the top (MYA, million years ago) estimates the times of their divergence. Roman numerals show subfamilies diagnosed on the basis of gene structure by comparisons of the positions of exons. Muscle subunits are labelled M and neuronal subunits are labelled N. (From Le Novère & Changeux, 1995. © Springer-Verlag New York Inc. 1995.).

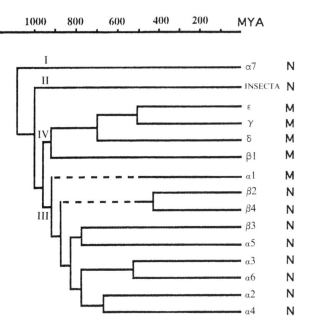

libraries made from brain mRNAs or libraries of genomic DNA. At least seven different α subunits exist (named α2 to α8, while α1 is the corresponding muscle subunit), and at least three different β subunits. Messenger RNAs coding for these can be localized in sections of nervous tissue by using autoradiography after *in situ* hybridization with appropriate radioactive antisense oligonucleotide probes; different subunits have different distributions in the brain (Wada *et al.*, 1988). It seems likely that neuronal receptors are usually made up of two α and three β subunits (Cooper *et al.*, 1991), but α7 can form functional homo-oligomer channels just on its own (Couturier *et al.*, 1990). A further subunit, α9, forms homomeric receptors that appear to mediate the cholinergic efferent nerve action on the sensory hair cells of the cochlea (Elgoyhen *et al.*, 1994).

Since nAChRs are found in nematodes and insects as well as in all the classes of vertebrates, they form an ancient group of molecules and the evolutionary divergence of their subunits may have started over a billion years ago. We may speculate that the original nAChR in some of the earliest multicellular animals was a homo-oligomer composed of identical subunits all coded for by a single gene. Gene duplication was probably the trigger for the subsequent divergence of the different subunits. Detailed comparison of over 60 sequences has led to the evolutionary tree shown in fig. 4.11. It suggests that the α7 subunit diverged from the rest at an early stage, and that the vertebrate muscle α subunits are perhaps more closely related to vertebrate neuronal subunits than to the other muscle subunits (Le Novère & Changeux, 1995).

Other neurotransmitter-gated channels

The nicotinic acetylcholine receptor is just one of a whole group of ion channels that are opened directly by the action of neurotransmitters. These channels are concerned with fast synaptic transmission, especially in the central nervous system. They include the anion-selective channels activated by γ-aminobutyric acid (GABA) and by glycine that produce inhibitory responses, the excitatory cation-selective channels operated by 5-hydroxytryptamine (5-HT) and by glutamate, and probably also the channels activated by ATP acting as a neurotransmitter.

It has been one of the exciting findings of recent years that some of these channels possess structural features in common with each other. The GABA, glycine and 5-HT_3 receptor channels show considerable sequence identity with the nAChR channel, hence they can be grouped together with it as a family. Their subunits are similar in size and have a common secondary structure, with four hydrophobic putative membrane-crossing segments, a large extracellular N-terminal section, and a large cytoplasmic section between M3 and M4. The various glutamate receptor channels have rather different sequences and a different transmembrane topology and so form a separate family. The P_{2X} receptor, responsive to ATP acting as a neurotransmitter, is different again and quite distinct in structure.

(GABA, 5-HT and glutamate can also produce much slower synaptic responses by acting via different receptors that do not contain intrinsic ion channels. GABA, for example, acts via *ionotropic* GABA_A receptors for fast responses, and via metabotropic GABA_B receptors for slow responses. *Metabotropic* glutamate receptors, GABA_B receptors, and 5HT_1, 5HT_2 and 5HT_4 receptors all produce their responses by activating G proteins. Each receptor is a single polypeptide chain with seven transmembrane α-helices and no intrinsic ion channel.)

GABA and glycine receptor channels

Fast inhibitory transmission in the mammalian central nervous system involves the opening of chloride-selective channels in GABA_A or glycine receptors. GABA_A receptors are widely distributed throughout the brain, whereas glycine receptors are largely restricted to the brain stem and spinal cord. There is good chemical evidence that the glycine receptor has five subunits, just like the acetylcholine receptor (Langosch *et al.*, 1988). The similarity in subunit secondary structure among members of the nAChR-related family suggests that they all have a similar pentameric arrangement with five subunits grouped round a central pore.

A remarkable feature of these channels is the diversity of subunit make-up that they show. GABA$_A$ receptors contain two main types of subunit, α and β, with additional γ and δ chains. A fifth type of chain, ρ (rho), is found in the retina. But there are at least six different varieties of the α chain, four of the β chain, three of the γ chain and two of the δ chain. The amino acid sequences show 70 to 80% identity between different members of the same subunit type, 30% to 40% between different types (Barnard, 1992; Macdonald & Olsen, 1994).

Complete GABA$_A$ receptors in the nervous system probably contain both α and β chains together with either γ or δ. Different combinations of subunits are found in different parts of the nervous system. Motor neurons in the spinal cord, for example, have GABA$_A$ receptors with the $\alpha2$, $\beta3$ and $\gamma2$ subunits, whereas many brain neurons show the $\alpha1$, $\beta2$ and $\gamma2$ combination.

Glycine receptor channels are composed of α and β subunits, probably forming a pentamer in the ratio $\alpha_3\beta_2$ (Langosch et al., 1988). There are at least four different types of α subunit, showing over 80% sequence identity with each other, and about 47% sequence identity with the β subunit (Kuhse et al., 1991). The antagonist strychnine binds to the α subunits.

The 5-HT-gated channel

The neurotransmitter 5-hydroxytryptamine (serotonin) exerts its varied actions by activating at least four different receptor types. Three of these produce relatively slow responses mediated via G proteins and second-messenger systems, but the 5HT$_3$ receptor is a ligand-gated ion channel and so produces fast synaptic responses (Derkach et al., 1989). Messenger RNA from a hybrid line of mouse/hamster brain-derived cells produces 5HT$_3$ receptors when injected into *Xenopus* oocytes. The receptor amino acid sequence shows 27% identity with that of the *Torpedo* nAChR α subunit, and has a generally similar structure, with the usual set of four transmembrane α-helices, as is shown in fig. 4.12 (Maricq et al., 1991). The expression in oocytes shows that a single subunit type is sufficient to produce functioning channels (probably as homopentamers, we may suppose), but it may well be that other subunits are involved as well *in vivo*.

Glutamate receptor channels

The great majority of the synapses for fast excitatory transmission in the vertebrate central nervous system use glutamate as the neurotransmitter to

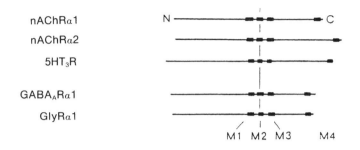

nAChRα1

nAChRα2

5HT₃R

GABAₐRα1

GlyRα1

M1 M2 M3   M4

Fig. 4.12. Subunit lengths in the nAChR-related receptor family. Black bars show the position of the putative membrane-crossing segments M1 to M4. The nAChR α1 subunit is 437 amino acid residues long and the rest are drawn to scale. (From Unwin, 1993b. Reproduced with permission from the *Cell* **72**/*Neuron* **10** supplement, copyright Cell Press.)

activate ionotropic glutamate receptor channels. Such glutamatergic transmission is involved in some of the long-lasting changes in synaptic efficiency thought to underlie the processes of learning and memory. High glutamate levels may cause the death of neurons, and these neurotoxic effects may be involved in neurodegenerative conditions such as Huntingdon's and Alzheimer's diseases (see Collingridge & Lester, 1989; Monaghan *et al.*, 1989; Westbrook & Jahr,1989; Nakanishi & Masu, 1994).

The three major types of ionotropic glutamate receptors are named AMPA, kainate and NMDA receptors according to the agonists that will activate them. AMPA receptors are activated by α-amino-3-hydroxy-5-methyl-4-isoxazolepropionate and also by quisqualate; they show rapid responses and are permeable to sodium and potassium ions when their channels are open. Kainate receptors are similar to them in general properties and show 35% to 40% amino acid sequence identity with them, so the two groups are sometimes put together as AMPA-kainate receptors. NMDA receptors are activated by *N*-methyl-D-aspartate; they are permeable to sodium, potassium and calcium ions when activated and are blocked by magnesium ions. Their subunits show 22% to 24% sequence identity with those of AMPA or kainate receptors (Sommer & Seeburg, 1992). The block by extracellular magnesium ions in NMDA receptor channels can be removed by depolarization, hence they are more likely to be open in a depolarized cell. This feature, and also their permeability to calcium ions, may be important aspects of their functioning (Nowak *et al.*, 1984; MacDermott *et al.*, 1986).

Molecular cloning shows that the protein chains making up glutamate receptor channels are much longer than those of the nAChR family (fig. 4.13). The diversity of glutamate subunits is considerable (Nakanishi, 1992; Seeburg, 1993; Hollmann & Heinemann, 1994). AMPA receptor subunits are named GluR1 to GluR4 (or GluR-A to D), kainate receptor subunits are named GluR5 to 7. KA1 and KA2 are high-affinity kainate receptors. The KBP (kainate binding proteins) chains from chick and frog are much shorter chains than the other subunits; they have not been shown to form channels. DGluR-I and DGluR-II have been isolated from the fruit fly *Drosophila* and

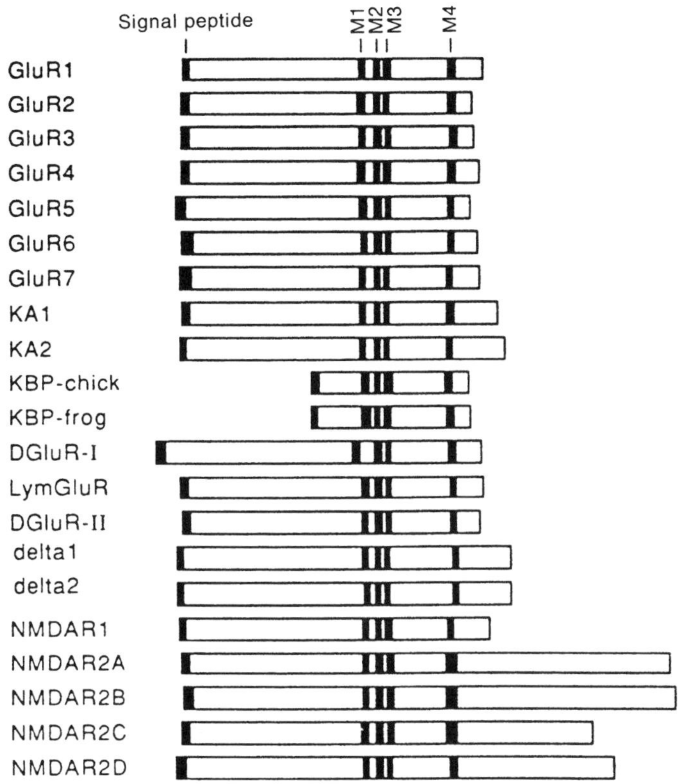

Fig. 4.13. Bar graph representations of 21 subunits of ionotropic glutamate receptors, aligned on their M2 segments. Putative membrane-crossing (or membrane-dipping for M2) segments are shown black. The GluR1 subunit is 889 amino acid residues long, and the rest are drawn to scale. (From Hollmann & Heinemann, 1994. Reproduced with permission from the *Annual Review of Neuroscience* **17**, © 1994 by Annual Reviews Inc.)

LymGluR comes from the pond snail *Lymnaea*. The delta1 and delta2 chains are found in the mammalian brain but have not been shown to form channels.

NMDA receptors are made up of NMDAR1 (also called NR1) chains, which exist in at least seven splice variants, and may or may not contain NMDAR2A to NMDAR2D chains (also called NR2A etc.). The four NMDAR2 subunits are rather larger in size (they have an extensive C-terminal region) and cannot form functional NMDA receptors without combining with NMDAR1 chains (Ishii *et al.*, 1993; McBain & Mayer, 1994). Heteromers of NMDAR1 with different NMDAR2 chains show differences in gating and sensitivity to magnesium block when expressed in cultured cells. Messenger RNAs coding for the different chains can be detected and localized in the brain by means of *in situ* hybridization experiments, using radioactive anti-sense oligonucleotide probes. The results show that NMDAR1 chains are found throughout the brain, but the different NMDAR2 chains are more localized: NMDAR2B is present in the rat cerebral cortex but not in the cerebellum, for example, whereas the reverse is true for NMDAR2C (Monyer *et al.*, 1992).

Diversity is further increased by alternative splicing. In the GluR1 to GluR4 subunits a section 38 residues long situated just before M4 is coded for by either one of two adjacent exons; these alternatives have been called the 'flip' and 'flop' segments. They alter the relative responses of the receptors to glutamate or AMPA and to kainate, and have different distributions in the brain (Sommer *et al.*, 1990). Alternative splicing of three exons in NMDAR1 subunits generates eight different variants, which are expressed differentially in different parts of the brain (Zukin & Bennett, 1995).

A remarkable further source of diversity occurs at a particular site in the M2 segment of the GluR2, GluR5 and GluR6 subunits (Sommer *et al.*, 1991). Here the genomic DNA codes for glutamine (Q) but the residue in the protein (as determined from cDNA made from mRNA) is often arginine (R). The position is thus called the Q/R site or (since asparagine occurs at the homologous position in NMDA receptors) the QRN site. The change appears to arise from the process of RNA-editing, whereby the mRNA sequence is altered by enzymic action so that the RNA codon for glutamine (CAG) is converted to that for arginine (CGG). With arginine at the site, the channel is not permeable to calcium ions, whereas with glutamine present it is (Hume *et al.*, 1991).

The precise way in which the subunit chain crosses the membrane has been the subject of many differing views in recent years, with disagreements about how many membrane-crossing segments there are and whether the C terminus is extracellular or cytoplasmic. The problem seems to have been resolved by using prolactin as a reporter in an epitope protection assay, as in the experiments described above for the nAChR (Bennett & Dingledine, 1995). The results show that the M2 segment does not cross the membrane from one side to the other, but loops into it so that both its ends are cytoplasmic (fig. 4.14). Confirmation of this model has been provided by introducing sites for glycosylation into the molecule at different points; only if these were extracellular would glycosylation occur. In GluR1 it was found that only M1, M3 and M4 were completely transmembrane, that the M3–M4 link was extracellular, and that the N terminus was extracellular and the C terminus cytoplasmic (Hollmann *et al.*, 1994). Similar results have been obtained for two kainate binding proteins from the goldfish (Wo & Oswald, 1994).

Some similarities have been detected between the M2 loop of glutamate channels and the H5 loop thought to line the pore of potassium channels (Wo & Oswald, 1995). It is as if the potassium channel H5 loop had been turned through 180° to enter from the cytoplasmic side of the membrane, and it is interesting that the NMDA channel is blocked by external magnesium ions whereas the inward rectifier potassium channel is blocked by internal magnesium.

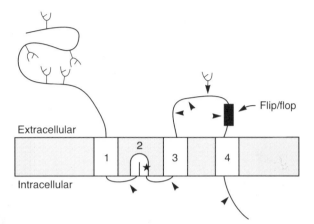

Fig. 4.14. The membrane-crossing topology of the GluR3 subunit as determined by the prolactin fusion protein method. The tree-like structures are N-linked glycosylation sites, so they must be extracellular; the tree with an arrow shows where a glycosylation site can be inserted, and so this position is probably extracellular. The M1, M3 and M4 transmembrane segments are shown. The star shows the Q/R site in the M2 membrane-dipping segment. Arrowheads show prolactin fusion sites, leading to the results that justify this model. (From Bennett & Dingledine, 1995. Reproduced with permission from *Neuron* **14**, copyright Cell Press.)

The P_{2X} ATP receptor channel

An external-ligand-gated channel of quite different design from that of the nAChR-related and glutamate receptor families has been detected in cDNA libraries prepared from smooth muscle mRNA (Valera *et al.*, 1994; Brake *et al.*, 1994). Extracellular ATP acts on smooth muscle cells via P_2 purinoceptors, of which the P_{2X} receptors are ion channels. The protein involved is 399 amino acid residues long. It appears to have two membrane-crossing segments separated by a large extracellular loop. An H5 region next to M2 shows considerable similarity to the H5 regions of various potassium channels. When expressed in *Xenopus* oocytes it forms channels that allow sodium and calcium ions to enter when they are activated by extracellular ATP.

It seems likely that the P_{2X} receptor is involved in fast neurotransmission with ATP as the transmitter substance (Edwards *et al.*, 1992; Surprenant *et al.*, 1995). Presumably the channel is made of a number of subunits surrounding a central pore.

Voltage-gated channels and their relatives

In their classic study of the action potential in squid giant axons, Hodgkin & Huxley (1952b) showed that the ionic current flow through the axon membrane was largely accounted for by changes in its conductance to sodium and potassium ions. Depolarization produced a rapid transient increase in sodium conductance and a slower maintained increase in potassium conductance. Hodgkin & Huxley's equations describing these changes were based on a model in which particular sites for ionic flow were opened as a result of membrane potential change. These sites we now call the voltage-gated sodium and potassium channels.

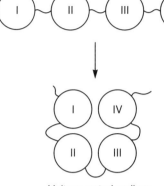

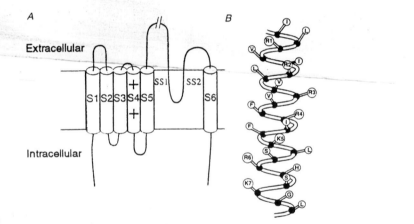

Fig. 4.15. Subunits and domains in voltage-gated channels. In voltage-gated potassium channels four separate protein chains are brought together as subunits to form the whole channel. In sodium and calcium channels the whole channel is made from a single long protein chain (encoded by a single mRNA molecule) that has four homologous but not identical domains.

Fig. 4.16. Characteristic structures in voltage-gated channels and their relatives. Each subunit or domain contains six transmembrane α-helices, shown in *A* as cylinders, with the rest of the nearby peptide chain drawn as a line connecting them. The membrane-associated segment SS1–SS2 (also called H5 or P) occurs between S5 and S6 and probably forms part of the lining of the pore. The S4 segment contains the positively charged amino acid residues arginine (R) or lysine (K) at every third position. This is shown for the *Shaker* potassium channel in *B*. (From Catterall, 1993. Reproduced from *Trends in Neurosciences*, with permission from Elsevier Trends Journals.)

Molecular studies in recent years have shown that they and similar channels form a superfamily of related proteins with a number of structural features in common. Not all members of the superfamily are gated solely by changes in membrane potential, and so partly for this reason the terms 'voltage-dependent' or 'voltage-sensitive' are sometimes used instead of 'voltage-gated' to describe them.

The channel pore in this superfamily is formed from the union of four similar subunits, or domains of a single protein chain in the case of sodium and calcium channels, as is shown in fig. 4.15. The four principal subunits or domains have a characteristic structure (fig. 4.16). In each of them there are six membrane-crossing segments, S1 to S6, which appear to be α-helices. There is also a section between S5 and S6, known as the SS1–SS2 loop or the H5 or P region, which probably (as we shall see in chapter 5) forms part of the lining of the pore. The S4 segment has a remarkable structure: every third

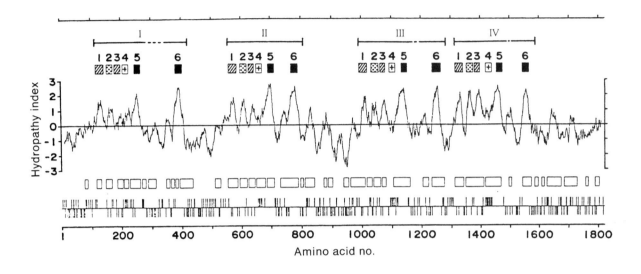

Fig. 4.17. Hydropathy profile of the electric eel voltage-gated sodium channel. The vertical axis shows the mean hydropathy index of 19 adjacent amino acid residues from $i - 9$ to $i + 9$. Positive peaks represent segments with a high proportion of hydrophobic residues that are interpreted as α-helices that cross the plasma membrane lipid bilayer. Homologous domains I to IV are shown, and the S1 to S6 segments within each one of them. The line of white boxes shows all sections of predicted α-helix or β-strand structure. The bottom line shows positions of the positively charged arginine and lysine residues as upward lines and the negatively charged aspartate and glutamate residues as downward lines. (From Noda *et al.*, 1984. Reprinted with permission from *Nature* **312**, p. 126, Copyright 1984 Macmillan Magazines Limited.)

amino acid residue is positively charged. It seems likely that the movement of these positive charges is in some way responsible for the sensitivity of the channel to changes in membrane potential.

In addition to the principal subunits that form the channel pore, there may be auxiliary subunits present. These may serve to promote the expression of the channel in the cell, or to modify the channel function in some way.

The voltage-gated sodium channel

Molecular investigations on voltage-gated channels began with the isolation of the sodium channel protein from the electric organs of the electric eel *Electrophorus*, utilizing its highly specific binding to the puffer fish poison tetrodotoxin. It is a large protein, with a molecular weight of about 260 000. The amino acid sequence was first determined by Numa and his colleagues at Kyoto University using cDNA prepared from electric organ messenger RNA (Noda *et al.*, 1984). They found that the protein chain is 1820 amino acid residues long and consists of four homologous domains that have very similar sequences. Hydropathy analysis (fig. 4.17) shows that within each domain there are six segments that probably form transmembrane α-helices; they are conventionally labelled S1 to S6.

In contrast to the nAChR and its relatives, there is no hydrophobic signal peptide at the N terminus of the protein chain, hence it seems likely that the N terminus itself is cytoplasmic. If we assume that the hydrophobic S1 to S6 segments are the only places where the chain crosses from one side of the membrane to the other, then the loops between the four homologous domains and the C-terminal segment will all be cytoplasmic also, giving the

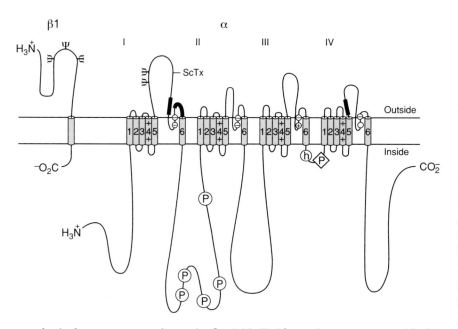

Fig. 4.18. Membrane topology of the rat brain voltage-gated sodium channel. The four domains in the α chain are labelled I to IV, and the S1 to S6 transmembrane segments are shown in each of them. The SS1–SS2 (H5 or P) segments are shown dipping into the lipid bilayer between S5 and S6; small circles in them represent residues involved in tetrodotoxin binding. N-linked glycosylation sites are shown as Ψ. Sites of phosphorylation (chapter 6) are shown as P in a circle for cAMP-dependent phosphorylation and P in a diamond for protein kinase C phosphorylation. The inactivation particle (chapter 6) is shown by h in a circle. Black rectangles and ScTx show sites involved in binding scorpion α-toxins. The principal α subunit contains the transmembrane pore; two auxiliary β subunits are present. The β2 chain (not shown in this diagram) is similar to the β1 chain in its topology. (From Catterall, 1992.)

topological arrangement shown in fig. 4.18. Evidence in agreement with this comes from the positions of phosphorylation sites (which must be cytoplasmic), and of glycosylation sites and binding sites for scorpion toxins (which must both be on the extracellular side of the membrane), and also from antibody binding studies (see Catterall, 1988, 1992; Trimmer & Agnew, 1989).

It seems very reasonable to assume that the four domains of the protein chain are clustered together to form a cylindrical rosette with an aqueous pore down its central axis, as suggested in fig. 4.15. Evidence in favour of this view is that antibodies binding to parts of the S5–S6 link in domains I and IV interfere with the binding of scorpion α-toxins, suggesting that these two domains are adjacent to each other (Thomsen & Catterall, 1989). The domains are thus positioned in a tetrameric structure, in contrast to the pentameric structure of the nAChR and other transmitter-gated channels.

Screening of cDNA libraries from rat brain with probes from the electric eel sodium channel sequence produced sequences for the principal subunits of four distinct sodium channels, called types I, II, IIA (an alternatively spliced variant of type II) and III (Noda *et al.*, 1986; Kayano *et al.*, 1988). Two further types are expressed primarily in adult skeletal muscle (type μ1) and in heart and embryonic or denervated skeletal muscle (h1) (Trimmer *et al.*, 1989; Rogart *et al.*, 1989). All these sequences show a high degree of homology with each other. Two more homologous sequences were discovered in *Drosophila* (Ramaswami & Tanouye, 1989). Two further sequences, similar to each other but rather distinct from the other sodium channels (they have fewer positive charges in the domain IV S4 segment, for example), have been cloned from glial cell and heart cDNA libraries (Gautron *et al.*, 1992; George *et al.*, 1992).

In the sodium channels of rat brain the large principal α subunit is accompanied by two much smaller auxiliary subunits named $\beta 1$ and $\beta 2$. In channels from skeletal and cardiac muscle the $\beta 2$ subunit is absent. Functional voltage-gated sodium channels are produced when the mRNA for the mammalian α subunit alone is injected into *Xenopus* oocytes, but their expression is enhanced (i.e. more channels are produced) and their properties are altered (rates of activation and inactivation are higher) if the $\beta 1$ subunit is also present (Isom *et al.*, 1992). The β subunits thus serve an accessory modulating function.

Voltage-gated calcium channels

In many cells the voltage-gated sodium currents are supplemented, or even replaced, by calcium currents flowing through their own specific channels. Calcium channels are entirely responsible for voltage-gated depolarizations in most invertebrate muscles and in vertebrate smooth muscles, and serve to maintain the depolarization during the plateau of the action potential in vertebrate heart muscle. They are also crucially important in controlling the internal calcium ion concentration in many muscle cells and in secretory cells and at nerve terminals. Since calcium ions act as intracellular messengers for a wide variety of cellular processes, this means that calcium channels are intimately involved in intracellular control systems in addition to any effect they may have on the cell membrane potential.

Voltage-gated calcium channels have been purified from the transverse tubules of rabbit skeletal muscle. They consist of a large principal subunit, called $\alpha 1$, which contains the channel pore, together with auxiliary $\alpha 2$, β, γ and δ subunits. Molecular cloning shows that the $\alpha 1$ subunit has a structure very similar to that of the sodium channel, with four homologous domains each with six transmembrane segments (Tanabe *et al.*, 1987). Its amino acid sequence shows some homology with that of the voltage-gated sodium channel: 29% of the positions are occupied by the same residue.

Calcium channels are found in brain, muscle, and a wide variety of other tissues. They show considerable diversity (Tsien *et al.*, 1988; Snutch & Reiner, 1992; Spedding & Paoletti, 1992; McCleskey, 1994). This first became evident from physiological measurements and from differential sensitivity of the channels to different pharmacological agents, and then by the discovery of their molecular structures. These different approaches have led to alternative nomenclatures. In classification by their properties, T type channels have a lower threshold than the other types, L type channels are blocked by dihydropyridines, P type channels are blocked by ω-Aga-IVA, a peptide component of funnel-web spider venom, and N type channels are blocked by

Table 4.4. *Different types of mammalian voltage-gated calcium channels*

Channel type	α1 cDNA type	Agents causing block	Primary tissue
L	S	DHP	Skeletal muscle
L	C	DHP	Brain, heart, lung
L	D	DHP	Brain, heart, pancreas
N	B	ω-Ctx-GVIA	Brain
P	A	ω-Aga-IVA	Brain, heart
Q	A	None specific	Brain
R	E	None	Brain
T (= B)	E	Ni^{2+}	Brain

Abbreviations: DHP, 1,4-dihydropyridine (nifedipine) and other dihydropyridines; ω-Aga-IVA, a peptide toxin from funnel-web spider (*Agelenopsis aperta*) venom (Mintz *et al.*, 1992); ω-Ctx-GVIA, a toxin from the cone shell *Conus geographus*. Simplified after McCleskey (1994), Hofmann *et al.* (1994) and Dolphin (1995).

ω-conotoxin. Q and R channels have also been described (Zhang *et al.*, 1994). There has been much interest in how these different types are utilized in different cell activities, but the situation is complex. An example of this is in the calcium channels of neuromuscular presynaptic nerve terminals; these are N type in frogs and birds but P type in mammals (Uchitel *et al.*, 1992).

Cloning of calcium channel α1 subunits has distinguished at least five different types of sequence (A to E) from mammalian brain, in addition to the skeletal muscle form (Snutch & Reiner, 1992; Soong *et al.*, 1993). The C form of the rat brain α1 subunit exists in two isoforms as a result of alternative splicing (Snutch *et al.*, 1991). Table 4.4 shows the relations between the functional classes and the molecular sequence types.

Voltage-gated potassium channels

We have seen how the combination of fish electric organs and channel-specific neurotoxins proved highly productive for biochemists at work on the structure of the nAChR channel and the sodium channel. Determination of voltage-gated potassium channel sequences had to follow a different track, however, since tissues correspondingly rich in them were not to be found. Nor were neurotoxins of specificity binding affinity comparable to that for α-bungarotoxin or tetrodotoxin available until quite recently.

Shaker is a behavioural mutant of the fruit fly *Drosophila*. It is easily detected since the flies shake their legs when anaesthetized with ether, and has been localized at a particular point on the X chromosome. Voltage clamp studies on flight muscle cells show that flies with mutations at the *Shaker* locus have

Fig. 4.19. Alternative splicing of *Shaker* potassium channels. The core region, containing segments S1 to S5 and H5, is common to all transcripts. This is combined with any one of five different N-terminal regions and with either of two different S6 and C-terminal regions. Terminology derived from this model gives names such as *Shaker* A1 or D2 etc. for particular splice variants; other terminologies exist. The scale bar is 20 amino acid residues long. (From Pongs, 1992.)

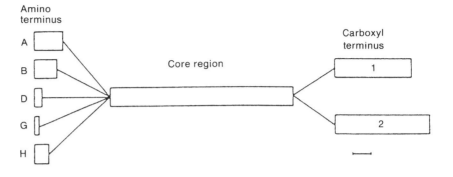

modified potassium currents (Salkoff, 1983). Cloning of the genomic DNA from the *Shaker* locus enabled the amino acid sequence of the putative potassium channel protein to be deduced (Papazian *et al.*, 1987; Tempel *et al.*, 1987). It was 616 amino acid residues long and had a hydrophobicity profile just like one of the four domains in the sodium and calcium channels, and suggesting a pattern of transmembrane folding like that in fig. 4.16*A*.

By analogy with the other voltage-gated channels we would expect the complete potassium channel to be a tetramer of the individual subunits. The first evidence that potassium channels are multimers came from experiments in which mRNA coding for subunits with different properties was injected into *Xenopus* oocytes: channels with hybrid properties were produced (Isacoff *et al.*, 1990; Ruppersberg *et al.*, 1990). MacKinnon (1991) used a charybdotoxin-resistant mutant expressed in oocytes with wild-type subunits to investigate this further. By measuring inhibition of the channels at high toxin concentrations, he could estimate the fraction of channels containing only mutants. This is related to the subunit number by the binomial expansion, and the results indicated that the number was four. Confirmation of this has since been provided by biochemical isolation of tetramers assembled in microsomal membranes *in vitro* (Shen *et al.*, 1993), and by the manufacture of an artificial tetramer (a protein chain containing four subunits linked together) with properties just like those of a normal potassium channel (Liman *et al.*, 1992).

A striking feature of voltage-gated potassium channels is the degree of diversity that they show. This arises in three different ways. Firstly, the *Shaker* gene in *Drosophila* contains at least 23 exons, and alternative splicing of these leads to 10 different channel variants. There are five different N-terminal regions and two different C termini, but the central section containing most of the membrane-crossing region is invariant (fig. 4.19). Particular splice variants can be indicated by the nomenclature *Shaker* A1, G2, etc. (Stocker *et al.*, 1990).

A second source of diversity is the presence of sister genes similar to, but not identical with, *Shaker*. They are known as *Shab*, *Shaw* and *Shal*, and occur at

Table 4.5. *Homologies and nomenclature of some* Drosophila *and mammalian voltage-gated potassium channel genes*

Drosophila		Mammal		
		Mouse	Rat	Human
Shaker	Kv1.1	MBK1 MK1	RCK1, RBK1 RMK1, RK1	HBK1, HK1
	Kv1.2	MK2	RCK5, BK2 RK2, NGK1	HBK4
	Kv1.3	MK3	RCK3, RGK5 KV3	HGK5, HPCN
	Kv1.4		RCK4, RHK1 RK3	HBK4, HK2 hPCN2
	Kv1.5		RCK7, KV1 RK4	hPCN1
	Kv1.6		RCK2, KV2	HBK2
	Kv1.7	MK4		
	Kv1.8		RCK9	
Shab	Kv2.1	M*Shab*	DRK1	DHK1
	Kv2.2		cdrk	
Shaw	Kv3.1	NGK2	KV4, Raw2 Raw2a	
	Kv3.2		Rk*ShIII*A, Raw1	
	Kv3.3	MK5		KCNC3
	Kv3.4	MK6	Raw3	KCNC4
Shal	Kv4.1	M*Shal*	R*Shal*	
	Kv4.2		RK5	

Different laboratories have used different ways of naming the various mammalian channel genes; some of these are listed in the table. The Kv1.1 etc. notation was introduced by Chandy *et al.* (1991), and is now generally used. The other names can be converted to this system: DHK1, for example, becomes human Kv2.1 or hKv2.1. Partly after Pongs (1992), with information from Strong *et al.* (1993).

particular loci on chromosomes 2 and 3; the proteins of any two of them show about 40% identity in their amino acid sequences. Homologous genes occur in mice and other mammals. The mammalian genes also fall into four subfamilies that are homologous with the four *Drosophila* genes, as indicated in table 4.5 (Butler *et al.*, 1989; Wei *et al.*, 1990; Salkoff *et al.*, 1992). The mouse versions of *Shaker* (called mKv1) show on average 70% identity with *Shaker* but only about 40% with the mouse versions of the other genes. The mouse genes usually do not possess introns in their coding regions and so cannot display alternative splicing. Perhaps as a result of this there are numerous slightly different copies of the genes present in mice; thus there are at least 12 varieties of the mouse version of *Shaker*, called Kv1.1 to Kv1.12. Almost identical sequences have been determined for rat (rKv1.1 etc.) and human (hKv1.1 etc.) potassium channels. Exons and alternative splicing do occur in the rat Kv3 family.

Further diversity arises because individual potassium channels can be heteromultimers, assembled from subunits that are not identical with one another. Oocytes injected with mRNAs for just one subunit produce channels that must be homomultimers. Those expressing more than one subunit can produce channels with hybrid properties, as long as the subunits all come from the same subfamily. However, subunits from different subfamilies will not combine to form channels (Covarrubias *et al.*, 1991). Heteromultimers have been demonstrated in brain nerve cells from mice (mKv1.1 and mKv1.2) and rats (rKv1.1 and rKv1.4); in each case the two subunits could be immunoprecipitated together by antibodies that recognize only one of them (Sheng *et al.*, 1993; Wang *et al.*, 1993). The ability to form heteromultimers allows even further diversity of potassium channel function.

The assembly of potassium channel subunits into tetramers is dependent upon a particular section of the polypeptide chain in the cytoplasmic region between the N terminus and the S1 segment; it has been called the T1 domain (Li *et al.*, 1992a; Shen *et al.*, 1993; Shen & Pfaffinger, 1995). *Shaker* B fragments will not bind to each other if residues 83 to 196 are missing. Normally *Shaker* subunits will not combine with DRK1 (Kv2.1) subunits to form heterotetramers, but they will with a chimera in which the N-terminal domain of *Shaker* B is combined with the rest of DRK1. So the T1 domain specifies just which heterotetramers can form, as well as being essential to tetramer formation itself.

What is the purpose of all this diversity? The various members of the subfamilies have slightly different properties, especially in their rates of activation and inactivation. Perhaps cells need to fine-tune their potassium conductances according to their differing roles in the functioning of the body (Salkoff *et al.*, 1992).

Calcium-activated potassium channels

Not all potassium channels are opened solely by changes in membrane potential. An important group found in many cells contains those that are opened by an increase in the intracellular calcium ion concentration, such as might be brought about by calcium inflow through voltage-gated calcium channels or calcium release from intracellular stores (Meech & Standen, 1975; Higashida & Brown, 1986; Latorre *et al.*, 1989). They can be subdivided according to the size of their single channel conductances. High conductance K(Ca) channels (maxi-K or BK channels) have large single channel conductances, in the range 100 to 250 pS. They are voltage dependent as well as calcium dependent, and are blocked by the scorpion toxin charybdotoxin. Small conductance K(Ca) channels (SK channels) have much lower values, down to 6 pS. They show little or no voltage dependence. Channels with conductances in the range 18 to 50 pS are sometimes known as intermediate conductance (IK) channels.

Mutations at the *slowpoke* (*slo*) locus in *Drosophila* abolish a calcium-activated potassium current in fly flight muscles and larval neurons. Mutant flies have action potentials that do not repolarize as fast as usual. Cloning of genomic DNA from the *slo* locus, and of cDNA isolated from a fly head cDNA library, enabled the *slo* gene product to be determined (Atkinson *et al.*, 1991; Adelman *et al.*, 1992). This was a protein about 1200 amino acid residues long, with considerable similarity to the *Shaker*-like potassium channels for the N-terminal third of its sequence, with the usual S1 to S6 and H5 segments present. The central third of the chain was derived from a number of different exons, with considerable alternative splicing leading to a variety of different channel types. The calcium binding site is presumably in the C-terminal third of the sequence.

A similar gene, *mSlo*, encodes for BK ('maxi') calcium-activated potassium channels in mice (Butler *et al.*, 1993). It produces channels with a conductance of 272 pS in symmetrical isotonic potassium solutions. As with the *Drosophila* channel, alternative splicing leads to a number of different variants. Production of chimeras between the mouse and fly channels show that the permeability and gating characteristics are determined by the N-terminal third of the molecule, whereas the calcium sensitivity depends on the C-terminal third (Wei *et al.*, 1994).

Cyclic-nucleotide-gated channels

Vertebrate photoreceptor cells respond to light by the closing of cation-selective channels in the plasma membrane. These channels are responsible for the flow of sodium and calcium ions into the rod outer segments in the dark. They are opened by cytoplasmic cyclic guanosine monophosphate (cGMP). Light absorption by the visual pigment leads indirectly to the activation of the enzyme cGMP phosphodiesterase, which breaks down cGMP and so closes the channels (see Yau & Baylor, 1989; Stryer, 1991).

The amino acid sequence of the rod channel protein has been determined by cDNA cloning. It has much in common with the sequences of *Shaker*-like potassium channels. There are six membrane-crossing segments and an H5 region, plus a region soon after the S6 segment that shows some sequence similarity with the cyclic nucleotide binding domains in cGMP- and cAMP-dependent protein kinase, and so is probably the cGMP binding site. Expression in *Xenopus* oocytes produced channels that would open in the presence of internal cGMP (Kaupp *et al.*, 1989). A similar protein, which will not form channels on its own, seems to combine with it to produce hetero-oligomers that have properties very similar to the native channels (Chen *et al.*, 1993).

Olfactory and visual transduction have much in common. Odorants increase the intracellular concentration of cyclic adenosine monophosphate (cAMP) in the cilia of olfactory sensory neurons, and this opens channels to produce depolarization (Nakamura & Gold, 1987). To see whether these channels had molecular structures homologous with those in the retina, a rat olfactory epithelium cDNA library was screened at low stringency with a retinal channel cDNA probe. This led to the isolation of a cDNA sequence that was very similar to the corresponding retinal sequence, and which coded for a cation-selective channel (Dhallan *et al.*, 1990).

There are sufficient similarities between the S1 to S6 segment sequences of these channels and the corresponding ones of voltage-gated channels to show that they must be related to each other in evolution, and so we can regard them as members of the same superfamily (Jan & Jan, 1990, 1992). Similar channels seem to be quite widely distributed in the tissues of the body, and may be concerned with the regulation of calcium entry by cAMP and cGMP (Biel *et al.*, 1994; Yau, 1994).

Further cyclic-nucleotide-gated channel proteins that seem to be members of this superfamily have been found in insects and in plants. The *eag* (*ether-à-go-go*) locus in *Drosophila* encodes a polypeptide that has the same general structure as those produced by the *Shaker* group, but not much sequence similarity with them except in the S4 and S5–S6 linker sequences (Warmke *et al.*, 1991). It encodes a voltage-gated channel that is permeable to potassium and calcium ions and is modulated by internal cAMP (Brüggemann *et al.*, 1993). Remarkably, two potassium channel peptides with S1 to S6 segments and a cyclic nucleotide binding site, called AKT1 and KAT1, have been isolated from the flowering plant *Arabidopsis* (Anderson *et al.*, 1992; Sentenac *et al.*, 1992). It seems likely that these act as inward rectifiers and are concerned with uptake of potassium ions into plants (Schroeder *et al.*, 1994).

Channels with two membrane-crossing segments per subunit

A number of channels with two membrane-crossing segments, M1 and M2, in each subunit have been cloned recently. They form a rather heterogeneous group and may not all be related to each other. Some of them show appreciable sequence similarity, whereas others do not. We have already briefly looked at the P_{2X} receptor, a neurotransmitter-gated channel that could be included in this group.

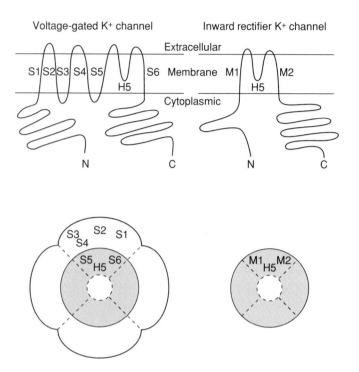

Fig. 4.20. Membrane topology proposed for the inward rectifier potassium channel IRK1 (right) compared with that of voltage-gated potassium channels (left). The lower diagrams show how four subunits might come together to form a channel. The inward rectifier channel, according to this model, corresponds to the inner core of the voltage-gated channel. (From Kubo *et al.*, 1993a. Reprinted with permission from *Nature* **362**, p. 132, Copyright 1993 Macmillan Magazines Limited.)

Inward rectifier channels

Not all potassium channels are gated by depolarization. Inward rectifier channels open at membrane potentials near to, or more negative than, the resting potential, and allow potassium ions to flow inwards but not outwards. Cloning studies show that their subunits are smaller than those of the voltage-gated channels: they possess just two membrane-spanning segments, together with an H5 pore-lining segment that seems to be homologous with those of voltage-gated potassium channels. There is no segment similar to the S4 segment of voltage-gated channels. It seems reasonable to suppose that the channels are homotetramers, made from four identical subunits (fig. 4.20).

One of these channels was isolated by cloning from rat kidney outer medulla and thus called ROMK1 (Ho *et al.*, 1993). It was a peptide with only 391 amino acid residues and so appreciably smaller than the subunits of voltage-gated channels. When expressed in *Xenopus* oocytes the channels required internal ATP and showed some inward rectification. A splice variant of it (ROMK2) has since been isolated (Zhou *et al.*, 1994). Another channel of the same general structure, and with 40% amino-acid identity, was isolated from mouse macrophages and named IRK1 (Kubo *et al.*, 1993a). Related channels (IRK2 and IRK3) have been isolated from brain cDNA libraries (Morishige *et al.*, 1994; Takahashi *et al.*, 1994). The rectification depends upon the presence of magnesium or polyamine ions on the inside of the membrane; presumably they block the inside of the pore when current flow is outwards (see chapter 6).

Knowledge of the amino acid sequences for ROMK1 and IRK1 encouraged Kubo and his colleagues to look for a similar channel with a very distinguished history. One of the classical experiments on chemical transmission at synapses was Loewi's demonstration in 1921 that stimulation of the vagus causes slowing of the heart beat by releasing a chemical substance, later shown to be acetylcholine (Loewi, 1936). The acetylcholine acts on muscarinic receptors in the sinoatrial node and the atrium, and these receptors activate G protein molecules, which then open inward rectifier potassium channels (see chapter 6). A clone apparently coding for channels of this type was isolated by screening a rat heart cDNA library with oligonucleotide probes coding for short sequences of IRK1. The new protein was called GIRK1. It has two transmembrane segments (M1 and M2) and a putative H5 pore-lining region, and its amino acid sequence shows 43% identity with IRK1 and 39% with ROMK1 (Kubo et al., 1993b). When expressed in Xenopus oocytes it is opened by the $\beta\gamma$ subunits of cytoplasmic G proteins (Reuveny et al., 1994).

There is now evidence, however, that the GIRK1 subunit is not the only component of the muscarinically activated channel. A second protein, homologous with other inward rectifier subunits and called CIR (cardiac inward rectifier), has been identified as associated with GIRK1 (Krapivinsky et al., 1995). This will form channels when expressed on its own in oocytes in the presence of the m2 muscarinic acetylcholine receptor, but shows only a low sensitivity to acetylcholine; GIRK1 is similar in these respects. But if both channels are expressed together with the m2 receptor in the same oocyte, channels that are highly sensitive to acetylcholine (which activates G proteins via the m2 receptors) are produced. Hence it seems likely that the atrial muscarinically activated potassium channel is a heteromultimer of GIRK1 and CIR.

Further inward rectifier potassium channels have been identified by screening a rat brain cDNA library with primers similar to the H5 region; they have been called BIR9 and BIR10 (Bond et al., 1994). An ATP-sensitive inward rectifier channel, uK_{ATP}-1, has been isolated from a rat pancreatic islet cDNA library. It shares 43% to 46% sequence identity with other members of the inward rectifier family. When expressed in oocytes, it closes in the presence of ATP and is activated by diazoxide, a drug which opens ATP-sensitive channels. It is widely expressed in the tissues of the body, including pancreatic β cells (compare fig. 1.2), and skeletal and heart muscle (Inagaki et al., 1995).

These various inward rectifier potassium channels have been brought into a unified classification scheme comparable to that for voltage-gated potassium channels (Doupnik et al., 1995). The ROMK group, for example, is included in the Kir1 subfamily, with ROMK1 and ROMK2 as the splice variants

Table 4.6. *A classification of inward rectifier potassium channel clones in mammals*

	Standardized nomenclature	Channel clone
Kir1 subfamily	Kir1.1a	ROMK1
	Kir1.1b	ROMK2
Kir2 subfamily	Kir2.1	IRK1
	Kir2.2	IRK2
	Kir2.3	IRK3
Kir3 subfamily	Kir3.1	GIRK1
	Kir3.2	GIRK2
	Kir3.3	GIRK3
	Kir3.4	CIR
Kir4 subfamily	Kir4.1	BIR10
Kir5 subfamily	Kir5.1	BIR9
Kir6 subfamily	Kir6.1	$uK_{ATP}-1$

Simplified after Doupnik *et al.* (1995), with the addition of the Kir6 subfamily for the channel described by Inagaki *et al.* (1995).

Kir1.1a and Kir1.1b. Channels are put in the same subfamily if their sequence identities are high; different subfamilies show sequence identities of 40% or more. Details are shown in table 4.6.

A pH-sensitive potassium channel with two transmembrane segments has been cloned from rabbit kidney collecting tubule cells, and called RACTK1 (Suzuki *et al.*, 1994). The H5 region was similar to that of the other inward rectifier potassium channels (Sutcliffe & Stanfield, 1994), but the rest of the molecule showed little sequence similarity with them. The channel is probably concerned with the movement of potassium ions into the urine.

Amiloride-sensitive sodium channels

Sodium reabsorption in the distal parts of the kidney tubules occurs via channels that are specifically permeable to sodium ions and that are blocked by the diuretic agent amiloride. Amiloride-sensitive sodium channels also occur in other epithelial cells, in such organs as the colon, the bladder and the lung. The channels are localized to the apical membrane of the cells, and the net transport is driven by a sodium potassium ATPase located on the basal membrane. They tend to be regulated by hormones such as vasopressin and oxytocin (Benos *et al.*, 1995).

Three homologous subunits of these channels have been isolated from a rat colon library by using the *Xenopus* oocyte expression system (Canessa *et al.*, 1993, 1994). The α, β and γ peptides are 698, 638 and 650 amino acid residues long, respectively. They each have M1 and M2 membrane-crossing segments separated by a long extracellular loop. All three subunits are necessary for full expression of the amiloride-sensitive currents in the oocyte, although a small response can be produced by the α subunit alone. This suggests that the channel is a multimer, perhaps normally with a stoichiometry of $\alpha_2\beta\gamma$

A subunit homologous with the rat colon α subunit (669 residues long, 81% identity) has been isolated from human lung. It seems to be important in the removal of pulmonary fluid when the baby switches to air-breathing at birth (Voilley *et al.*, 1994).

Calcium release channels

We now look at two channels found in intracellular membranes, both of them large in size and both concerned with calcium movements within the cell. Sequence analysis shows considerable similarities between them.

The ryanodine receptor

Ryanodine is a plant alkaloid with insecticidal properties. It affects intracellular calcium release during activation in skeletal muscle, and the protein to which it binds is known as the ryanodine receptor. Electron micrographs of skeletal muscle show the presence of structures crossing the gap between the T tubules and the sarcoplasmic reticulum. They are known as 'feet', they are grouped in fours, and they seem to be involved in connecting excitation in the T tubule release of calcium from the sarcoplasmic reticulum (Franzini-Armstrong, 1980). Isolation of the ryanodine receptor from skeletal muscle shows that it is a large protein that forms tetrameric particles just like the 'feet', so the ryanodine receptor and the calcium-release channel are one and the same (Inui *et al.*, 1987).

Cloning and sequencing of the skeletal muscle ryanodine receptor (RyR1) shows that it is a very large molecule. Each protein chain contains 5032 or perhaps 5037 amino acid residues. One model of the chain has four membrane-crossing α-helix segments in the C-terminal region (Takeshima *et al.*, 1989), another has 10 of them, or perhaps 12 if two segments near the middle of the peptide chain act in this way (Zorzato *et al.*, 1990). Since one ryanodine molecule is bound for each four monomers, it seems likely that the complete channel is composed of four subunits and has a molecular weight of over two million. Probably the C-terminal regions from the four subunits come together to make the pore in the sarcoplasmic reticulum membrane. Most of the mass of the molecule is in the 'foot' region. Electron micrograph studies suggest that the foot region of each subunit is in close contact with a

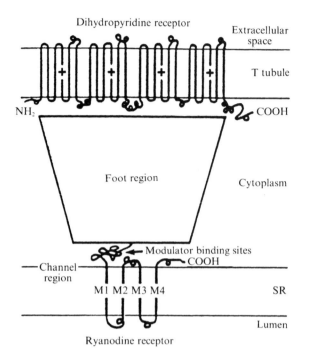

Fig. 4.21. Proposed membrane topology of the skeletal muscle ryanodine receptor RyR1. Only one of the four identical subunits forming the channel is shown. An alternative model proposes 10 or 12 transmembrane segments in the sarcoplasmic reticulum (SR) membrane. Each subunit is closely associated with a voltage-gated calcium channel (dihydropyridine receptor) in the T tubule membrane, which probably acts primarily as a voltage sensor. Only the α_1 subunit of the dihydropyridine receptor is shown here. (From Takeshima *et al.*, 1989. Reprinted with permission from *Nature* **339**, p. 445, Copyright 1989 Macmillan Magazines Limited.)

dihydropyridine receptor (a voltage-gated calcium channel) in the T tubule plasma membrane, as is shown in fig. 4.21. It is likely that the dihydropyridine receptor acts primarily as a voltage sensor, linking depolarization of the muscle plasma membrane to opening of the calcium-release channel. When the calcium-release channel opens, calcium ions pour out of the sarcoplasmic reticulum into the cytoplasm and act as the trigger for muscular contraction.

Other ryanodine receptor isoforms have been cloned (Sorrentino & Volpe, 1993). The RyR2 receptor occurs in cardiac muscle; it is probably activated by calcium ions entering via the voltage-gated calcium channels rather than by direct interaction between the two proteins. Each RyR2 subunit is 4968 amino acid residues long and shows 66% sequence identity with RyR1 (Nakai *et al.*, 1990). The RyR3 receptor has been isolated from rabbit brain. It is 4872 residues long and is 67% and 70%, respectively, identical with RyR1 and RyR2 (Hakamata *et al.*, 1992). Unlike the other two isoforms, RyR3 channel opening is not stimulated by caffeine.

The IP₃ Receptor

Inositol 1,4,5-trisphosphate, abbreviated as $InsP_3$ or IP_3, is an important second messenger in many cellular operations It is produced in a signal transduction cascade initiated by some messenger molecule originating outside the

Fig. 4.22. Schematic diagram to show the transmembrane topology of the IP$_3$ receptor; just two of the four subunits are shown. ATP binding sites and sites of phosphorylation (P) in the regulatory domain are shown. The shaded regions may be omitted by alternative splicing. (From Mikoshiba, 1993. Reproduced from *Trends in Pharmacological Sciences*, with permission from Elsevier Trends Journals.)

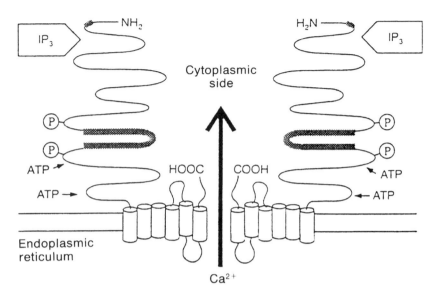

cell, and acts by controlling intracellular calcium ion concentrations (Berridge, 1993). A hormone or neurotransmitter combines with its appropriate G-protein-linked receptor, and the activated G protein then itself activates the enzyme phospholipase C-β. Alternatively a growth factor acts on a tyrosine kinase receptor so as to activate phospholipase C-γ. Whichever phospholipase C is activated then hydrolyses the membrane lipid phosphatidylinositol 4,5-bisphosphate to give IP$_3$ and diacylglycerol. The IP$_3$ combines with calcium release channels (IP$_3$ receptors) in the endoplasmic reticulum membrane so as to open them and allow calcium ions to flow out into the cytoplasm.

At least three different IP$_3$ receptor isoforms are known. The most widespread, IP$_3$R1, was first characterized as a protein called P$_{400}$ that was present in the normal cerebellum but virtually absent in cerebella from mutant mice with no Purkinje cells. It was later isolated by cloning from mouse cerebellum cDNA libraries (Furuichi *et al.*, 1989; see also Mikoshiba, 1993; Furuichi *et al.*, 1994). The protein chain is typically 2749 amino acid residues long, and four of them are used to make one receptor. Receptor subtypes occur as a result of alternative splicing.

Each protein chain in IP$_3$R1 consists of three domains (fig. 4.22). The channel domain at the C-terminal end contains six or eight putative transmembrane segments and shows appreciable similarity to the corresponding part of the ryanodine receptor. Removal of this domain produces a soluble protein which binds IP$_3$ but does not form homotetramers, and removal of all or parts of the N-terminal domain abolish IP$_3$ binding (Mignery & Sudhof, 1990; Mignery *et al.*, 1990). A regulatory or coupling domain connects the other two domains; it contains sites for binding of ATP (which increases the likelihood of channel opening) and for phosphorylation.

The other IP_3 receptor isoforms, known as IP_3R2 and IP_3R3, are similar in structure. Sequence homology between the three types is highest in the IP_3-binding domain (72% to 77%) and the channel domain (66% to 75%) (Furuichi *et al.*, 1994).

Background chloride channels

Since chloride is the commonest anion in most extracellular solutions, it is likely to be the commonest ion flowing through a channel that is selectively permeable to anions, even if the channel permeability to some other anions is somewhat higher. An anion-selective channel, therefore, is likely to be called a chloride channel unless there is good reason to do otherwise.

We have already discussed the neurotransmitter-gated chloride channels activated by GABA and glycine. Here we consider some chloride-selective channels that open without being gated by combination with an external ligand.

The CLC group of channels

The cells in the electric organs of the electric ray *Torpedo* are asymmetrical. On the upper, innervated face are large numbers of nAChR channels that allow sodium ions to pour into the cell when they are opened by nervous activity. The non-innervated lower face has to have a high conductance so as to allow the current to flow readily, and this is brought about by a high density of chloride channels (White & Miller, 1979).

Jentsch and his colleagues (1990) in Hamburg used an ingenious cloning method to determine the amino acid sequence of the chloride channel protein. Messenger RNA from the electric organ would produce chloride channels when injected into *Xenopus* oocytes. A cDNA library made from this mRNA provided cDNA clones that could be injected into the oocyte in single-stranded form so as to hybridize with particular species of the mRNA. Thus the cDNA coding for the chloride channel would hybridize with the specific chloride channel mRNA so as to inactivate it, and this would reduce the amount of chloride channel activity in the oocyte. In this way they were able to isolate the cDNA coding for the chloride channel protein, sequence it and so determine the primary structure of the protein. The protein was 805 amino acid residues long and showed no homologies with any other protein known at the time. It became known as CLC-0. (The name is more correctly written as ClC-0, but since this is too easily misread as CIC-0 or C1C-0, we shall use the capitalized form here.)

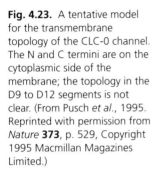

Fig. 4.23. A tentative model for the transmembrane topology of the CLC-0 channel. The N and C termini are on the cytoplasmic side of the membrane; the topology in the D9 to D12 segments is not clear. (From Pusch *et al.*, 1995. Reprinted with permission from *Nature* **373**, p. 529, Copyright 1995 Macmillan Magazines Limited.)

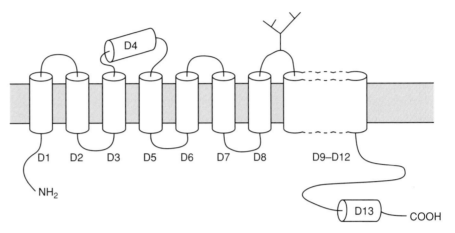

Do similar channels exist in mammals? Skeletal muscle has a high resting chloride conductance, so the Hamburg group used homology screening of a rat muscle cDNA library to look for proteins homologous with the *Torpedo* channel. They found one, called CLC-1, that was 944 residues long, showing 54% sequence identity with the *Torpedo* channel, and which produced chloride currents similar to those of skeletal muscle when expressed in *Xenopus* oocytes (Steinmeyer *et al.*, 1992). Mutations in this channel produce myotonia (congenital muscle stiffness) and decreased muscle chloride conductance in mice and humans (Koch *et al.*, 1992; Gronemeier *et al.*, 1994).

CLC-2 is another member of this group. It is about 50% homologous with CLC-0 and CLC-1 and is widely expressed in most cells of the mammalian body (Thiemann *et al.*, 1992). It may be involved in cell volume regulation. Pairs of further CLC channels have been found in kidneys from rats (rCLC-K1 and rCLC-K2) and humans (hCLC-Ka and hCLC-Kb); they show 80% to 90% homology with each other and about 40% with the other CLC channels (Kieferle *et al.*, 1994). They are presumably involved in the extensive movements of sodium chloride in the kidney tubules.

All these various CLC channel proteins have hydropathy plots with 13 hydrophobic peaks, conventionally labelled D1 to D13. First models of the membrane topology postulated 12 or 13 transmembrane segments. However, more recent work has shown that the C terminus is cytoplasmic (Gründer *et al.*, 1992), that N-linked glycosylation occurs between D8 and D9, indicating that this section must be on the outside of the plasma membrane (Kieferle *et al.*, 1994), and that the C-terminal end of D12 is on the inside (Pusch *et al.*, 1995). It looks as though not all of the D1 to D13 segments actually cross the membrane. Figure 4.23 shows one view of how the protein chain might be arranged.

How many subunits make a CLC channel? Experiments with mutants of human CLC-1 causing dominant myotonia congenita (Thomsen's disease) have suggested some answers (Steinmeyer *et al.*, 1994). Two mutant sequences were used, a severe one, P480L, and one producing a milder myotonia, G230E. Different quantities of mutant mRNA were injected into *Xenopus* oocytes at the same time as normal (wild-type) mRNA. P480L mRNA caused considerable reduction in the chloride conductance of the oocyte at quite low concentrations, suggesting that just one mutant subunit in a multimer is sufficient to make a chloride channel non-functional. The effect with G230E was not so pronounced, so perhaps in this case two mutant subunits per multimer are required to prevent channel function. The results overall could best be explained if each channel is normally a homo-oligomer, most likely with four subunits.

The CFTR channel

Cystic fibrosis is a hereditary disease that afflicts about one in 2500 children of north European descent. Sufferers have thick mucous secretions that block the smaller airways of the lung and the ducts of the pancreas. This leads to inflammation and infection of the lung and pancreas, then progressive destruction of both organs and eventual death. Affected individuals show abnormalities of electrolyte transport, such that excessive salt loss occurs via the sweat glands. In 1983 it was found that the essential feature of the disease is a failure of chloride transport (Knowles *et al.*, 1983; Quinton, 1983), and further work showed that the defect lies with a chloride channel that is normally activated by cAMP-regulated protein kinase A (Quinton, 1990; Super, 1992).

The gene product responsible for the cystic fibrosis defect was identified by Tsui, Riordan and Collins and their collaborators in 1989; the work was a triumph of molecular genetics (Kerem *et al.*, 1989; Riordan *et al.*, 1989; Rommens *et al.*, 1989; Santis, 1995). First, DNA from the cystic fibrosis locus at band q31 on chromosome 7 was cloned. Then probes from this were used to screen cDNA libraries from sweat gland epithelia and other tissues. A number of overlapping cDNA clones were obtained, and from these the amino acid sequence of the gene product was determined. It was named the cystic fibrosis transmembrane conductance regulator, or CFTR.

The CFTR protein is 1480 amino acid residues long. Hydropathy analysis suggests that there are two membrane-spanning domains, MSD1 and MSD2, each with six putative transmembrane segments (fig. 4.24). Each of these is followed by a hydrophilic region which, by comparison with other known sequences, would be expected to bind ATP; they are called the nucleotide-binding domains, NBD1 and NBD2. In the middle of the chain is a region called the R (regulatory) domain, which has numbers of positively and

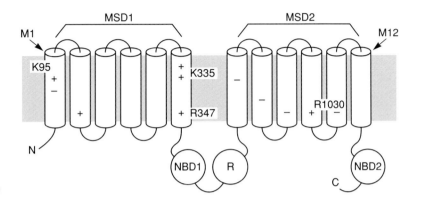

Fig. 4.24. Proposed domain structure for the CFTR. The 12 putative membrane-crossing segments, M1 to M12, are grouped into two domains MSD1 and MSD2. NBD1 and NBD2 are the nucleotide (ATP) binding domains, and R is the regulatory region. Charged residues in the MSDs are shown by plus (+) and minus (−) signs. The N and C termini are on the cytoplasmic side of the membrane. (From Welsh *et al.*, 1992. Reproduced with permission from *Neuron* **8**, copyright Cell Press.)

negatively charged amino acids and also sites where phosphorylation could occur.

The CFTR is similar in structure to a number of transporter proteins that together form the ATP-binding cassette (ABC) superfamily. Many of these are found in bacterial cells and are concerned with the membrane transport of a variety of different molecules, including various sugars and amino acids (Higgins, 1992). One of them is the human P-glycoprotein MDR1, associated with multiple drug resistance, which acts as a pump extruding drugs from the cell (Gros & Buschman, 1993; Zaman *et al.*, 1994). Another is the sulphonyl-urea receptor, which has been cloned from pancreatic β cells and may act as the ATP sensor of ATP-sensitive potassium channels (Aguilar-Bryan *et al.*, 1995).

The CFTR's name reflects some initial doubt as to whether it really formed a chloride channel. It showed no homology with other known channels, and other members of the ABC superfamily acted as carriers rather than channels. Perhaps it just acted as a regulator of some other protein that formed the actual chloride channel. But it was soon found that the CFTR could be expressed in various cells, which then showed single channel chloride currents when injected with cAMP and protein kinase A. An ingenious experiment by Anderson and his colleagues (1991b) produced good evidence that it really is a channel. They prepared mutants in which positively charged residues in the MSDs were changed to negatively charged ones. In two cases, K95D and K335E, the permeability sequence to different anions was changed, from $Br^- > Cl^- > I^-$ to $I^- > Br^- > Cl^-$. This suggests quite strongly that M1 and M6, where these two mutant residues are situated, are involved in forming the channel. Further evidence was provided by Bear and her colleagues (1992). They prepared a very pure sample of the CFTR protein after expressing it in an insect cell line and incorporated it into a planar lipid bilayer. Chloride channels would open in the presence of protein kinase A and cAMP.

Fig. 4.25. The hexameric nature of rat heart muscle gap junction channels, as seen by electron microscopic image analysis of crystalline arrays stained with uranyl acetate. The centre-to-centre distance between adjacent channels is 85 Å. (From Yeager & Gilula, 1992.)

Gap junction channels

Adjacent cells of multicellular animals may communicate with each other via gap junctions. These were first detected as regions where the plasma membranes of two cells were separated by 20 to 30 Å instead of the usual 200 Å or so, and were given the name 'gap junction' to distinguish them from 'tight junctions' where the two membranes are in contact with no space between them (Revel & Karnovsky, 1967). Permeability studies show that the channels are not particularly selective, and will readily allow ions and small molecules to pass through them when they are open, as we shall see in the next chapter.

Electron microscopy and image reconstruction shows that gap junction channels are hexameric structures (fig. 4.25). The individual subunits are called connexins, and the groups of six of them are called connexons. A complete gap junction channel is formed of two connexons from adjacent membranes (fig. 4.26). The cytoplasmic compartments of the two cells are thus connected by the channel pore passing through the middle. Gap junction channels are often found in close arrays, as for example in liver cells and in the intercalated discs of heart muscle.

At least 16 different connexins have been cloned from vertebrate tissues. They range in size from 225 to 510 amino acid residues long and show considerable sequence similarities. They are commonly named by reference to their molecular weight as predicted from their amino acid sequence. For example, rat connexin 32 (Cx32) has a molecular weight of 32 007 (Paul, 1986) and rat connexin 43 (Cx43) has one of 43 036 (Beyer *et al.*, 1987; Bennett *et al.*, 1991).

Fig. 4.26. The structure of gap junctions, based on X-ray diffraction studies on mouse liver cells. Notice how two connexin hexamers (connexons), one in each membrane, join to form a channel connecting the cytoplasmic compartments of the two cells. (From Makowski *et al.*, 1977. Reproduced from the *Journal of Cell Biology* **74**, p. 643, by copyright permission of The Rockefeller University Press.)

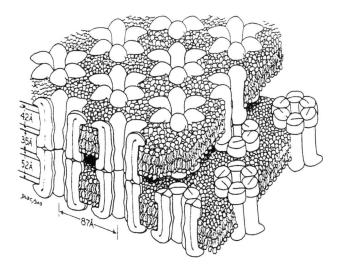

Fig. 4.27. The membrane topology of connexins, based on rat liver connexin 32. There are four transmembrane segments (TM1 to TM4), two outer loops (OL-1 and 2) on the extracellular side of the membrane, and the N and C termini and the inner loop (IL) on the cytoplasmic side. Black circles show hydrophobic residues. For the hydrophilic residues plus (+) and minus (−) signs show charged residues, P shows proline, and in the outer loops the conserved cysteine (C) and glycine (G) residues are shown. (From Peracchia *et al* 1994.)

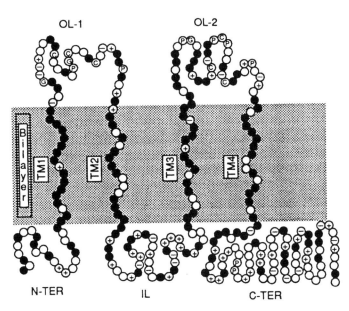

Cytoplasm

Analysis of connexin amino acid sequences suggests that there are always four transmembrane segments, probably α-helical in structure. Experiments with antibodies to particular parts of the chain have been used to determine the membrane topology (Milks *et al.*, 1988). The N and C termini are cytoplasmic, and so also is the inner loop between TM2 and TM3, and there are two outer loops on the extracellular side of the membrane (fig. 4.27). The

N-terminal region, the TM segments and the two outer loops show considerable similarity in all connexins, but the inner loop and the C-terminal region vary from one connexin to another both in sequence and in length. The two outer loops each contain three cysteine residues, and also between them at least five prolines and three glycines. All of these residues are always in the same place in the chain, suggesting (since prolines and glycines are often involved in bends and cysteines form covalent disulphide links) that the folding pattern of the outer loops is stable and constant.

In all connexins the third transmembrane segment contains some polar residues and also one acidic and one basic residue; hence it may be that it forms the lining of the channel pore. The connexin outer loops in one connexon must engage with those in the connexon forming the other half of the channel. The second outer loop is longer than the first, largely because it contains a group of hydrophobic residues; these hydrophobic regions might bond to each other. Both outer loops contain conserved negatively charged residues (aspartate and glutamate), and these might link to calcium ions to form bridges between the two connexons. A model of how these interactions might occur has been proposed by Peracchia and his colleagues (1994). They suggest that one of the two connexons in a channel is rotated by 30° (half a connexin) with respect to the other so that each connexin in one half of the channel is connected via its outer loops to two in the other half. Such an arrangement might well make the intercellular channel more rigid than it otherwise would be.

There is no evidence that different connexins can be combined in the same connexon, but functional channels can sometimes be formed from two different types of connexons. Connexins can be expressed in *Xenopus* oocytes, so that two oocytes can be put side by side in contact and gap junction channels then form between them. White and his colleagues (1994) found that heterotypic channels would form between Cx46 and either Cx43 or Cx50 from rat lens, but not between Cx43 and Cx50.

Some other channels and channel-like proteins

The MIP family

The major intrinsic protein of the mammalian lens is a 26 kDa protein found extensively in the membranes of lens fibre cells, and known as MIP, MIP26 or MP26. It was at first thought to be involved in lens gap junctions, but later work showed that this is not so. Reconstitution in lipid bilayers showed that MIP26 formed high conductance ion channels with some selectivity for anions (Ehring *et al.*, 1990). The protein is 263 amino acid residues long with six putative membrane-crossing α-helices, and the channels are probably homotetramers (Gorin *et al.*, 1984; Reizer *et al.*, 1993).

MIP26 is just one of a family of at least 18 members with a surprisingly large variety of functions. They include the glycerol uptake facilitator of *E. coli*, nodulin 26 from the symbiotic root nodules of soybean, tonoplast intrinsic protein from various plant vacuole membranes, big brain protein from *Drosophila* (concerned with neural development), and the mammalian water channels CHIP28 (*ch*annel-forming *i*ntegral *p*rotein of weight 28 000) and WCH-CD (Reizer *et al.*, 1993; Chepelinsky, 1994). In each case the amino acid chain appears to consist of two homologous repeats, each with three or four membrane-crossing segments and always with an NPA (asparagine-proline-alanine) section between the last two of them. The first models suggested that there were three α-helices per repeat, but this would imply that the membrane topology was reversed in the two repeats, which seems rather unlikely if the complete chain arose by gene duplication. An alternative model assumes that there are four membrane-crossing segments (of which only the third and fourth are α-helices), in which case the two repeats would have the same orientation with respect to the membrane (Wistow *et al.*, 1991).

CHIP28 occurs in red blood cells, kidney proximal tubules and elsewhere in the mammalian body. It can be expressed in *Xenopus* oocytes by injecting them with the appropriate mRNA. If such oocytes are placed in a hypo-osmotic solution (normally no problem for a freshwater frog) they swell up and burst within five minutes, showing that water is entering through the CHIP28 channels. However, there is no change in membrane conductance when these channels are expressed in oocytes, suggesting that they are not permeable to ions (Preston *et al.*, 1992; Agre *et al.*, 1993). A similar channel (WCH-CD) occurs in the collecting ducts of the kidney (Fushimi *et al.*, 1993). These water channels have been called aquaporins (Agre *et al.*, 1993; Knepper, 1994); they are of course quite different from the porins of bacterial outer membranes.

MinK, phospholemman and I_{Cln}

By cloning mRNA from rat kidney and using *Xenopus* oocytes as an assay system, Takumi and his colleagues (1988) found a membrane protein that induces a slowly activating potassium current on depolarization. The protein was only 130 amino acid residues long, with just one putative membrane-crossing segment, an extracellular N terminus and a cytoplasmic C terminus. This small size lead to the name minK or min K (miniature potassium channel), but it is also known as the I_{sK} or IsK protein, in reference to the current it induces. A very similar protein occurs in heart muscle, and appears to underly the slow component I_{Ks} of the delayed rectifier potassium current

(Zhang *et al.*, 1994). There is some doubt as to whether minK forms channels directly, perhaps by aggregation of a number of monomers (Takumi, 1993; Varnum *et al.*, 1995), or whether it serves to activate channels (including perhaps chloride channels) from some other source (Attali *et al.*, 1993). There is some evidence that in *Xenopus* oocytes it may form channels by combination with some other endogenous protein (Blumenthal & Kaczmarek, 1994).

Phospholemman is a cardiac membrane protein 72 amino acid residues long, with a single putative transmembrane α-helix. When phospholemman mRNA is injected into *Xenopus* oocytes, they develop a chloride-selective current that is activated by hyperpolarization (Moorman *et al.*, 1992). This could mean that phospholemman forms channels directly, but there is also the possibility that it serves to activate chloride channels that are already present in the oocytes (see Pusch & Jentsch, 1994).

Another protein that activates chloride channels when expressed in *Xenopus* oocytes is I_{Cln}. It is 235 residues long and has no α-helices long enough to span the membrane. A model with four β-sheet strands, forming a dimer with an eight-strand β-barrel, was initially suggested (Paulmichl *et al.*, 1992). However, most of the protein is found in the cytoplasm, suggesting that it is regulatory in function rather than forming the channel itself (Krapivinsky *et al.*, 1994).

Porins

Gram-negative bacteria such as *E. coli* have two membranes bounding their cells. There is an outer membrane in addition to the plasma membrane, and the two are separated by the periplasmic space containing the cell wall. The outer membrane is unusual in that its outer leaflet is composed of lipopolysaccharides instead of the usual phospholipids. Porins are channels in the outer membrane. They are not very selective, have a high conductance to ions and permit the passage of small molecules with molecular weights up to about 600. They protect the inner plasma membrane of the bacterial cell from a number of harmful compounds whose molecules are too large to pass through them.

Porins have molecular weights in the range 30 000 to 50 000 and usually form trimeric structures. Unlike most membrane channels, porins can form crystals, and this means that their structure can be determined by X-ray diffraction (Weiss *et al.*, 1990, 1991; Cowan *et al.*, 1992; Schulz, 1993). The results show that each monomer contains 16 β sections arranged antiparallel so as to form a 16-stranded β-barrel, as is shown in fig. 4.28. Loops between the β-sheet strands are short on the periplasmic side of the membrane, much longer on the external side. One of the long loops folds into the barrel and so reduces the size of the pore somewhat. The monomers are arranged in trimeric structures so that there are three similar pores in each one.

Fig. 4.28. The form of the protein chain in one of the three subunits of the porin from the photosynthetic bacterium *Rhodobacter capsulatus*. β-Strands (arrows) and α-helices are shown. In this picture the top of the porin faces the external medium and the bottom faces the periplasmic space. (From Weiss *et al.*, 1991. Reprinted with permission from Molecular architecture and electrostatic properties of a bacterial porin. *Science* **254**, pp. 1627–30. Copyright 1991 American Association for the Advancement of Science.)

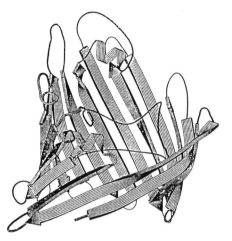

The unusual structure of porins, with β-sheet strands crossing the membrane rather than α-helices, may be related to their position in the cell. They have to pass through the inner plasma membrane and then the periplasmic space before being inserted into the outer membrane. Perhaps they would be retained in the inner membrane if they contained more α-helices, with disastrous results for its permeability (Nikaido, 1994).

Voltage-dependent anion channel

A voltage-dependent anion channel (VDAC) has been extracted from the mitochondrial outer membrane in a wide variety of eukaryotic organisms, including protozoa, fungi, plants, insects, fish and mammals. After reconstitution in bilayer lipid membranes the open channels are of high conductance (650 pS), slightly selective for anions over cations, and somewhat voltage dependent. The amino acid sequence suggests that the channel is a β-barrel structure with 12 to 19 β-sheet strands (Manella, 1992; Peng *et al.*, 1992). This invites comparison with the structure of bacterial porins, especially since mitochondria may have evolved from symbiotic bacteria, but there is no similarity between the two channel types in their amino acid sequences.

Gramicidin

Gramicidin is an peptide antibiotic produced by *Bacillus brevis* and effective against Gram-positive bacteria. It is a linear chain 15 amino acid residues long, but unusual in that the amino acids in the chain are alternately D- and L-isomers. The primary sequence of gramicidin A is as follows:

HCO − L-Val − Gly − L-Ala − D-Leu − L-Ala − D-Val − L-Val − D-Val − L-Trp −
D-Leu − L-Trp − D-Leu − L-Trp − D-Leu − L-Trp − NHCH$_2$CH$_2$OH

The tryptophan at position 11 is replaced by phenylalanine in gramicidin B and by tyrosine in gramicidin C, and the valine at position 1 may be replaced by isoleucine. The natural mixture contains about 80% gramicidin A; it is sometimes called gramicidin D. (Gramicidin S is a cyclic peptide quite different in structure.) The amino acid residues are all non-polar, the N terminus is formylated and the C terminus has ethanolamine tacked onto it, so the molecule is highly non-polar and therefore very lipid soluble.

Gramicidin readily enters lipid bilayers to form ion channels permeable to monovalent cations, and indeed gramicidin channels were the first to have their single channel currents measured, by Hladky & Haydon in 1970 (see fig. 3.10). The alternating D- and L-amino acids allow the formation of a single-stranded right-handed β-helix with 6.3 residues per turn, in which all the side-chains point away from the axis and the interior of the helix is an open cylindrical pore. This particular arrangement occurs only when the molecules are embedded in the bilayer membrane; in solution dimers with intertwining helices form (Wallace, 1990; Busath, 1993).

Gramicidin channels in membranes are also dimers: two monomers are placed head-to-head with their N termini in near contact so that the whole channel is a symmetrical structure, as is shown in fig. 4.29 (Arseniev *et al.*, 1985). It seems very likely that channels form by the coming together of monomers each floating in their own half of the lipid bilayer. Some nice evidence for this has been produced by O'Connell and her colleagues (1990). They used an analogue of gramicidin A in which the valine in position 1 was replaced by trifluorovaline. Channel conductance for this analogue was lower than that for the normal compound. But, if the trifluorovaline analogue was introduced on just one side of the lipid bilayer and the normal gramicidin on the other, then channel conductances were very largely intermediate in size. This suggests that most of the channels contained one of each gramicidin type, and hence that gramicidin molecules introduced to one of the lipid leaflets do not readily cross over the membrane into the other leaflet.

Because of their relatively simple structure, gramicidin channels have been much used as models for transmembrane ion permeation. Interactions with ions and water molecules in the channel probably occur mainly at the peptide bond carbonyl groups, which are exposed on the inside of the β-helix.

Alamethicin

Alamethicin is an antibiotic produced by the fungus *Trichoderma viride*. It is a peptide with 20 amino acid residues, of which eight are the unusual α-amino-

Fig. 4.29. Molecular model of the gramicidin A channel. The two monomers are placed head-to-head with their N termini in near contact. (From Urry *et al.*, 1988.)

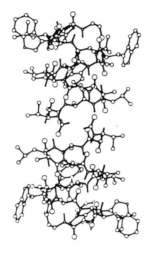

isobutyric acid (or α-methylalanine), whose presence tends to promote helix formation. The C-terminal residue is the alcohol derived from phenylalanine, phenylalaninol, and the N-terminal residue, α-amino-isobutyric acid, is acetylated. X-ray diffraction studies of alamethicin crystals show that each molecule is largely a single α-helix with a bend at the proline residue at position 14 (Fox & Richards, 1982). In line with the proline residue are the polar residues Gln-7 and Glu-18, and also an exposed carbonyl group on Gly-11; this gives an amphipathic character to the helix, with a polar stripe down one side.

Alamethicin can be incorporated into lipid bilayers, leading to an increase in their ionic conductance, especially in the presence of a transmembrane potential (Gordon & Haydon, 1975). Single channel measurements show fluctuations between different conductance levels, with the striking feature that the steps between one level and the next become successively larger (Sansom, 1991). One interpretation of this is that channels form by the aggregation of four or more alamethicin monomers and the conductance changes up or down as monomers are added to or subtracted from such a bundle.

The voltage sensitivity of alamethicin channel formation has given rise to much speculation. An important feature here is that an α-helix has an appreciable dipole moment associated with it. Each peptide bond acts as a small dipole (electrons are drawn towards the oxygen atom in the $C=O$ bond and towards the nitrogen atom in the $N-H$ bond), and these are all aligned in the α-helix, so that the N-terminal end is positive to the C-terminal end. Perhaps the potential across the membrane is necessary to pull the monomers into the membrane or to orient them all with their N-termini on the negative side (see Sansom, 1991, 1993; Woolley & Wallace, 1992; Cafiso, 1994).

A number of natural and synthetic analogues of alamethicin occur, many of them useful as antibiotics. They are all peptides rich in α-amino-isobutyric acid (Aib) and often with an alcohol at the C terminus, so they have been given the name peptaibols. They form channels in lipid bilayers, and presumably their antibiotic action results from their forming channels in the bacterial (or in some cases protozoal) plasma membrane.

Influenza virus M_2 protein

The free influenza A virus consists of ribonucleoprotein surrounded by a lipid envelope. The integral membrane protein M_2 is a minor component of this membrane but becomes expressed abundantly in the plasma membrane of virus-infected cells (Lamb *et al.*, 1985). It is 97 amino acid residues long, with a 23-residue extracellular N-terminal segment and a 54-residue cytoplasmic C terminus, separated by a hydrophobic membrane-crossing segment that is probably an α-helix. It occurs as tetramers or higher order oligomers in which some of the subunits are linked by disulphide bonds.

Injection of M_2 mRNA into *Xenopus* oocytes produces an inward current, largely of sodium ions, that can be blocked by the antiviral drug amantadine (Pinto *et al.*, 1992). Molecular modelling suggests that the ion channel could be formed from a tetramer with its four α-helices surrounding a central pore (Sansom & Kerr, 1993).

Other channels formed by venoms and toxins

Gramicidin, alamethicin and the influenza virus M_2 protein are all examples of channel-forming peptides produced by one organism that have a deleterious effect on another organism. Peptides of this type enter the plasma membrane of the host cell and alter its permeability, with consequent effects upon the ionic composition of the cell. Examples from bacteria include diphtheria toxin (Silverman *et al.*, 1994), aerolysin (Buckley, 1992; Parker *et al.*, 1994), tetanus toxin (Rauch *et al.*, 1990), *Staphylococcus* δ-toxin (Sansom, 1991) and the colicins (Parker *et al.*, 1989; Slatin *et al.*, 1994). Animal venoms that work in this way include mellitin, a 26-residue peptide from bee venom (Dempsey, 1990), and margainin, a toxin from *Xenopus* skin (Duclohier *et al.*, 1989).

Channels formed from synthetic peptides

Most of the channels examined in this chapter have α-helices crossing the membrane, and in many cases the transmembrane pore seems to be lined or partly lined by them. A number of investigators have therefore made synthetic α-helices and applied them to membranes to see whether they will form channels. Thus Lear and his colleagues (1988) made three similar peptides from leucine (a highly hydrophobic and helix-forming residue) and serine (the smallest uncharged polar residue). Their sequences were made so that the helix would be amphiphilic, with all the serine residues on one side. This would promote aggregations in which the polar sides of the helices would line the walls of the transmembrane pore. One of their compounds was a 21-residue peptide containing 12 leucines and 9 serines in the sequence $H_2N-(Leu\text{-}Ser\text{-}Ser\text{-}Leu\text{-}Leu\text{-}Ser\text{-}Leu\text{-})_3-CONH_2$, written more briefly as $(LSSLLSL)_3$. They found that this would form well-defined ion channels permeable to monovalent cations when it was incorporated into bilayers. However, a similar but shorter peptide, $(LSSLLSL)_2$, did not form channels, presumably because it was too short to cross the membrane. A peptide with fewer serines, $(LSLLLSL)_3$, would not form ion channels permeable to metallic cations but would conduct protons.

Another approach is to synthesize peptides with sequences similar to those in the putative membrane-crossing segments of real channels and see whether they will form channels. Thus Montal and his colleagues have synthesized peptides corresponding to the S3 segment of voltage-gated sodium channels and the M2 segment of the nAChR and found that in each case they will form ion channels in lipid bilayers (Oiki *et al.*, 1988a,b; Montal, 1990). But not all α-helices will do this. Neither the M1 segment from the nAChR nor a segment containing the M2 amino acid residues in randomized order will produce channels. Analogous results were obtained with the CFTR; peptides with sequences corresponding to the M2 and M6 segments will form channels, but those of M1, M3, M4 or M5 will not, and neither will scrambled sequences of M2 and M6 (Oblatt-Montal *et al.*, 1994).

Differences in single channel conductances suggest that these synthetic nAChR M2 channels may be formed from different numbers of monomers, perhaps four in some cases, five or even six in others. To overcome the uncertainties produced by this effect, Montal and his colleagues made M2 tetramers and pentamers by binding the segments to appropriate short peptides. They found that such multimers would form channels with constant open-channel conductances, 26 pS (in 0.5 M KCl) for the M2δ tetramer and 40 pS for the pentamer. The pentamer figure compares well with the 45 pS single channel conductance recorded from purified nAChR channels under similar conditions (Oblatt-Montal *et al.*, 1993a,b).

5 Permeability and selectivity

In this chapter we consider questions about how easy it is for different ions to pass through open ion channels. *Permeability* refers to the rate at which any particular ion species will pass through an open channel under standard conditions. *Selectivity* compares the permeabilities of the open channel to different ion species.

Permeability: theoretical approaches

We have seen in chapter 2 that ions move through channels down their electrochemical gradients. An ion will tend to move across the membrane away from the side with the higher concentration of its own ion species and towards the side whose electrical potential is opposite in sign to its own electrical charge. Net ionic movements will thus reflect the combined effect of the concentration gradient and the electrical field. Knowledge of the electrochemical gradient enables us to predict the *direction* in which the ions will flow, but it cannot tell us what the *rate* of that flow will be.

A relatively simple way of thinking about ion movement through channels is to assume that the channel provides an aqueous medium not greatly different from the aqueous solutions on each side of the membrane, and that each ion moves independently of every other ion. The ions would then move in the channel as if they were in free solution. We can call this approach the independent electrodiffusion model. It originated in the 1940s before channel concepts were established, but some aspects are still useful in describing membrane permeability in terms of channel behaviour.

An alternative view is to consider the channel as providing sites where the permeant ion can bind to the channel wall. Each binding site constitutes an energy well and is surrounded on each side by an energy barrier. An ion moving through the channel then has to cross a series of these energy barriers and energy wells as it passes from the aqueous solution on one side of the membrane, via one or more binding sites inside the channel, and finally out to the other side. This approach usually utilizes the absolute reaction rate analysis developed by Henry Eyring and his colleagues.

Neither of these two approaches is wholly satisfactory. Each will account for some, but not all, of the features of ion channels. The electrodiffusion model is useful in defining the permeabilities of channels; the Eyring rate theory analysis may give us fruitful ideas about mechanisms of permeation.

There are other ways of modelling channel permeability. One method is to consider the ion channel as a continuum with the ions subject to diffusive processes, but without assuming that ion movements are independent of one another (Levitt, 1986). More heroically, one can try to calculate all the movements of all the ions and water molecules passing through the channel on a sub-picosecond time scale. This procedure is called molecular dynamics (see Polymeropoulos & Brickmann, 1985; Karplus & Petsko, 1990). It requires a precise knowledge of the positions of the various atoms in the lining of the channel pore, a good understanding of the forces between them, and a vast amount of computer time. The method has been used to calculate the movement of water molecules and ions in the gramicidin channel (Chiu *et al.*, 1989; Poxleitner *et al.*, 1993; Roux & Karplus, 1994), but any extension to more typical channels must wait until we have much more detailed information about their structure.

One difficulty with complex theoretical analyses is that as they get more complicated, so they become less subject to adequate testing procedures. Simple theories are attractive because they can more readily be related to experimental results, but the consequence of this is that they are more readily disproved. But when the simple theories have been shown not to fit the facts, there is no alternative but to look for more complex ones. We shall not examine these more complex theories here; they are beyond the scope of this book and they have not yet met with general acceptance in the field. Nevertheless, their development may well be necessary for further understanding of what happens in channels. Those interested in pursuing these ideas further should look at the treatments by Läuger (1982), Levitt (1986), Cooper *et al.* (1988), Dani & Levitt (1990), Partenskii & Jordan (1992), Chiu *et al.* (1993) and Skinner *et al.* (1993).

The independent electrodiffusion model

Goldman in 1943 produced a model of ion movement through the plasma membrane that did not include the concept of channels, but nevertheless has been very influential in later thinking about them. He assumed that the membrane was homogeneous in nature and that ions would cross it independently of one another, moving at a rate proportional to that in free solution. He assumed that the electrical gradient across the membrane was linear, hence his analysis is sometimes known as the *constant field theory*. His approach was taken up by Hodgkin & Katz in 1949, in analysing the properties of squid axon membranes. The equations that they produced are known as the Goldman–Hodgkin–Katz or GHK equations, and have been much used since then in the description of channel properties. Let us have a look at them.

First consider the diffusion of potassium ions down a concentration gradient from the inside (i) of the cell to the external medium (o) through a thin membrane. We assume initially that there is no potential difference across the membrane. The rate of movement of ions outward through the membrane is the efflux J_o, measured in moles per second per unit area of membrane. The rate of movement in the opposite direction, inwards through the membrane, is the influx J_i. The net flux J_K is the difference between J_o and J_i. The net flux will be proportional to the concentration gradient:

$$J_K = P_K \left([K]_i - [K]_o \right) \tag{5.1}$$

where the constant of proportionality, P_K, is an important quantity called the *permeability* or *permeability coefficient*. Look at the units in equation 5.1: if J_K is measured in mol cm^{-2} s^{-1} and the concentrations in mol cm^{-3}, then (since the units on one side of an equation must balance those on the other) P_K must be measured in cm s^{-1}, the dimensions of a velocity.

It is assumed in the constant-field model that ions will partition themselves between the homogeneous membrane and the aqueous solutions each side in a constant ratio that we can call the partition coefficient β. Rates of diffusion through the membrane will be proportional to a diffusion coefficient D of the ion in the membrane phase (see equation 2.8) and inversely proportional to its thickness a. Then P_K is equal to $\beta D_K / a$.

If there is a potential difference E between the two compartments, we need something more complicated than equation 5.1 to describe the net flux. Goldman showed that, provided the assumptions of the constant-field model apply

$$J_K = P_K \frac{E z F}{RT} \left[\frac{[K]_i - [K]_o \exp \left(-E z F / RT \right)}{1 - \exp \left(-E z F / RT \right)} \right] \tag{5.2}$$

We can convert the net flux into units of electrical current by multiplying by Faraday's constant F and the charge on the ion z, so that the current per unit area carried by potassium ions, I_K, is then

$$I_K = P_K \frac{E z^2 F^2}{RT} \left[\frac{[K]_i - [K]_o \exp \left(-E z F / RT \right)}{1 - \exp \left(-E z F / RT \right)} \right] \tag{5.3}$$

This equation is sometimes known as the *GHK current equation* for potassium ions. It implies that the relation between current density I_K and voltage E is not linear, in other words there is some rectification present. At large positive values of E, the slope of the current voltage curve approaches $[K]_i$, whereas at large negative voltages it approaches $[K]_o$.

Equations similar to 5.3 can be written for sodium and chloride currents. These expressions can then be summed to give the total ionic current through the membrane. If the system is in a steady state so that there is no net current flowing through the membrane (as at the resting potential, or at the reversal potential for some induced change), then we get the *GHK voltage equation*:

$$E = \frac{RT}{F} \ln \left[\frac{P_K[K]_o + P_{Na}[Na]_o + P_{Cl}[Cl]_i}{P_K[K]_i + P_{Na}[Na]_i + P_{Cl}[Cl]_o} \right] \tag{5.4}$$

where P_K, P_{Na} and P_{Cl} are the permeability coefficients for the potassium, sodium and chloride ions. There are other versions of this equation that take account of the roles of calcium and other ions, with some extra complications when monovalent and divalent ions are considered in the same expression (see Lewis, 1979).

Describing selectivity with the GHK equations

Many ion channels are remarkable for their selectivity, as is indeed suggested by their names. But do sodium channels let only sodium ions through, and do potassium channels let only potassium ions through? Perhaps, on the contrary, such channels may be slightly permeable to ions other than the ones they are primarily selective for. One way of testing this idea is to see how the ionic concentration gradients affect the relations between current and voltage for the open channels.

Consider a channel that is permeable only to sodium ions, and suppose the external sodium ion concentration is nine times the internal concentration. Then the sodium equilibrium potential E_{Na} is given by the Nernst equation for sodium ions:

$$E_{Na} = \frac{RT}{F} \ln \frac{[Na]_o}{[Na]_i}$$

which works out close to $+55$ mV. If the membrane potential is negative, or less positive than $+55$ mV, then the flow of sodium ions will be inward when the channels open. If the membrane potential is more positive than $+55$ mV then the sodium ion current flow will be outward. In other words, the current reverses in sign as we pass through $+55$ mV, which is thus (as we have seen in chapter 2) known as the reversal potential for this current.

But suppose the sodium channels will allow some potassium ions to flow through. At $+55$ mV the sodium ions are in equilibrium and so there is no net sodium ion flow in either direction. But the potassium ions are not in equilibrium (E_K is likely to be about -80 mV, for example), and so they will be flowing outwards. In order to get a net zero current flow (i.e. in order to reach the reversal potential), we must balance the small outward flow of potassium ions with a small inward flow of sodium ions, and to do this we must set the membrane potential at a value less positive than $+55$ mV. So the reversal potential will be different from the value of E_{Na}. To put the argument the other way round, if the reversal potential is not equal to the equilibrium potential for an ion that we know flows through the open channels, then some other ion must also be passing through them.

In a situation like this we need some means of describing in numerical terms the relation between the permeabilities for the different ions. Equation 5.4 is particularly useful in this respect. Suppose, for example, we wish to describe the relative permeabilities of sodium channels to sodium and potassium ions. Neglecting chloride ions (a safe assumption for cation-selective channels), equation 5.4 becomes

$$E = \frac{RT}{F} \ln\left[\frac{[\mathrm{Na}]_o + (P_K/P_{Na})[\mathrm{K}]_o}{[\mathrm{Na}]_i + (P_K/P_{Na})[\mathrm{K}]_i} \right]$$

A similar expression can be used to describe the reversal potential of currents through potassium channels, which may have some low permeability to sodium ions:

$$E = \frac{RT}{F} \ln\left[\frac{[\mathrm{K}]_o + (P_{Na}/P_K)[\mathrm{Na}]_o}{[\mathrm{K}]_i + (P_{Na}/P_K)[\mathrm{Na}]_i} \right] \tag{5.5}$$

Notice that we can work out the *ratio* of the permeability coefficients using this type of equation even if we do not know their absolute values. Such ratios are robust and useful as empirical descriptions of channel properties.

These equations can be applied to single channels. Instead of the permeability coefficients relating to a particular area of membrane (P_K for a square centimetre, for example), we use the lower case p_K etc. to denote the single channel permeability. The units for p_K are now $cm^3\,s^{-1}$, with the dimensions of a volume flow rather than a velocity. Figure 5.1 shows equation 5.5 applied to the reversal potential for currents through ATP-sensitive potassium channels at different external potassium ion concentrations. Here $p_{Na}/p_K = 0.015$, implying that the channel permeability for potassium ions is 65 times that for sodium ions.

A more rigorous way of estimating relative permeabilities is to substitute a test ion for a standard one and see how this affects the reversal potential. Suppose we wish to see which of the alkali metal ions will pass through sodium channels. With just sodium chloride in the external solution, and if chloride permeability is negligible, equation 5.4 becomes

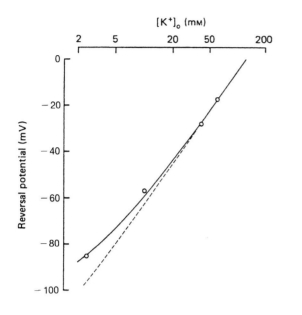

Fig. 5.1. Reversal potentials for ATP-sensitive potassium channel unitary currents, determined from patch clamp measurements at different external potassium ion concentrations. The dashed line is the Nernst equation (equations 2.9 and 2.10) for potassium ions. The full line is the Goldman–Hodgkin–Katz (GHK) voltage equation (equation 5.5) with $[K]_i = 120$ mM and $p_{Na}/p_K = 0.015$. Notice the logarithmic scale for $[K]_o$. (From Spruce *et al.*, 1987a.)

$$E_{\mathrm{rev,Na}} = \frac{RT}{F} \ln\left(\frac{P_{\mathrm{Na}}[\mathrm{Na}]_o}{P_{\mathrm{Na}}[\mathrm{Na}]_i + P_{\mathrm{K}}[\mathrm{K}]_i}\right)$$

When the sodium is replaced by the test ion X in the external solution, we have

$$E_{\mathrm{rev,X}} = \frac{RT}{F} \ln\left(\frac{P_{\mathrm{X}}[\mathrm{X}]_o}{P_{\mathrm{Na}}[\mathrm{Na}]_i + P_{\mathrm{K}}[\mathrm{K}]_i}\right)$$

whence
$$P_{\mathrm{X}}/P_{\mathrm{Na}} = \frac{[\mathrm{Na}]_o}{[\mathrm{X}]_o} \exp\left(\Delta E_{\mathrm{rev}} F/RT\right) \tag{5.6}$$

where ΔE_{rev} is the change in reversal potential, $(E_{\mathrm{rev,X}} - E_{\mathrm{rev,Na}})$.

Hille (1972) used this method to investigate the selectivity of the sodium channels of frog nerve fibres. With lithium replacing sodium in the external solution, the reversal potential changed by only -1.6 mV. This gives a value of 0.93 for the ratio $P_{\mathrm{Li}}/P_{\mathrm{Na}}$, so the permeability to lithium ions is only slightly less than that to sodium ions. With potassium substituted for sodium, however, the change in reversal potential was -59 mV, giving a $P_{\mathrm{K}}/P_{\mathrm{Na}}$ ratio of 0.086, so the permeability of the channel to potassium ions is only one twelfth of that to sodium ions. Permeabilities to rubidium, caesium, magnesium and calcium ions were too small to measure.

Permeability ratios for various channels

The voltage-gated channels found in nerve and muscle cells are highly selective. In this group the calcium channels are the most selective, then potassium

Table 5.1. *Relative permeabilities to different cations in three voltage-gated channels*

Ion X	Na channel P_X/P_{Na}	K channel P_X/P_K	Ca channel P_X/P_{Ca}
Li	0.93	0.018	0.0024
Na	1	0.010	0.0008
K	0.086	1	0.0003
Rb	<0.012	0.91	
Cs	<0.016	0.07	0.0002
Ca	Too small to measure		1
Sr			0.67
Ba			0.40
Tl	0.33	2.30	
NH$_4$	0.16	0.13	

Data for the sodium and potassium channels from frog node (Hille, 1971, 1972, 1973), and for the calcium channel from guinea-pig heart muscle cells (Tsien *et al.*, 1987).

channels, then sodium channels. Thus P_{Ca} is typically over 1000-fold greater than P_{Na} in calcium channels and P_K is about a 100-fold greater than P_{Na} in delayed rectifier potassium channels, whereas P_{Na} is only 12-fold greater than P_K in voltage-gated sodium channels. Further details are given in table 5.1.

The sarcoplasmic reticulum potassium channel is absolutely selective for monovalent cations, but not so selective for potassium ions over sodium ions as the voltage-gated potassium channels of nerve axon membranes are. In channels from frog muscle sarcoplasmic reticulum, P_X/P_K values were in the sequence K$^+$ 1, Rb$^+$ 0.59, Na$^+$ 0.45, Cs$^+$ 0.37 and Li$^+$ 0.30 (Wang & Best, 1994).

The cGMP-gated channels that carry the 'light-sensitive' sodium current in retinal photoreceptor cells are preferentially permeable to calcium ions over sodium ions. P_X/P_{Na} values are Ca^{2+} 12, Mg^{2+} 2.5, Na$^+$ 1 (Nakatani & Yau, 1988). However, since the external sodium concentration is much greater than that of calcium, sodium ions carry about 80% of the current through these channels. A feature of these channels is that their conductances are much lower in the presence of divalent cations, which probably enter the channel and cause partial block. Thus, the unit conductance falls from about 25 pS in zero calcium to 0.1 pS in physiological calcium ion concentrations (Yau & Baylor, 1989).

It is firmly established that nicotinic acetylcholine receptor (nAChR) channels have five subunits, and it seems very likely that other neurotransmitter-gated channels have a similar pentameric structure. We might expect these channels to have larger pores than the tetrameric voltage-gated channels, and therefore to be somewhat less selective. In accordance with this idea,

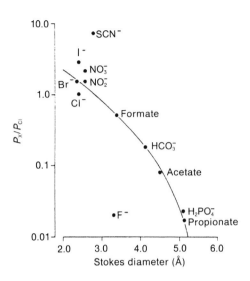

Fig. 5.2. Selective permeability of GABA-activated channels in cultured spinal neurons. Permeability ratios P_X/P_{Cl} were obtained, as in equation 5.6, by measuring the change in reversal potential after substituting the ion X for chloride, the normal permeant ion. With some exceptions (especially fluoride) the permeability ratio diminishes as the size of the ion increases. The Stokes diameters are derived from conductivity measurements, hence they are estimates of hydrated radii rather than crystal radii. (From Bormann *et al.*, 1987.)

neurotransmitter-gated channels tend to be highly selective for either cations or anions, but less selective within these groups. Thus, the nAChR channels from vertebrate skeletal muscle are selective for cations over anions (they are effectively impermeable to anions), and somewhat selective for monovalent cations over divalent cations, but only slightly selective for sodium ions over potassium ions (Adams *et al.*, 1980). Glutamate AMPA receptor channels show a similar spectrum. But NMDA glutamate receptor channels, and also some neuronal nAChR channels, are rather more permeable to calcium than to sodium or potassium ions (MacDermott *et al.*, 1986; Vernino *et al.*, 1992).

It is common for anion-selective channels to be known as chloride channels because chloride is the physiologically predominant anion, and so in the living cell chloride ions are the main carriers of current through them. Reversal potential experiments may show that other small anions are equally or more permeant. In the channels activated by the inhibitory neurotransmitters GABA or glycine the permeability sequence is $SCN^- > I^- > Br^- > Cl^- \gg F^-$, as is shown in fig. 5.2 (Bormann *et al.*, 1987). These particular channels are effectively impermeable to cations.

An exceptional example of an anion-selective channel that shows some cation permeability has been described (Franciolini & Nonner, 1987). This is a neuronal chloride channel that has a P_{Cl}/P_{Na} ratio of 5 to 1. However, the cations can only pass through it if accompanied by anions, and vice versa. Thus, sodium ions fail to carry current if the permeant chloride ions are replaced by the impermeant sulphate. Perhaps the sodium ion can screen a negative charge in the wall of the channel pore, so preventing it from excluding chloride ions.

Gap junction channels allow both cations and anions to pass through, and this is probably related to the relatively large diameter of the transmembrane pore. Relative values of P_K, P_{Na} and P_{Cl} have been determined as 1, 0.81 and 0.69 (Neyton & Trautmann, 1985). Bacterial porins, however, show some selectivity for either cations or anions, and at least one is highly cation selective (Bishop & Lea, 1994; Nikaido, 1994).

Permeability and conductance

The permeability p_X of a channel selective for the particular ion X is related to its unitary conductance γ. According to the constant field model, the relation between the two parameters at the reversal potential E_{rev} is given by

$$\gamma = p_X \frac{E_{rev} z^3 F^3}{(RT)^2} \left(\frac{[X]_o [X]_i}{[X]_o - [X]_i} \right) \qquad (5.7)$$

If the channel is permeable to more than one ion, then this equation can be modified to take account of that. For example, for a potassium channel that is also permeable to sodium ions, with a permeability ratio $\alpha = p_{Na}/p_K$,

$$\gamma = p_K \frac{E_{rev} z^3 F^3}{(RT)^2} \left[\frac{([K]_o + \alpha[Na]_o)[K]_i}{([K]_o + \alpha[Na]_o) - [K]_i} \right] \qquad (5.8)$$

Equation 5.8 was used by Spruce and his colleagues (1987a) to calculate a value for p_K from measurements of γ in ATP-sensitive potassium channels. When $[K]_o$ was 2.5 mM, γ was 14.8 pS. Using a value of 0.015 for α (as in fig. 5.1), p_K worked out at 2.6×10^{-13} cm^3 s^{-1}. However, at higher external potassium concentrations, the values of γ were lower than this value of p_K predicted, which suggests that the constant field theory has its limitations.

Another example where relations between the conductance and the permeability coefficient do not follow the predictions of the constant-field theory occurs in the GABA receptor channels illustrated in fig. 5.2. The P_{SCN}/P_{Cl} ratio shown there is 7.3, but the conductance γ in thiocyanate solutions is actually smaller than it is in chloride solutions, 22 pS as opposed to 30 pS. Clearly we need a theoretical framework that will accommodate facts like this, and we shall return to this particular example later. Let us first look at some other situations where the constant-field model cannot adequately account for the facts.

Interdependence of ion movements

Investigations such as those described in the previous section show the usefulness of the constant-field theory in describing relative permeabilities. However, ion channels often show more complicated behaviour than is predicted

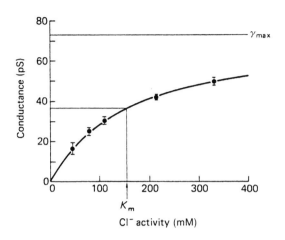

Fig. 5.3. Single channel conductance of GABA-activated channels in cultured spinal neurons at different chloride concentrations, showing saturation. Measurements were made with outside-out patches and equal chloride concentrations on each side of the membrane. The curve is drawn according to equation 5.9. (From Bormann et al., 1987.)

by the theory and the electrodiffusion model from which it is derived. These complexities arise because the movements of ions through the channels are not independent of one another.

Saturation

If each individual ion moved independently of every other ion in the system, then its movement would not be affected by the concentration of the ions present. Hence the net rate at which ions moved through an open channel would increase steadily with increasing concentration. So the conductance of an open channel would also increase continually as the concentration is raised. This does not happen: instead the conductance shows *saturation*, i.e. it rises at first quite rapidly with increasing concentration, then more slowly, and finally it reaches a maximum. An example of this effect is shown in fig. 5.3.

What does the phenomenon of saturation imply? Clearly there is an upper limit on the rate at which ions can pass through the channel. The most likely explanation is that each ion combines with one or more particular binding sites in the channel pore during its passage through it, and these binding sites can each bind only one ion at a time.

The form of the curve in fig. 5.3 is similar to those obtained in enzyme reactions when the rate of the reaction is plotted against the concentration of the substrate. The enzymic curves are described by the Michaelis–Menten relation, and this also fits the conductance curves, so that

$$\gamma = \frac{\gamma_{max}}{1 + K_m/[X]} \tag{5.9}$$

where γ is the single channel conductance, γ_{max} is its maximal value, K_m is the Michaelis constant (the concentration at which the conductance is half its maximum value) and [X] is the concentration of the ion. In order to apply this equation, it is necessary to have equal concentrations of X on each side of the membrane. The K_m value for the chloride conductance of the GABA receptor channel in fig 5.3 is 155 mM.

In some channels the conductance does not merely saturate as the permeant ion concentration is increased, it actually goes through a maximum and then decreases. This suggests that the channel pore can be occupied by more than one ion at a time, that it becomes full of permeant ions at the higher concentrations, and that full pores are less permeable than empty or partially empty ones. Perhaps it is more difficult for an ion to move to the next binding site in the channel if that site is already occupied by another ion.

The long-pore effect

Another example of departure from independence was discovered by Hodgkin & Keynes in 1955; indeed their results led them to one of the earliest explicit uses of the channel concept in its modern sense. Their experiments were concerned with the movement of potassium ions across the squid axon membrane in axons that had been treated with dinitrophenol to eliminate the sodium–potassium pump. Using radioactive potassium ions they were able to measure both inward and outward fluxes at different membrane potentials.

Ussing (1949) had shown that, if the movement of each ion is independent of its neighbours, then the ratio between the efflux J_o and the influx J_i should be given by

$$\frac{J_o}{J_i} = \exp\left[\frac{(E - E_K)zF}{RT}\right] \qquad (5.10)$$

This can be written as

$$\log_{10}\frac{J_o}{J_i} = \frac{E - E_K}{58}$$

(where E and E_K are measured in millivolts), which implies that the flux ratio should change ten-fold for every 58 mV change in membrane potential. What Hodgkin & Keynes found was that the relation was much steeper than this; a ten-fold change occurred with just a 23 mV change in membrane potential (fig. 5.4). This could be described by modifying equation 5.10 so as to introduce a factor n into the exponent, with $n = 2.5$:

$$\frac{J_o}{J_i} = \exp\left[\frac{n(E - E_K)zF}{RT}\right] \qquad (5.11)$$

What does this mean? Hodgkin & Keynes suggested that the potassium ions must move through a long narrow pore in single file. If there are n ions

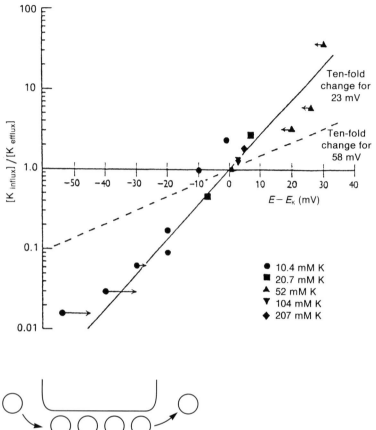

Fig. 5.4. The effect of membrane potential on the potassium ion flux ratio in cuttlefish giant axons, providing evidence for the long-pore effect. The dashed line is equation 5.10, the full line is equation 5.11 with $n = 2.5$. (From Hodgkin & Keynes, 1955.)

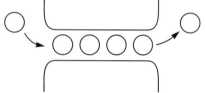

Fig. 5.5. The long-pore effect. The narrow pore region of the channel is occupied by four ions in this diagram. Three ions have to leave the channel to the right before the ion at the left end can leave to the right. Hence the movement of any one ion is closely linked to the movement of the others in the pore. If there is an overall movement from left to right (as a result of ions moving down the electrochemical gradient), then the chances of an ion moving against that flow will be much less than is predicted by an independent electrodiffusion model.

in the pore then an individual ion can only move from one side of the membrane to the other if it is nudged n times in succession from the same side by other ions (fig. 5.5). This will happen much more often if the ion is moving in the same direction as the majority of the other ions, although movements in the opposite direction may still occur occasionally. The phenomenon is known as the long-pore effect.

The anomalous mole-fraction effect

The occupancy of channels by several ions also has consequences if current is carried by a mixture of two ion species, one of which is more permeant than the other. If ions move through channels independently of one another, as is assumed by the independent electrodiffusion model, then the rate of flow of

Fig. 5.6. The anomalous mole-fraction effect in GABA-activated channels of cultured spinal neurons. The single channel conductance was measured using outside-out patches in solutions containing different proportions of chloride and thiocyanate ions. Solutions were the same on each side of the membrane, with total Cl⁻ and SCN⁻ concentrations equal to 145 mM. Notice that the conductance passes through a minimum at about 16% thiocyanate. The continuous line shows the relation predicted by a rate theory model with a single binding site; the dashed line represents that predicted by a two-site model. (From Bormann *et al.*, 1987.)

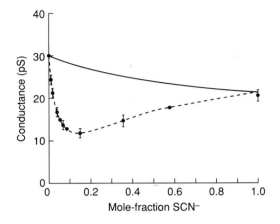

a mixture of the two permeant ions should be intermediate between the rates for their pure solutions. But experiments show that this is not always so. Figure 5.6 shows an example for GABA-activated chloride channels. The single channel conductance is 30 pS in chloride solutions, and 22 pS in thiocyanate solutions, but when both ions are present the conductance falls to a minimum of 12 pS at about 16% thiocyanate. Evidently thiocyanate interferes with the movement of chloride ions.

The most likely explanation for this effect is that for some of the time the open channel is occupied by at least two ions simultaneously. Suppose there are two binding sites for the ions, *A* and *B*, and that any ion passing through will have to bind first with *A* and then with *B*. Suppose thiocyanate binds more strongly to them than chloride does. Then a chloride ion on site *A* will have more difficulty in displacing a thiocyanate ion from site *B* than a thiocyanate ion would. The chloride ion might well leave the channel again, repelled by the thiocyanate ion, rather than continue its passage. The result is that the rate at which a mixture of the two ions can pass through the channel is less than the rate of either ion on its own.

Block by ions

There are cases in which ions get into channels but then find themselves unable to move all the way through. Such ions will block the movement of permeant ions through the channel. This blockage is likely to be affected by the membrane voltage field, which may draw the blocking ion into the pore or repel it from it. There may be competition between the blocking ion and the permeant ions; often blockage from one side of the membrane is reduced by raising the concentration of permeant ions on the other side.

Some of these blocking effects occur under physiological conditions, generating rectification in the relation between the current through a channel and the membrane potential: the conductance is lower when the current flows in one direction rather than in the other. Magnesium ions may generate such rectification. Currents through NMDA receptor channels show outward rectification (outward currents are larger than inward currents) because extracellular magnesium ions enter the pore from outside to block inward currents (Nowak *et al.*, 1984). Inward rectifier potassium channels, in contrast, are blocked by intracellular magnesium or polyamine ions so that potassium ions cannot readily move out through them (Vandenberg, 1987; Matsuda, 1988; Fakler *et al.*, 1995). Voltage-gated potassium channels may show some blockage by internal magnesium and sodium ions (Forsythe *et al.*, 1992). Sodium channels can be partially blocked by protons (Woodhull, 1973).

Study of channel blockage by various ions has been used to probe the nature of the pore through which ions move. We shall look at some aspects of this in the following section and at others in chapter 7.

Binding site models

Binding site models postulate that ions pass through channels by combining with one or more particular sites in the channel pore. Each site can bind only one ion at a time, and there are energy barriers separating these various stages in the ion's journey through the channel.

The simplest situation would be when there is a single binding site in the channel. We can represent the passage through the channel by the following scheme, where B is the binding site and X is the permeant ion:

$$X_o + B \underset{k_{-1}}{\overset{[X]_o k_1}{\rightleftharpoons}} XB \underset{[X]_i k_{-2}}{\overset{k_2}{\rightleftharpoons}} X_i + B \qquad (5.12)$$

In passing from free solution on the outside of the membrane to the binding site in the channel, the ion has to cross an energy barrier, and it has to cross another one when passing from the binding site to the aqueous solution inside the membrane. We can draw a hypothetical energy profile to show this (fig. 5.7); the binding site is an energy well between two peaks. The voltage across the membrane will determine how easily the ion climbs the energy barriers, and the transition rate constants k_1, k_{-1} etc. will be voltage dependent.

Rate theory models

Absolute reaction rate theory was developed by Henry Eyring to relate the values of the rate constants in a chemical reaction to the energy barriers separating the different stages (Glasstone *et al.*, 1941; Eyring *et al.*, 1949). It has

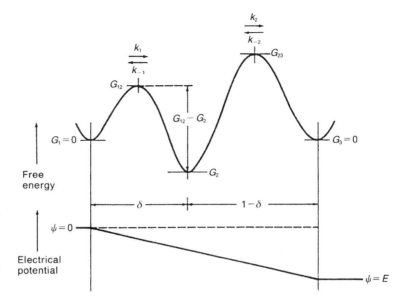

Fig. 5.7. Energy barrier model
for a channel with a single
binding site. The curve shows
how the chemical free energy
of an ion varies with its
position in the channel.
Heights of the barriers and the
well are measured with respect
to the external surface G_1. The
position of the well G_2 is given
by the fraction δ of the
electrical field at which it
occurs. It is common to
position the barriers midway
between the wells they
separate. The rate constants for
hopping from one minimum to
another (e.g. k_1 for the
transition from G_1 to G_2) are
shown. Barrier heights and well
depths will be different for
different ions. (From Hille,
1992a.)

been used to make useful models of a number of different ion channels. It
can provide explanations for some of the channel properties described in the
previous section, and has proved productive in focussing ideas on the nature
of permeation.

A transition rate constant in a rate theory analysis is related to the product
of two components: the height of the energy barrier that has to be sur-
mounted, and the fraction of the voltage field that acts on the ion as it climbs
it. The first component is equal to $\exp(-\Delta G^{\ddagger}/RT)$, where $-\Delta G^{\ddagger}$ is the
change of energy (J mol^{-1}) between the foot and the peak of the barrier. For
the rate constant k_1 in scheme 5.12 and fig. 5.7, $-\Delta G^{\ddagger}/RT$ is equal to
$G_{12}-G_1$, where the G values are expressed in RT units (RT is 2.436 kJ mol^{-1}
at 20 °C). The second component is equal to $\exp(-z\delta EF/2RT)$; here δ gives
the position of the energy well expressed as a fraction of the whole electrical
field across the membrane and we assume that the peak of the energy barrier
occurs halfway there at an electrical distance $\delta/2$ from the start. So the complete expression for k_1 is

$$k_1 = \nu \times \exp[-(G_{12} - G_1) - z\delta EF/2RT] \tag{5.13}$$

The constant ν (nu) is called the frequency factor, generally taken to be kT/h,
where k is the Boltzmann constant, T is the absolute temperature and h is the
Planck constant. kT/h is 6.1×10^{12} s^{-1} at 20 °C. We can write similar equations for the other rate constants in the model:

$$k_{-1} = \nu \times \exp[-(G_{12} - G_2) + z\delta EF/2RT] \tag{5.14}$$
$$k_2 = \nu \times \exp[-(G_{23} - G_2) - z(1 - \delta)EF/2RT]$$

and $\qquad k_{-2} = \nu \times \exp[-(G_{23} - G_3) - \chi(1 - \delta)EF/2RT]$

The net flux through the channel can be obtained by calculating the unidirectional fluxes over one of the two barriers, taking into account the probability that the binding site well is occupied. An ion can enter a well only if it is not already occupied by another ion, and clearly an ion can only leave a well if it is there already, i.e. if the well is occupied. So we need to know the probabilities for these two situations. The probability p_U that the well will be unoccupied at any particular instant is given by the sum of the rate constants for leaving the well divided by the sum of all the rate constants:

$$p_U = \frac{k_{-1} + k_2}{[X]_o k_1 + k_2 + k_{-1} + [X]_i k_{-2}}$$

Similarly the probability p_O that it is occupied is given by the sum of the rate constants for entering the well divided by the sum of all the rate constants:

$$p_O = \frac{[X]_i k_{-2} + [X]_o k_1}{[X]_o k_1 + k_2 + k_{-1} + [X]_i k_{-2}}$$

The efflux over the outer barrier is then $k_{-1} p_O$, and the influx is $[X]_o k_1 p_U$. The difference between the two gives the net flux j_X:

$$j_X = \frac{[X]_i k_{-2} k_{-1} - [X]_o k_1 k_2}{[X]_o k_1 + k_2 + k_{-1} + [X]_i k_{-2}}$$

The unitary current will then be given by the net flux, measured in ions s^{-1}, multiplied by the elementary charge e_0 and the charge number of the ion z.

Few channels actually behave in such a simple way that their properties can be described by a single site model. The treatment does, however, allow us to lay down certain ground rules in thinking about selectivity and permeability:

(1)   The ease with which an ion gets into a channel (or whether it does so at all) will be determined by the height of the energy barrier it has to traverse in order to get into the pore from the extracellular or intracellular solution. Part of the process contributing to the barrier is simply diffusional access into the channel. Also the ion may have to lose part or all of its hydration shell, breaking the ion–dipole bonds that hold the water molecules in place. A narrow pore diameter may produce a high energy barrier, since diffusional access will be restricted and there may be little or no room for any hydration shell.

(2)   Energy barriers determine the selectivity sequences (as in table 5.1) given by measurements of the change in reversal potential as one ion is substituted by another. The barriers will be high for the less permeant ions.

(3)   The speed with which an ion moves through the channel will also be

determined by the depth of the well. A cation may meet a negative charge in the pore wall to which it is attracted electrostatically. Different ions may be attracted more or less strongly; the stronger the attraction to the binding site, the deeper the well. The deeper well, the slower are the rate constants for leaving it, and so the slower the overall permeation process. It is this feature of the model that can account for some of the apparently paradoxical behaviour of channels: with a deep well the relative permeability for an ion may be high while its conductance is low, a feature that cannot be explained by simple electrodiffusion models

It is important to realise that the energy profile of the channel – the pattern of energy wells and barriers – cannot be determined directly and independently of the ion flow measurements that it is used to explain. Nevertheless, rate theory models have been very useful in thinking about the nature of permeation. The suggestions that channels include one or more binding sites for the permeant ions, and that one or more ions can occupy the channels at any particular time, are now generally accepted. It seems clear that all channels are not the same in these respects. Some channels behave as though they contain a single ion binding site in the pore, others as though they have multiple binding sites. Those with multiple binding sites may or may not behave as though they contain more than one ion at once. Let us look at some examples.

The voltage-gated sodium channel

Certain results of the simple model can be used in modelling the effect of blocking ions. An instructive and influential analysis was that carried out by Woodhull (1973) on the blockage of sodium channels in frog nerve by protons. She found that an increase in the hydrogen ion concentration reduces the sodium permeability of the nerve and that this reduction is dependent on the membrane potential; it is less at more positive membrane potentials.

The results could be explained if there were a negatively charged binding site for sodium ions in the pore at some distance from its mouth, and if this could be temporarily occupied by a hydrogen ion. Hydrogen ions could access or leave the site from the external side of the channel only. Using the model shown in fig. 5.7, the binding site would be the energy well at G_2, and the inner energy barrier (G_{23} in fig. 5.7) would be too high for protons to surmount.

We can write the reaction between hydrogen ions and the binding site as follows:

$$H_0^+ + B^- \underset{k_{-1}}{\overset{[H^+]_o k_1}{\rightleftharpoons}} HB \tag{5.15}$$

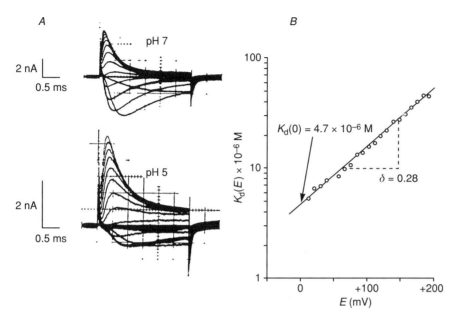

Fig. 5.8. Block of voltage-gated sodium channels by protons. *A* shows families of sodium currents from voltage clamped frog nerve at low (pH 7) and high (pH 5) hydrogen ion concentrations. The inward currents at lower membrane potentials are reduced much more at pH 5 than are the outward currents at the more positive potentials, so the partial block at pH 5 is voltage-dependent. The dissociation constant $K_d(E)$ is calculated from these currents using equation 5.19. *B* shows how $K_d(E)$ varies with membrane potential; the slope of the line gives the position of the binding site in the electric field as the electrical distance δ. (From Woodhull, 1973. Reproduced from the *Journal of General Physiology* 1973, **61**, pp. 687–708, by copyright permission of The Rockefeller University Press.)

We can assume that the energy barriers do not themselves change with membrane potential, so we can simplify equations 5.13 and 5.14 to give

$$k_1 = a_1 \times \exp(-z\delta EF/2RT) \tag{5.16}$$

and
$$k_{-1} = a_{-1} \times \exp(z\delta EF/2RT) \tag{5.17}$$

so that the rate constants are exponential functions of membrane potential. The dissociation constant K_d for scheme 5.15 is k_{-1}/k_1. Substituting equations 5.16 and 5.17 in this expression we get

$$K_d(E) = \frac{a_{-1}}{a_1} \exp\left(\frac{z\delta EF}{RT}\right) \tag{5.18}$$

Here $K_d(E)$ is the voltage-dependent dissociation constant; it increases e-fold (i.e. by a factor of 2.718) for every $25/z\delta$ mV increase in E. As we might expect, extracellular hydrogen ions are attracted into the channel at negative membrane potentials and driven out at positive ones. The affinity therefore appears greater at negative potentials and falls as the membrane is depolarized. In the absence of a membrane potential the dissociation constant is equal to a_{-1}/a_1, and we could write this as $K_d(0)$. The dissociation constant can be measured experimentally at any given voltage by comparing the macroscopic sodium currents at pH 7 and pH 5. The current will be proportional to the fraction of unoccupied channels (y), which is given by a modification of equation 5.9:

$$I_{Na} \propto y = \frac{K_d(E)}{K_d(E) + [H^+]_o}$$

Fig. 5.9. Hille's energy barrier model for the voltage-gated sodium channel of frog nerve (left), and its molecular interpretation (right). The deepest well (position 2) determines current amplitude (it is deeper for blocking ions such as thallium ions and protons); it might be associated with a carboxylate oxygen atom. The highest barrier (position 23) determines the selectivity sequence; it is postulated that this is the narrowest part of the pore, where the sodium ion is in contact with the fewest water molecules. (From Hille, 1975. Reproduced from *The Journal of General Physiology* 1975, **66**, pp. 535–60, by copyright permission of The Rockefeller University Press.)

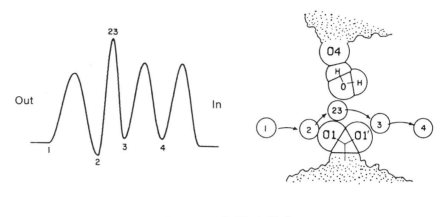

So
$$\frac{I_{\text{Na}-\text{pH5}}}{I_{\text{Na}-\text{pH7}}} = \frac{K_\text{d}(E) + 10^{-7}}{K_\text{d}(E) + 10^{-5}} \tag{5.19}$$

from which $K_\text{d}(E)$ can readily be calculated.

The relation between $K_\text{d}(E)$ and voltage is shown in fig. 5.8. $K_\text{d}(0)$ is 4.7×10^{-6} M and δ is 0.28. The simplest interpretation of this is that hydrogen ions move through 0.28 of the membrane potential field to reach their blocking site. This fraction is known as the *electrical distance* of the binding site through the membrane; it will not necessarily be the actual distance. The blocking site is a negative charge, the anion of a weak acid (such as the carboxyl oxygen atom in the side-chain of a glutamate or aspartate residue) with a pK_a of $-\log_{10}(4.7 \times 10^{-6})$ or 5.3.

Let us now look at a more complete model of the sodium channel that was developed by Hille (1975), partly as a development of Woodhull's work. It was one of the first attempts, long before the amino acid sequence of any channel was known, to relate permeation to structure. The channel model has three binding sites, although only one of them can be occupied at any one time, so its energy profile has four energy barriers and three wells, as is shown in fig. 5.9.

Permeability ratios, measured from reversal potentials, are determined in Hille's model by the height of the highest barrier, 23 in fig. 5.9. He suggested that this occurs at the narrowest part of the channel, where the permeant ion is in contact with the fewest water molecules. The height of this barrier was $9RT$ for sodium ions, $9.1RT$ for lithium ions, but $11.7RT$ for potassium ions and more than $14RT$ for rubidium and caesium ions. These heights predict the relative permeabilities shown in table 5.1.

The current through the channels is determined by the deepest well, placed where Woodhull (1973) estimated the site for blockage by hydrogen ions to be. Here, it is suggested, lies the negative charge of a carboxylic acid oxygen atom, and immediately beyond it is the narrowest part of the channel and the highest energy barrier. The well had a depth of $-1.0RT$ for sodium ions, with an apparent dissociation constant of 370 mM (equation 5.9). For ions that block the channel, the well was deepened to $-3.0RT$ for thallous ions and $-12.6RT$ for protons.

The acetylcholine receptor channel

The behaviour of the nAChR channel has been modelled by a single binding site between two energy barriers (Lewis & Stevens, 1979; Dani, 1989). It is one of the few channels to which this simple model has been successfully applied. Evidence for the binding site comes from measures of channel conductance at different sodium ion concentrations; it shows saturation in accordance with equation 5.9. Evidence that there is only one binding site comes from the findings that there is no long-pore effect and no anomalous mole-fraction effect.

GABA receptor chloride channels

We have already described some experiments by Bormann and his colleagues (1987) on GABA receptor channels, with the results shown in figs. 5.2, 5.3 and 5.6. Much of the behaviour of these channels can be described by a rate theory model with two positively charged binding sites, such as might be formed by the $-NH_3^+$ groups of basic amino acids. The model energy profile (fig. 5.10) has wells at these two binding sites and three energy barriers, with the outer and inner barriers equal to each other and a little higher than the middle one. The model also includes a factor for mutual repulsion between permeant ions when both wells are occupied by them, with the repulsion by thiocyanate ions greater than that by chloride.

In order to account for the result (shown in fig. 5.2) that P_{SCN}/P_{Cl} is 7.3, the barriers are higher for chloride ions ($9.95RT$ for the outer and inner barriers, $7.1RT$ for the middle one) than for thiocyanate ($8.35RT$ and $6.0RT$). In order to account for the lower unitary conductance in thiocyanate (22.5 pS, fig. 5.6) than for chloride (30 pS), the wells are deeper for thiocyanate ($-6.4RT$) than for chloride ions ($-3.15RT$). This also accounts for the K_m value (fig. 5.3) being lower for thiocyanate (57 mM) than for chloride (155 mM). Triumphantly, the model also explains the anomalous mole-fraction effect (fig. 5.6): thiocyanate binds more tightly than chloride so that it is less easy for chloride to expel a thiocyanate ion than for thiocyanate to expel a chloride ion.

Potassium channels as multi-ion pores

The first evidence that potassium channels may contain more than one ion was provided by Hodgkin & Keynes (1955). As described above (equation 5.11), they found that the potassium fluxes in delayed rectifier channels could

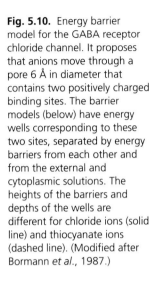

Fig. 5.10. Energy barrier model for the GABA receptor chloride channel. It proposes that anions move through a pore 6 Å in diameter that contains two positively charged binding sites. The barrier models (below) have energy wells corresponding to these two sites, separated by energy barriers from each other and from the external and cytoplasmic solutions. The heights of the barriers and depths of the wells are different for chloride ions (solid line) and thiocyanate ions (dashed line). (Modified after Bormann *et al.*, 1987.)

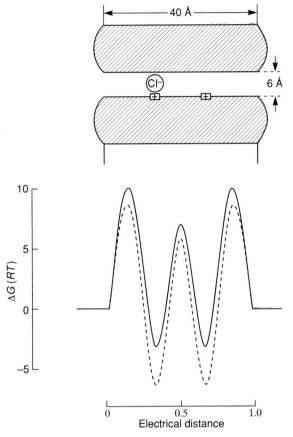

not be accounted for by the simple assumptions of the electrodiffusion model with independence of ion movements. The number of ions in the channel, n in equation 5.11, had to be more than one. Similar results have been obtained for the inward rectifier channels of frog muscle and the calcium-activated potassium channels of human red blood cells (Spalding *et al.*, 1981; Vestergaard-Bogind *et al.*, 1985).

There is further evidence that potassium channels are multi-ion pores. Anomalous mole-fraction effects occur in various different potassium channels when thallous or rubidium ions are present with potassium ions (Hagiwara *et al.*, 1977; Eisenman *et al.*, 1986; Wagoner & Oxford, 1987). Caesium ions block potassium channels in a voltage-dependent manner, with an apparent value of $z\delta$ (as in equation 5.13, for example) often greater than 1.0, which seems impossible for a univalent ion (Hagiwara *et al.*, 1976; Gay & Stanfield, 1977; Adelman & French, 1978).

Hille & Schwarz (1978) used rate theory models where the pore could be occupied by two or three ions at once to model the behaviour of potassium

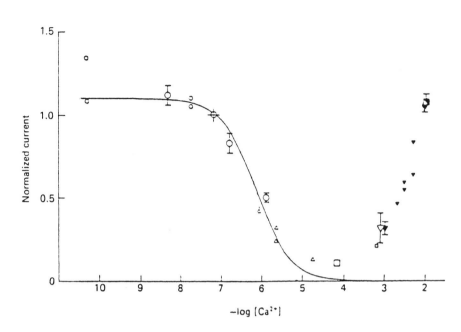

Fig. 5.11. Ion flow through the voltage-gated calcium channels of frog muscle at different external calcium ion concentrations. The points show peak inward currents when the muscle fibres were depolarized under voltage clamp, with tetrodotoxin present to block the sodium channels. Current is carried by sodium ions when pCa is below 5 (10 μM) and by calcium ions when it is above 3 (1 mM). The curve is calculated by assuming that sodium ion flow through a channel is blocked if it contains a single calcium ion. Different symbols show currents in the presence of different chelators. (From Almers *et al.*, 1984.)

channels. The energy barriers for hopping into and out of the pore in their model were higher than those for hopping between sites within the pore, and there was repulsion between ions occupying these sites. These models would simulate all the experimental evidence. The high value of $z\delta$ in the caesium block experiments could be explained by permeant ions having to vacate binding sites to make way for the blocking ion; the voltage then moves more than one ion, raising the effective value of z. Occupancy of sites towards the inside of the pore by permeant ions will tend to repel blocking ions occupying other sites, which explains how adding permeant ions inside reduces block from the outside. The model also predicts the reduction in conductance that occurs at very high potassium concentrations – the ionic traffic jam that occurs when all the sites are occupied.

Voltage-gated calcium channels

Voltage-gated calcium channels are highly selective; under normal conditions they are completely impermeable to monovalent cations (Almers *et al.*, 1984). Calcium channels are permeable to strontium and barium ions, but other divalent cations usually act as blocking agents.

If the external calcium ion concentration is greatly reduced, calcium channels become permeable to both sodium and potassium ions. Figure 5.11 shows how the current through the calcium channels in frog muscle varies with calcium ion concentration. The current falls as the calcium ion

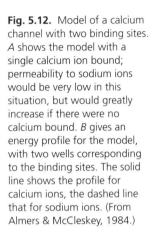

Fig. 5.12. Model of a calcium channel with two binding sites. *A* shows the model with a single calcium ion bound; permeability to sodium ions would be very low in this situation, but would greatly increase if there were no calcium bound. *B* gives an energy profile for the model, with two wells corresponding to the binding sites. The solid line shows the profile for calcium ions, the dashed line that for sodium ions. (From Almers & McCleskey, 1984.)

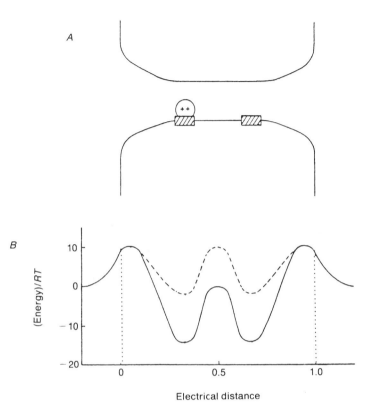

concentration is reduced from 10 mM (pCa = 2), reaching a minimum between 1 mM and 10 μM (pCa = 3 to 5). This implies that the calcium channels are highly selective, since other ions cannot substitute for calcium as its concentration is reduced. But then the current rises again as the calcium ion concentration is further reduced below 10 μM, reaching a maximum at 0.01 μM and below. Since there are so few calcium ions about, the current must be carried by some other ion. In this experiment sodium ions were the only other cations present, but in other experiments potassium, lithium, rubidium and caesium ions would all carry inward current in the absence of calcium ions.

Clearly, then, the selectivity of the calcium channel is dependent upon the presence of calcium ions. Almers & McCleskey (1984) produced a convincing model of how this feature might arise (fig. 5.12). They suggested that each calcium channel has two binding sites with a high affinity for calcium ions. If neither of these is occupied, monovalent ions can easily pass through the channel. If one site is occupied, then sodium and other ions are repelled and so cannot pass through the channel; the flow of calcium ions will also be low. But if both sites are occupied by calcium ions, the electrostatic repulsion between the ions will greatly increase the probability of one of them leaving the pore. Hence the rate of flow of calcium ions through the membrane is roughly proportional to the number of doubly occupied channels. We shall return to these ideas later.

The characteristics of the calcium channel are accountable for by a rate theory model with two wells (the binding sites) and three barriers, as is shown in fig. 5.12B. The depth of the wells is much greater for calcium ions than for sodium ions, but the barrier heights are just the same for the two ions. Calcium channels are often described as selecting for calcium ions by their high affinity for them; the wells in the energy profile are deeper for calcium ions. Other channels generally select by rejection of ions that do not fit; the barriers are higher for the less permeant ions.

The size of the selectivity filter

Most conceptual models of ion channels assume that there is some particular part of the transmembrane pore where selectivity for one ion over another is highest. It is known as the *selectivity filter*, a term introduced by Hille in 1971. The narrowest region of the pore will be the place where the permeant ion comes into most intimate contact with the channel protein; hence it seems likely that the selectivity filter will be here. We can imagine that it is here that permeant ions are let through and impermeant ions (if they have got that far through the channel) are finally stopped.

One way of thinking about the selectivity filter is to imagine that it has a particular size and shape, so that only ions that are small enough can pass through. Hille (1971) used this concept in his investigation of the selectivity of the voltage-gated sodium channels of frog nerve axons. He substituted a range of different organic cations for sodium, and measured their relative permeabilities from the changes in reversal potential, using equation 5.6. Some examples of his results are shown in fig. 5.13 and table 5.2.

Hille found that the relative permeability becomes smaller as the size of the cations is increased. The largest ion to pass through the channel was aminoguanidinium, with a permeability ratio (P_X/P_{Na}) of 0.06, whereas the much smaller hydroxylamine was almost as permeable as sodium, with a permeability ratio of 0.94. Impermeant ions included such relatively large ions as tetraethylammonium, but also quite small ones such as methylamine. Indeed, all ions containing methyl groups, including methyl derivatives of permeant ions, were impermeant.

The aminoguanidinium ion is a planar molecule with outer dimensions about $3.7 \times 5.9 \times 7.6$ Å; if we think of it as a postal package that has to be posted through a mail box, then the mouth of the mail box must be at least 3.7×5.9 Å, it would seem. But if this is the size of the selectivity filter, then why does it not

Fig. 5.13. (right) Ionic selectivity in the voltage-gated sodium channel of frog nerve. Each family of records shows the currents (I_m) through the sodium channels in a node, produced by a set of different clamped depolarizations. Potassium channels were blocked with tetraethylammonium ions. The sodium curves (left) were obtained in normal physiological saline. The other families of records show responses when the sodium in the external solution was substituted by the less permeant ions ammonium, guanidinium and potassium, or by the impermeant tetramethylammonium (TMA). Inward currents (downward deflections) would be carried by the external cation, outward currents would be carried mainly by internal sodium ions. Reversal potentials were determined from these curves, and permeability ratios P_X/P_{Na} (tables 5.1. and 5.2) calculated from equation 5.6. (From Hille, 1971, 1972 and 1992a. Reproduced from *The Journal of General Physiology* 1971, **58**, pp. 599–619, and 1972, **59**, pp. 637–58, by copyright permission of The Rockefeller University Press.)

Table 5.2. *Permeabilities of the frog nerve sodium channel to various organic cations*

Ion	Formula	Permeability P_X/P_{Na}
Hydroxylamine	NH_3OH^+	0.94
Hydrazine	$H_2NNH_3^+$	0.59
Ammonium	NH_4^+	0.16
Formamidine	$HN = CHNH_3^+$	0.14
Guanidine	$(NH_2)_2C = NH_2^+$	0.13
Aminoguanidine	$(NH_2)_2C = NHNH_2^+$	0.06
Methylamine	$CH_3NH_3^+$	
Methylhydrazine	$CH_3NHNH_3^+$	
Methylguanidine	$CH_3NHC(NH_2) = NH_2^+$	
Tetramethylammonium	$(CH_3)_4N^+$	
Tetraethylammonium	$(C_2H_5)_4N^+$	
Ethanolamine	$HOCH_2CH_2NH_3^+$	
Choline	$(CH_3)_3NCH_2CH_2OH^+$	
Biguanide	$H_2NC(=NH)NHC(=NH)NH_3^+$	

Data selected from Hille (1971). For the lower eight ions in the table the permeability was too low to be measured. Notice that none of the permeant ions contain CH_3- groups.

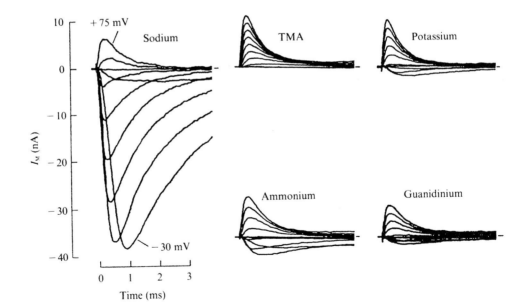

let through ions with methyl groups, whose diameter of 3.8 Å ought to allow them to just squeeze through? Hille suggested that the crucial feature is that the permeant ions all contain $-NH_2$ or $-OH$ groups, whose hydrogen atoms can form hydrogen bonds with adjacent oxygen atoms in the channel wall. Hydrogen bond formation allows the atoms involved to move closer to each other than would otherwise be the case; in ice, for example, there is an effective overlap of the hydrogen atom of one water molecule with the oxygen atom of the next by 0.84 Å (Pauling, 1960). So if the hydrogen atoms in the $-NH_2$ or $-OH$ groups of the permeant organic cations can form hydrogen bonds with oxygen atoms lining the walls of the selectivity filter, then the size of the filter may be smaller than the apparent size of the permeant ion. Hille suggested that the filter dimensions are about 3×5 Å, which would exclude methyl groups since they cannot form hydrogen bonds.

A selectivity filter of these dimensions would let through a sodium ion plus one water molecule in the transverse plane (there may well also be other water molecules in front of and behind the permeating ion in the pore). This water molecule would form hydrogen bonds with oxygen atoms lining the walls of the filter. The permeability of the channel to potassium ions is much less than to sodium ions, and the channel is impermeable to rubidium and caesium ions (table 5.1). These larger ions are too big to fit with a water molecule into the 3×5 Å filter.

None of the organic cations that pass through sodium channels can readily pass through voltage-gated potassium channels. Thus, although the crystal radius of potassium ions (1.33 Å) is greater than that of sodium ions (0.95 Å), potassium channel selectivity filters are likely to be smaller than those of sodium channels. Rubidium ions (crystal radius 1.48 Å) can pass through the potassium channel whereas caesium ions (radius 1.69 Å) cannot. Hence it seems reasonable to suggest that the radius of the selectivity filter is at least as large as the crystal radius of the rubidium ion and somewhat smaller than that of the caesium ion, suggesting a diameter between 3.0 and 3.3 Å for a circular cross-section (Hille, 1973).

The sodium ion is smaller than the potassium ion, so why does it not pass readily through the potassium channel? A possible explanation is as follows. We have seen in chapter 2 that an ion in aqueous solution is surrounded by a hydration shell of water molecules. A cation passing through the selectivity filter might replace its ion–dipole bonds to the oxygen atoms of these water molecules by similar links to the oxygen atoms in carbonyl groups of the amino acid residues lining the pore of the channel. The lining of the channel pore would thus be able to take the place of the hydration shell. If these oxygen atoms are held rigidly in position, ions smaller than potassium would be unable to make this replacement (Bezanilla & Armstrong, 1972). This idea is illustrated in fig. 5.14.

Fig. 5.14. Schematic representation of potassium and sodium ions in water and in the pore of a hypothetical potassium channel. The potassium ion bonds to all four oxygen atoms seen in the pore cross-section, whereas the sodium ion bonds to only two, with the result that its potential energy in the pore is much higher than it is in water and hence it will not readily enter the pore. (From Armstrong, 1975.)

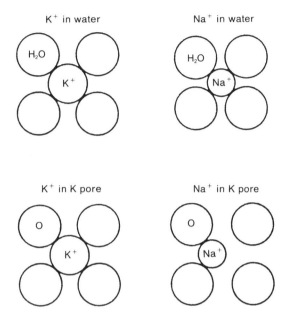

Experiments on the permeability of neurotransmitter-gated channels suggest that the selectivity filters here are considerably wider. Measurements by Bormann and his colleagues (1987) on GABA-activated channels in cultured neurons show that permeability to different ions falls with increasing ion size, reaching zero at a diameter of about 5.6 Å, as is shown in fig. 5.2. Similar experiments on the nAChR channel suggest that its selectivity filter is about 7 Å in diameter (Dwyer *et al.*, 1980).

The permeation pathways of gap junction channels are much larger. By injecting fluorescent compounds of various sizes into cells and looking for their appearance in neighbouring cells, it has been found in most cases that molecules up to about 1000 daltons in size can pass through them. This implies that the gap junction pores are about 16 Å in diameter (Schwartzmann *et al.*, 1981). Structural studies suggest that the pore is cylindrical in form and about 15 Å in diameter (Unwin & Ennis, 1984). Selectivity appears to be simply on the basis of size, and most ions of physiological importance will pass through quite readily.

Gap junction channels from different sources may have different conductances, suggesting that they may have different pore sizes (Spray, 1990). In osteoclasts channels made from connexin 43 would let lucifer yellow and similar dye compounds through, but channels made from connexin 45 would not, although both types would permit the flow of small ions (Steinberg *et al.*, 1994). There is no evidence that selectivity is associated with a particular

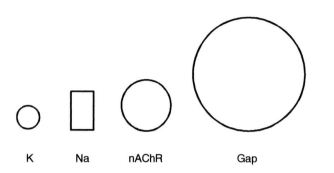

K   Na   nAChR   Gap

Fig. 5.15. Hypothetical cross-sections of the selectivity filter in four different channels, based on their permeabilities to ions of different sizes. Diameters of those with circular cross-sections are 3 Å for the voltage-gated potassium channel from frog nerve, 7 Å for the nicotinic acetylcholine receptor (nAChR) channel from frog muscle, and 15 Å for the gap junction channel. The voltage-gated sodium channel from frog nerve is shown with a 3 × 5 Å selectivity filter. (Based partly on Dwyer *et al.*, 1980.)

narrow region of the pore; most gap junction models assume that the pore is effectively a tube of constant diameter.

Figure 5.15 compares the estimated pore cross-section of gap junctions with the proposed selectivity filters of the nAChR and voltage-gated sodium and potassium channels, as suggested by Hille and his colleagues. It is notable that channels made from four subunits or domains (voltage-gated channels and potassium channels, for example) have smaller pores and higher selectivity than those made from five subunits (neurotransmitter-gated channels), and these in turn have smaller pores and are more selective than gap junction channels, which are made of six subunits. However, we should note that the cyclic-nucleotide-gated channels, which belong to the voltage-gated channel superfamily and so presumably are tetrameric, have selectivities and pore cross-sections nearer to those of the pentameric neurotransmitter-gated channels (Goulding *et al.*, 1993).

The porins of bacterial outer membranes have pore sizes about 11 Å in diameter, as judged by experiments on the movement of oligosaccharide sugars through them (Nikaido, 1994).

Water in the channel pore

Water molecules are about 3 Å in diameter so they are small enough to fit through the selectivity filters of ion channels. They can probably form hydrogen bonds with the polar groups of some of the amino acids lining the pore, hence it is generally assumed that they can pass through open channels. Experimental measurements on gramicidin channels in lipid bilayers have shown that the rate of water movement is comparable to the rate of ion movement, probably because water molecules and ions pass through together in single file (see Finkelstein, 1984).

Molecular dynamics simulations by have given us a graphic picture of water molecule movement in gramicidin channels (fig. 5.16; Chiu *et al.*,1989; Poxleitner *et al.*, 1993). They suggest that the water molecules in the channel

Fig. 5.16. Molecular dynamics simulations of water molecules and potassium ions in a gramicidin channel. Each picture shows a snapshot of the channel contents at one instant; the molecules would all move slightly in the next picosecond. *A* shows the pore filled with water molecules only. The water molecules are linked by hydrogen bonds. One water molecule (arrowed) acts as a double acceptor, so that the orientation of the others is different on each side of it. White circles in *A* show four carbonyl groups at the mouths of the pore. In *B* and *C* a potassium ion (large white sphere) is present, causing orientation of the adjacent water molecules, and in *D* three potassium ions are present. Arrows show double acceptor water molecules as in *A*, except for the left hand arrow in *D*, which shows a break in the chain of hydrogen bonds. (From Poxleitner *et al.*, 1993. Reproduced with permission from *Zeitschrift für Naturforschung*.)

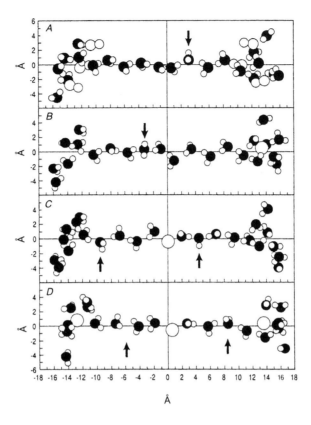

pore form hydrogen bonds with each other and also with the adjacent carbonyl groups in the gramicidin molecule. These hydrogen bonds would tend to promote chains of water molecules oriented in much the same direction, perhaps reversing in direction at some point, as is shown in fig. 5.16*A*. This orientation is enhanced if there is an ion in the pore.

Streaming potentials

The relation between movements of water molecules and ions can tell us something about the length of the narrow pore region in channels. If this part of the pore contains n water molecules and one ion, arranged in single file so that they cannot pass by one another, then for every ion that crosses the membrane from one side to the other, n water molecules will have to cross also.

The value of n can be determined from measurements of the *streaming potential*. Figure 5.17 shows in principle how this is done and how the potential arises. A membrane containing a channel selective for cations separates two compartments containing solutions with equal concentrations of an electrolyte such as potassium chloride. At zero membrane potential the

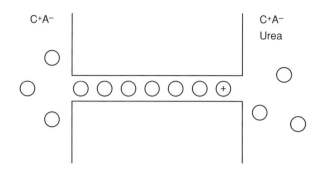

Fig. 5.17. The origin of the streaming potential. The diagram shows the narrow part of a cation-selective channel pore containing six water molecules and a cation. Solutions on each side of the membrane contain equal concentrations of a salt C^+A^- so the system is initially in equilibrium and there is no potential across the membrane. Addition of urea to the right hand side produces an osmotic gradient so that water moves through the pore from left to right. Cations are carried with the flow so that a membrane potential arises with the right hand side positive. The system reaches equilibrium when the electrochemical force on the ion balances the sum of the osmotic forces on the water molecules in the pore. This sum is proportional to the number of water molecules in the pore.

system is in equilibrium and there is no flow of current through the open channel. Now we add some impermeant non-electrolyte such as urea to the right hand compartment. This upsets the osmotic equilibrium so that each water molecule in the channel is subject to a force driving it from left to right. Hence water molecules will move through the channel carrying potassium ions with them. Consequently there will be a build-up of potassium ions in the right hand compartment and hence a potential across the membrane. This potential will tend to drive the potassium ions back from right to left, so that we rapidly reach an equilibrium position in which there is no more net flow of water and ions. The membrane potential at this point is called the streaming potential.

At this equilibrium position, the leftwards electrical force on a potassium ion in the channel will precisely balance the rightwards osmotic force exerted on all the water molecules in the channel. The more water molecules there are in the channel, the larger the osmotic force, and hence the larger the electrical force needed to balance it.

The electrical force on the cation at equilibrium is equal to $V_s z F/N_A \delta x$ (where V_s is the streaming potential, N_A is Avogadro's number and δx is the length of the pore), while the osmotic force on the water molecules in the channel is equal to $n \Delta \pi M_w / N_A \delta x$ (where $\Delta \pi$ is the osmotic pressure gradient and M_w is the partial molar volume of water). Since these two forces balance each other, we can write

$$V_s z F = n \Delta \pi M_w$$

and so the streaming potential is given by

$$V_s = n \Delta \pi M_w / z F$$

For a univalent cation at room temperature,

$$V_s = 0.46 \, \Delta \pi n \qquad\qquad (5.20)$$

where $\Delta \pi$ is measured as osmolal and V_s is given in millivolts (Rosenberg & Finkelstein, 1978).

Fig. 5.18. Streaming potential in the sarcoplasmic reticulum potassium channel reconstituted into an artificial lipid bilayer. White circles show single channel currents measured at different membrane potentials. Black circles show the shift produced by addition of 1.9 M urea to one side. The arrow marks the zero-current potential after addition of valinomycin, and the streaming potential of 2.3 mV is the difference between this and the reversal potential in the presence of urea. (From Miller, 1982a.)

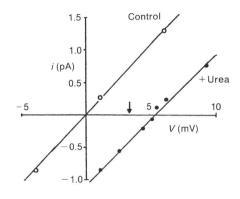

There is an extra complication that we need to consider. Addition of urea alters the activity of the cation in the right hand compartment, so that part of the potential difference is attributable to this and not to the streaming potential. The size of the correction factor can be calculated if suitable data on activity coefficients are available. For potassium ions it can be measured directly by introducing valinomycin into the membrane; this is a potassium carrier that does not involve any water movement, so the membrane potential observed is due to the potassium ion activity gradient alone.

Figure 5.18 shows the results of a streaming potential experiment on the sarcoplasmic reticulum potassium channel in a lipid bilayer (Miller, 1982a). With 250 mM potassium each side, the current voltage curve for single–channel currents was linear with a reversal potential at 0 mV. After 1.9 M urea was added to one side of the membrane, the current–voltage curve was shifted to the right, with a reversal potential at 5.5 mV. Then valinomycin was added, resulting in a huge increase in membrane conductance and a reversal potential at about 3.2 mV. The difference between these two figures, 2.3 mV, is the streaming potential. Applying equation 5.20 we get a value of 2.6 for n, implying that there are two or three water molecules for each ion in the single-file region of the channel pore. The length of this region therefore can be no greater than 10 Å (three water molecules and a potassium ion) and perhaps as short as 6 Å, provided we assume that there is usually only one ion in it at any one time,

Similar experiments have been done on some other channels. n is 2 to 4 in the calcium-activated potassium channel of muscle, giving a single-file region 6 to 12 Å long (Alcayaga *et al.*, 1989). For the gramicidin channel, n is 6 to 7 (Rosenberg & Finkelstein, 1978), and for a potassium channel from the alga *Chara* it is 9, suggesting a single-file region 28 Å long (Pottosin, 1992).

The nAChR channel has a selectivity filter which is about 7 Å in diameter (fig. 5.15). In a pore this wide, water molecules could easily bypass any of the cations that normally pass through the channel, and therefore we would not expect to see any streaming potential associated with their movement. A large enough cation, however, could sweep water molecules before it and so should produce a streaming potential. Dani (1989) used triethanolammonium ions for this purpose, and obtained streaming potentials of 2.4 mV. This suggests that five or six water molecules moved with each triethanolammonium ion. Since the water molecules will be four abreast in the narrow region, it must be about 4 to 5 Å long.

All these results fit well with the models of channels that propose a narrow region in the middle of the pore bounded by wider vestibules or mouths at its inner and outer ends. This narrow region will contain the selectivity filter, and it is longer in some channels than in others. In calculating the length of the narrow region, we have assumed that it contains only one ion. n is the ratio of water molecules to ions in the narrow region, so if there is good evidence to suppose that more than one ion is present, then the calculated length of the narrow region would be proportionately greater.

The molecular basis of selectivity

With the discovery of the amino acid sequences of channel proteins that began in the 1980s, hope was high that it would be possible to see how these sequences determine the selectivity of the various channels. First it is necessary to find out which parts of the molecule form the lining of the channel pore. Inspection of the sequences involved may suggest how selectivity arises, and these ideas may be tested by making mutant channels in which crucial amino acid residues are replaced by others. Much progress has indeed been made in this area (see Sather *et al.*, 1994). Let us have a look at some examples.

Voltage-gated potassium channels

It has become clear in recent years that the pore of voltage-gated potassium channels is probably lined at least in part by the H5 segment (also called the SS1–SS2 or P region) of the S5–S6 link. This stretch of 21 amino acid residues probably forms a hairpin loop so that the proline residues at each end are on the outer side of the membrane and the methionine and threonine residues in the middle are towards the inner side, as is shown in fig. 5.19 (Bogusz *et al.*, 1992; Brown *et al.*, 1993). A similar but not identical structure forms the selectivity filter of inward rectifier potassium channels. The inner part of the pore of voltage-gated potassium channels appears to be lined with the cytoplasmic S4–S5 loop and the cytoplasmic end of the S6 segment (Isacoff *et al.*, 1991; Choi *et al.*, 1993; Slesinger *et al.*, 1993).

Convincing evidence that the H5 segment forms the lining of the most

Fig. 5.19. The P region (H5 or SS1–SS2 loop) in two voltage-gated potassium channels, *Shaker* and Kv2. Amino acid residues are shown by the single letter code, and numbered from 0 to 20, beginning at the homologous proline residues P430 in *Shaker* and P361 in Kv2. For the *Shaker* channel (*A*), residues in circles are those that line the pore as determined by the method of individual cysteine substitution and subsequent block by silver ions (data from Lü & Miller, 1995). Mutations at D1 and T19 (cross-hatched) affect block by external tetraethylammonium (TEA) ions, whereas mutations at T11 (shaded) affect block by internal TEA (see chapter 7). For the Kv2 channel (*B*), residues in circles indicate those that combine with sulphydryl agents after individual mutagenesis to cysteine; extracellular sulphydryl agents combined with residues in white circles, whereas agents applied to the cytoplasmic side of the membrane combined with residues in shaded circles (data from Kürz *et al.*, 1995; Pascaul *et al.*, 1995). (Courtesy of Dr M.J. Sutcliffe.)

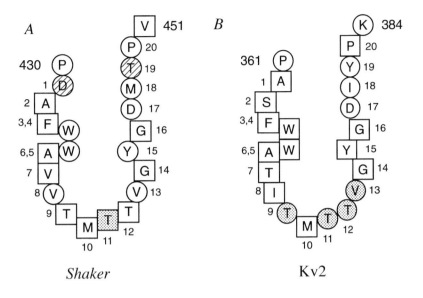

Shaker Kv2

selective part of the channel pore comes from a neat experiment in molecular engineering. NGK2 (Kv3.1) and DRK1 (Kv2.1) are two voltage-gated potassium channels from different mammalian subfamilies. The single channel conductance of NGK2 is almost three-fold that of DRK1. Hartmann and her colleagues (1991) replaced a section of DRK1 that contained most of the H5 segment with the corresponding section of NGK2. They found that the resulting chimera had the conductance properties of NGK2, even though most of its amino acid sequence was still DRK1.

The loops from the four subunits that make up the whole channel must come together to form a cohesive structure. It was initially proposed that the limbs of each loop were held together by hydrogen bonds as in an anti-parallel β-sheet, the whole arrangement forming a β-barrel, which is a fairly rigid structure (Bogusz *et al.*, 1992; Durell & Guy, 1992). If this is the case, the side-chains of the amino acid residues in the sequence would point alternately into and away from the pore. This hypothesis has been tested by mutating each H5 amino acid residue in turn to cysteine, making channels with one mutant per tetramer, and then looking at the channel conductance in the presence of silver ions. These react with the sulphydryl group of cysteine to form a covalent S—Ag link. If the silver atom is immobilized in the pore in this way the channel will be blocked, but this can only occur if the cysteine side-chain projects into the pore. In fact the residues that were unambiguously projecting into the pore were 0, 1, 4, 5, 8, 13, 15, and 17 to 20 (the numbering starts with 0 as the proline P430 of the *Shaker* sequence, as shown in fig. 5.19*A*), which is not compatible with a β-barrel structure (Lü & Miller, 1995).

Related experiments on Kv2.1 channels using larger sulphydryl reagents, which cannot penetrate the narrowest part of the pore, suggest that residues 0, 17 to 19 and 21 are exposed at the outer mouth of the pore, whereas 9 and 11 to 13 are accessible at the inner mouth (Kürz *et al.*, 1995; Pascual *et al.*, 1995); the numbering here starts with proline P361 as 0 (fig. 5.19*B*).

Most members of the voltage-gated and inward rectifier potassium channel gene families have a similar, perhaps consensus sequence within their H5 selectivity filters: Thr/Ser-X-X-Thr-X-Gly-Tyr-Gly (residues 9 to 16 in the H5 sequence, see fig. 5.19). In voltage-gated channels removal of the Tyr-Gly pair, which lines the narrowest part of the pore, eliminates the ability to select potassium over sodium ions, so the tyrosine residue may be very important in conferring selectivity (Heginbotham *et al.*, 1992). In general, potassium channels contain large numbers of the amino acids whose side-chains contain an aromatic ring, tyrosine, phenylalanine and tryptophan. It has been suggested that the electrons in the π molecular orbital, generated at the face of the aromatic rings of these side-chains by resonance of their double bonds, may provide negative sites that coordinate with potassium ions, and that these interactions are selective for potassium ions over sodium and other alkali metal ions (Heginbotham & MacKinnon, 1992; Kumpf & Dougherty, 1993).

Other residues also affect selectivity. The chimeric channel referred to above, with the NGK2 H5 segment, has a higher permeability to potassium ions than to rubidium ions, whereas this is less pronounced for DRK1. One difference between the two is that DRK1 has a valine residue at position 374 (V13 in fig. 5.19*B*) whereas the chimera has a leucine (L). The mutational change L374V in the chimera produced a reversion of the permeability changes so that the channel was now more permeable to rubidium ions (Kirsch *et al.*, 1992). Further substitutions at this point showed that hydrophobic residues (valine and isoleucine) favoured rubidium whereas polar residues (threonine and serine) favoured potassium (Taglialatela *et al.*, 1993).

Further information about the structure of the pore comes from experiments with blocking agents, especially tetraethylammonium and other quaternary ammonium ions and, more recently, charybdotoxin and related scorpion toxins. We shall look at these in chapter 7.

Sodium and calcium voltage-gated channels

The pores of sodium and calcium voltage-gated channels are probably also lined by their H5 segments. The H5 amino acid sequences are similar in the two channels but rather different from those in potassium channels. Calcium channels always have a glutamic acid residue at the same position in the four H5 repeats, about halfway along the second limb of the hairpin. This must produce a ring of four negative charges in the pore. In sodium channels, however, the equivalent residues are different in the four repeats. Only the first

Fig. 5.20. The blocking action of calcium ions on lithium currents through voltage-gated calcium channels with modified H5 glutamate rings. At low external calcium ion concentrations, normal calcium channels (wild-type, WT) are permeable to monovalent cations such as sodium or lithium, but this permeability is reduced as the calcium ion concentration is increased, as in fig. 5.11. Substitution of glutamine for the H5 glutamate residues in domains I to IV of the channel reduces this effect; substitution of two glutamates (I + IV) reduces it further. The ordinates show lithium currents (I_{Li}) produced by depolarization of *Xenopus* oocytes in which the WT or mutant calcium channels were expressed. Arrows for the different mutants show calcium ion concentrations at which I_{Li} was reduced to 50% of its maximum value. (From Yang *et al.*, 1993. Reprinted with permission from *Nature* **366**, p. 159, Copyright 1993 Macmillan Magazines Limited.)

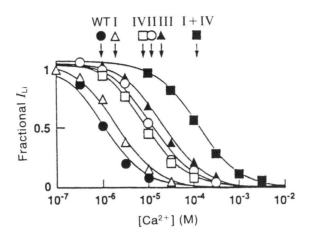

two of them (aspartate and glutamate) are negatively charged; the third (lysine) is positively charged and the fourth (alanine) is uncharged.

Does the ring of glutamate residues account for the selectivity of the calcium channel? Heinemann and his colleagues (1992) used point mutations to make the sodium channel more like the calcium channel: they replaced the lysine in the third repeat and/or the alanine in the fourth repeat with glutamic acid. Sodium currents through the mutant channels were very sensitive to block by calcium ions, and K1422E mutants showed appreciable calcium currents at higher external calcium ion concentrations. These exciting results suggest strongly that the ring of glutamic acid residues forms the selectivity filter for the calcium channel. Perhaps the equivalent ring forms the selectivity filter in the sodium channel.

A related approach is to use point mutations to alter the glutamate ring of the calcium channel. Yang and his colleagues (1993) found that substitution of glutamine or lysine for one of the glutamates reduces the blocking effects of low external calcium concentrations on monovalent cation currents (fig. 5.20), and reduces or even reverses the selectivity of the channel for divalent cations over monovalent ones. Substitutions in other parts of the H5 loop did not have these effects.

Since there is only one ring of glutamate residues, it looks as though there is no other part of the channel pore that could bind calcium ions. This is confirmed by the reduction in calcium block produced by the mutations; when the calcium concentration producing 50% block rises from 1 μM to 100 μM (fig. 5.20) or 1 mM (as occurs with lysine substitutions in domains II and III), then clearly there is no other high affinity calcium binding site in the channel pore. So the binding site model shown in fig. 5.12, with two widely separated high affinity sites in the channel, needs to be modified. Perhaps the glutamate ring can actually bind two calcium ions at once, as shown in fig. 5.21.

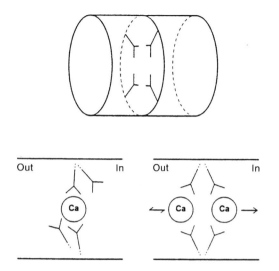

Fig. 5.21. How the ring of four glutamate residues in the calcium channel pore might bind one or two calcium ions. Forks represent negatively charged carboxyl groups. The upper diagram shows a strictly symmetrical arrangement, but some asymmetry in the system (as is suggested from the mutation experiments) would allow different glutamate residues to bind calcium at different levels. With a single ion in the ring, all four glutamates might be involved in high- affinity binding. Entry of a second calcium ion into the ring would reduce the binding affinity and so promote flow of ions through the pore. (From Yang *et al.*, 1993. Reprinted with permission from *Nature* **366**, p. 161, Copyright 1993 Macmillan Magazines Limited.)

A marked feature of these results was that the individual glutamate residues are not precisely equivalent to each other; substitution of the one in domain III has much larger effects than substitution of that in domain I, for example, as is evident in fig. 5.20. Perhaps the four glutamates occur at slightly different levels in the channel pore, and perhaps it is this irregularity that permits the ring to bind two calcium ions at once.

Cyclic-nucleotide-gated channels

The cyclic-nucleotide-gated (CNG) channels of vertebrate retinal (rod) and olfactory receptor cells have structures showing considerable similarity to those of the voltage-gated channel superfamily. Their subunits have six membrane-crossing α-helices, of which S4 shows the characteristic pattern, with every third residue carrying a positive charge, and there is an H5 region that probably forms part of the pore lining. The channels are permeable to monovalent cations, but not greatly selective between them. They are preferentially permeable to divalent cations, but such ions also cause partial block.

Goulding and his colleagues (1993) have performed some very informative experiments comparing the permeabilities of retinal and olfactory CNG channels. The single channel conductance of the olfactory channel is more than twice that of the retinal channel, and at negative membrane potentials there is a prominent subconductance level for the olfactory channel but not for the retinal channel. A chimeric channel made mostly from the retinal channel but containing the H5 sequence from the olfactory channel had the permeability properties of the olfactory channel.

The olfactory channel was more permeable to methyl-substituted

Fig. 5.22. Removal of two amino acid residues from the H5 region (here called the P region) of a potassium channel makes its permeability characteristics like those of cyclic-nucleotide-gated channels. Sequences of this region are shown for the potassium channel (*Shaker*), a retinal CNG channel (CNGC), and a *Shaker* mutant with the tyrosine (Y) and glycine (G) residues at positions 445 and 446 removed (Deletion). Voltage clamp currents from *Xenopus* oocytes in which the channels were expressed are shown, with depolarizations from −70 to +20 mV in 10 mV steps. For *Shaker* the currents are always outward, as expected for a potassium-selective channel. For the deletion mutant, currents are inward at potentials below −10 mV, suggesting that the channel is permeable to sodium and calcium ions as well. (From Heginbotham *et al.*, 1992.) A functional connection between the pores of distantly related ion channels as revealed by mutant K$^+$ channels. Reprinted with permission from *Science* **258**, pp. 1152–5. Copyright 1992 American Association for the Advancement of Science.)

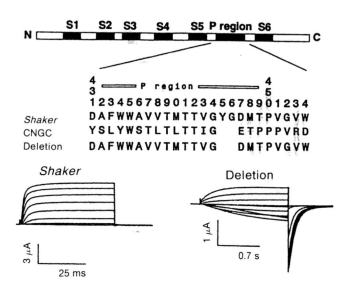

ammonium ions than the retinal channel, and the largest of these, trimethyl-ammonium, would pass through the olfactory channel but not through the retinal channel. This allowed the diameter of the narrowest part of the pore to be estimated; it was about 5.9 Å for the retinal channel and 6.4 Å for the olfactory channel. The results for the chimera were similar to those for the olfactory channel, giving further confirmation for the view that the H5 region lines at least part of the transmembrane pore and determines its selectivity.

The H5 region of CNG channels shows some homology with that in voltage-gated potassium channels. Heginbotham and her colleagues (1992) have identified the crucial difference between the two sections as two particular amino acid residues (tyrosine and glycine) that are present in the potassium channel and absent in the CNG channels. Deletion of the Tyr-Gly (YG) pair from the H5 region in a *Shaker* mutant produced channels with the permeability characteristics of CNG channels, as is shown in fig. 5.22.

The nicotinic acetylcholine receptor channel

There is good structural evidence that the membrane-crossing pore of the nicotinic acetylcholine receptor channel is lined by the M2 α-helices of its five subunits (Changeux *et al.*, 1992; Unwin 1993a). Their amino acid sequences are very similar to one another, so the pore is lined by rings of similar or identical residues, occurring at every third or fourth position along the length of the α-helices (figs. 5.23 and 5.24). Three rings of negatively charged residues seem to be particularly important in determining the selectivity of the channel. There is one on the extracellular end of the M2 region at position 20′

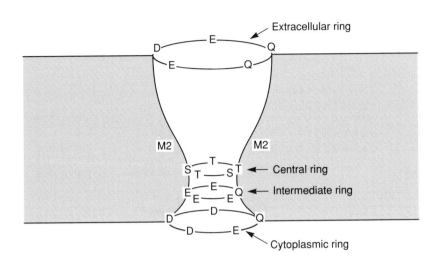

Fig. 5.23. Rings of similar amino acid residues in the M2 segments lining the membrane-crossing pore of the nicotinic acetylcholine receptor channel. Note the three anionic rings with negatively charged glutamate (E) and aspartate (D) residues, and the central ring with uncharged but polar serine (S) and threonine (T) residues. (From Imoto, 1993.)

in fig. 5.24, another one (the 'intermediate ring') just at the inner end (position −1′) of the M2 segment, and a third on the cytoplasmic side of the pore at position −4′.

Imoto and his colleagues have used point mutations to look at the importance of these rings for channel permeability (Imoto *et al.*, 1988). Reduction of the negative charge in any of these rings causes a corresponding reduction in single channel conductance. The effect is greater for changes in the intermediate ring, and furthermore changes here produce changes in the relative permeabilities to different cations (Konno *et al.*, 1991). This suggests that the intermediate ring may be part of the selectivity filter. Perhaps the extracellular and cytoplasmic rings serve to attract cations and repel anions (see Green & Andersen, 1991), whereas the intermediate ring acts as a binding site for cations.

Point mutations in another ring (the 'central ring' or 'hydroxyl ring', at position 2′) also affect single channel conductance (Imoto *et al.*, 1991; Villaroel & Sakmann, 1992). This central ring contains serine or threonine residues, both relatively small residues whose hydroxyl groups make them relatively polar. It also may be part of the selectivity filter. As a result of his high resolution electron image analysis of the nAChR, which we shall look at in the next chapter, Unwin (1995) suggests that a ring of leucine residues near the middle of the pore (position 9′) is the narrowest part of the pore in the closed channel and so acts as a hydrophobic barrier to ion flow, but that it moves out of the way when the channel opens. This movement would leave the central ring at 2′ as the narrowest part of the pore, so presumably it would form the selectivity filter.

The neuronal α7 acetylcholine receptor forms homo-oligomers (channels made of probably five identical subunits) when it is expressed in *Xenopus*

Fig. 5.24. Amino acid sequences of the M2 segments and adjacent regions in subunits of the nAChR-related neurotransmitter-gated channel family. The upper six sequences are from the mouse muscle nAChR (α1 to δ), chick neuronal α7 nAChR, and an ionotropic 5-hydroxytryptamine (5HT₃) receptor, all of which are cation selective; notice the negatively charged residues (D and E) at positions −4′, −1′ and 20′, the positively charged residues (K and R) at position 0, and the polar (S and T) residues at position 2′. The two lower sequences are from the α subunits of the GABA_A and glycine receptors, which are selectively permeable to anions; notice the positively charged R residues at position 19′ and the absence of negative charge at 20′, although the negative charge at −4′ remains. Notice also, in all the channels, the large hydrophobic L residue at 9′, and the positively charged K and R residues at 0′. (Redrawn from Lester, 1992. Reproduced with permission from the *Annual Review of Biophysics and Biomolecular Structure* **21**, © 1992 by Annual Reviews Inc.)

		−4′			−1′		2′								9′				13′							20′
nAChR	α1	D S G - E K M T L S I S V L L S L T V F L L V I V E
	β	D A G - E K M G L S I F A L L T L T V F L L L L A D
	γ	K A G G Q K C T V A T N V L L A Q T V F L L L L A D
	δ	D C G - E K T S V A I S V L L A Q S V F L L L I S K
	α7	D S G - E K I S L G I T V L L S L T V F M L L V A E
5HT₃R		D S G - E R V S F K I T L L L G Y S V F L I I V S D
GABA_AR α		E S V P A R T V F G V T T V L T M T T L S I S A R N
GlyR		D A A P A R V G L G I Y T V L Y M T T Q S S G S R A

oocytes, and these possess an appreciable permeability to calcium ions (Séguéla *et al.*, 1993). Mutation of the anionic intermediate ring so as to replace the glutamate residues by alanine greatly reduced the calcium permeability without affecting the permeability to sodium and potassium ions (Bertrand *et al.*, 1993).

A fine demonstration that selectivity is controlled by the nature of the amino acid residues lining the pore has been provided by mutagenesis experiments by Galzi and his colleagues (1992). They found that appropriate mutations in the M2 region of the neuronal α7 acetylcholine receptor changed the channel from cation-selective to anion-selective. The crucial changes were the introduction of a proline or alanine residue between −2′ and −1′, the replacement of the positively charged glutamate residue at −1′ by the neutral alanine, and the replacement of the hydrophobic valine residue at 13′ by the polar threonine. It is notable that such changes remove only one positive charge from the M2 region; neutralization of the charges at −4′ and 20′ was not a necessary component of conversion to anion-selectivity. This suggests that some of the effective mutations were concerned more with the precise positioning of the M2 segment within the channel than with a simple change in the overall charge count.

GABA and glycine receptor chloride channels

The general structure of the chloride channels opened by the inhibitory neurotransmitters GABA and glycine is similar to that of the nAChR channel. Homologies are strong in the M2 region as is shown in fig. 5.24. But there are notable differences. At position −1′ GABA and glycine receptors have a hydrophobic ring of alanine (A) residues, not the negatively charged glutamate (E) ring, and between this ring and position −2′ there is a ring of proline (P) residues inserted. Towards the outer end of the pore there is a ring of positively charged arginine (R) residues at 19′ instead of the negatively charged glutamate (E) residues at 20′ in the nAChR. Overall, then, in the section from −4′ to 20′ shown in fig. 5.24, the nAChR has three negative charges per subunit and one positive charge (an excess of ten negative charges per pore), whereas the GABA and glycine receptors have one negative charge and two positive charges (an excess of five positive charges per pore).

Glutamate receptor channels

When the various subunits in the ionotropic glutamate receptor family were first cloned it was commonly thought that there were four membrane-crossing segments and that the M2 segment might show some similarity to that of the nAChR. However, there is now good evidence that both ends of the M2 segment are on the cytoplasmic side of the membrane, so it appears to be a looped 'membrane-dipping' structure (fig. 4.14) rather than a membrane-spanning α-helix (Bennett & Dingledine, 1995; Wo & Oswald, 1995).

We have seen in chapter 4 that RNA editing can change a glutamine (Q) residue in the M2 segment to arginine (R) in certain AMPA-kainate receptors, and that this position is called the Q/R or QRN site. AMPA receptors containing subunits with arginine at the QRN site are not permeable to calcium ions, but where there is no arginine in this position they are (Hume *et al.*, 1991; Burnashev *et al.*, 1992a). With GluR6 kainate receptors, however, the situation is more complicated. If RNA editing occurs at two particular sites in the M1 segment, then calcium permeability is higher with arginine at the QRN position than with glutamine. If there is no RNA editing of the M1 segment then editing at the QRN site has little effect on calcium permeability (Köhler *et al.*, 1993). Clearly the nature of the pore in kainate receptors is rather different from that in AMPA receptors.

NMDA receptor subunits have asparagine (N) residues at the QRN position, and the channels are appreciably permeable to calcium ions. Replacement of this residue by glutamine in the NR1 subunit reduces the permeability of the channel to calcium ions. A similar substitution in the NR2 subunits increases the permeability to magnesium and reduces its blocking action, but has little effect on the calcium permeability (Burnashev *et al.*, 1992b).

6 Gating and modulation

Channels can be either open or shut. We can think that there is some structure or property of the channel that is concerned with the transition between these two states, and the word *gate* is used to describe this concept. When the gate is open the ions can flow through the channel, when it is shut they cannot. *Gating* is the process whereby the gate is opened and shut. There may be a number of different shut or open states, so the gating process may involve a number of different sequential or alternative transitions from one state of the channel to another. *Modulation* occurs when some substance or agent affects the gating of the channel in some way.

For ligand-gated channels the trigger event in gating is the binding of the neurotransmitter or internal messenger to one or more particular binding sites in the channel molecule. For voltage-gated channels, it is probably the movement of some internal sensor in response to a change in the electric field across the membrane. In each case a change in one part of the molecule produces an effect in a different part of it as the permeant pathway opens to permit the movement of ions. Gating is thus an allosteric process, involving a conformational change in the channel protein (see Perutz, 1989).

Single channel kinetics

What state changes does a channel undergo, and what are the rates of change between one state and another? Questions of this type form the subject matter of kinetics. The aim of a kinetic analysis is to describe the time course of the changes in channel properties in the hope that this will lead to ideas about their mechanism. The analysis is mathematical and can get quite complicated, so we will here provide no more than a taster. Much more thorough approaches are given by Colquhoun & Hawkes (1977, 1981,

1982, 1994, 1995) and others (Horn & Lange, 1983; Kienker, 1989; Ball & Rice, 1992).

What do we mean by the word 'state'? Channels can exist in two *conductive states*, open and closed (we neglect here the existence of subconductance states), and they change from one to the other during their functioning in the cell. Direct observation with the patch clamp technique can tell us which conductive state a channel is in. But there are also different *conformational states* of the channel protein, and a channel may pass through a number of these during the gating process. Ligand-gated channels, for example, can exist in different states according to whether they have bound one or more ligand molecules or not. Kinetic analysis may postulate the existence of different conformational states in order to explain the experimental data obtained from measurements on conductive states.

A two-state channel

Let us begin with a very simple situation. We assume that a channel exists in just two states, closed (C) and open (O). The channel can change from one state to the other at random. The changes are stochastic events – that is to say they occur at random in the time domain – and so we can describe their timing only in probabilistic terms.

When the channel is open there is a constant probability of it changing its state from O to C in a defined short period of time δt, irrespective of how long it has been in state O or how it arrived there. There might, for example, be a probability of 0.3 that the change will occur in the next 0.1 ms. This means that we can make statistical predictions about the change from O to C. Out of a large group of open channels, about 30% will have changed to state C after 0.1 ms, leaving 70% of them still in state O. In the next 0.1 ms, a further 30% of the remaining 70% will change, and so on. We can make similar predictions for a single channel over a period of time: out of a large number of occasions when the channel is open, about 30% will last only up to 0.1 ms, a further 30% of the remaining 70% will last up to 0.2 ms, and so on.

Processes with these characteristics, that the probability of a particular change in successive small time periods is constant, are examples of *Markov processes*. What happens to their constituent units in the future is unaffected by what has happened to them in the past. To be precise about it, in a Markov process if something goes through a set of states $x_1, x_2, \ldots, x_n$, then the probability of a further change to x_{n+1} is determined solely by the characteristics of x_n and is unaffected by the characteristics of x_{n-1} and all previous states. Radioactive decay is a simple example: any particular atom of a radioactive isotope may or may not disintegrate in the next 24 hours, but its chances of doing so are completely unaffected by how long it has been in existence already. The same applies to one of our two-state channels; when it is open the probability of it closing in the next millisecond is the same whether it has been open for several milliseconds already or for just a few microseconds.

The behaviour of our simple two-state channel can be described by the following scheme:

$$C \underset{k_{OC}}{\overset{k_{CO}}{\rightleftharpoons}} O \qquad (6.1)$$

Here k_{CO} and k_{OC} are *transition rate constants* for the opening and closing changes respectively. They are expressed in units of frequency, s^{-1}. Thus k_{CO} is the frequency of openings per unit closed time; a value of 20 s^{-1}, for example, means that for every second that the channel is closed there are on average 20 changes to the open state. Clearly this means that on average we have to include 20 closed periods in order to get a second of closed time, so the mean closed lifetime must be 0.05 s. In this simple two-state scheme, then, the mean open lifetime m_O and the mean closed lifetime m_C are given by

$$m_O = 1/k_{OC}$$

and
$$m_C = 1/k_{CO}$$

Notice that the mean lifetime for either state is the reciprocal of the rate constant for the change leading away from it. The higher the rate constant for the change from a particular state, the shorter the average time spent in that state.

The proportion of time that the channel spends in either state depends on both rate constants. Thus,

$$\text{proportion of time in the open state} = \frac{m_O}{m_O + m_C}$$

$$= \frac{k_{CO}}{k_{CO} + k_{OC}}$$

What about the variation in the durations of the two states? If we measure a large number of successive open times, for example, what sort of distribution do we see? Let us return to our model channel in which in the open state there is a probability of 0.3 that it will close in the next 0.1 ms. Out of 1000 open times, then, approximately 700 would remain open after 0.1 ms, 490 (70% of the remainder) would still be open at 0.2 ms, 343 at 0.3 ms, 240 at 0.4 ms and so on. In other words, the number of open times in any particular duration class falls exponentially as the duration increases.

This exponential distribution of open times can be written in terms of the rate constant for closure as follows:

$$f(t_{open}) = k_{OC} \exp(-k_{OC}t) \tag{6.2a}$$

And similarly the distribution of closed times in our two-state model is

$$f(t_{closed}) = k_{CO} \exp(-k_{CO}t) \tag{6.2b}$$

Each of these is a *probability density function* for the state it describes. Such functions have the property that the area under the curve between any two particular time values is the probability of observing a state lifetime in that duration range. Equations 6.2 can be written alternatively in terms of the mean state lifetimes m_O and m_C:

$$f(t_{open}) = m_O^{-1} \exp(-t/m_O) \tag{6.3a}$$

$$f(t_{closed}) = m_C^{-1} \exp(-t/m_C) \tag{6.3b}$$

A dwell time histogram derived from N experimental measurements of open or closed lifetimes (as in chapter 3 and figs. 3.18 and 3.19) has much in common with a probability density function. If the ordinate is divided by N then the height of any column in the histogram shows the proportion of events within that time range. This is the same as the probability of finding an event in that time range if any one of the N events were chosen at random. If our particular kinetic model happens to fit the facts for an actual channel, then the probability density function for a state should fit the appropriate dwell time histogram. For a simple two-state system it would be possible, therefore, to determine the transition rate constants directly from the dwell time histograms.

Multi-state channel kinetics

It turns out that the two-state model given in scheme 6.1 is too simple to describe the kinetic behaviour of most channels. Although there are usually only two conducting states, open and closed, each of these may include a number of different conformational states of the channel macromolecule. Hence more complex models have to be devised. Let us first consider a situation like scheme 6.1 but in which the open channel can be inactivated by converting to a third state I:

$$C \underset{k_{OC}}{\overset{k_{CO}}{\rightleftharpoons}} O \underset{k_{IO}}{\overset{k_{OI}}{\rightleftharpoons}} I \tag{6.4}$$

If the channel is in state O, it can shut by changing either to I or to C. Which of these is more likely depends on the relative size of the rate constants for the changes leading away from O. Thus if k_{OI} is greater than k_{OC} then the channel is more likely to move to I than to C, and vice versa.

The mean open lifetime m_O for scheme 6.4 is determined by two rate constants:

$$m_O = \frac{1}{k_{OC} + k_{OI}}$$

This is an example of a general rule: the mean lifetime in a particular state is the reciprocal of the sum of the rate constants for the transitions leading away from that state. An open dwell time histogram for scheme 6.4 would be fitted by equation 6.3a, which would give us an estimate of m_O but would not allow us to determine k_{OC} and k_{OI} separately.

Many channels appear to go through a series of separate states between the initial gating event and the opening. We shall see particular examples of this later. If there were two closed states and one open, for example, we would again have a scheme with four rate constants:

$$C_1 \underset{k_{-1}}{\overset{k_{+1}}{\rightleftharpoons}} C_2 \underset{k_{-2}}{\overset{k_{+2}}{\rightleftharpoons}} O \qquad (6.5)$$

The distribution of closed lifetimes in this scheme can be described by the sum of two exponentials:

$$f(t_{closed}) = a_1 \tau_1^{-1} \exp(-t/\tau_1) + a_2 \tau_2^{-1} \exp(-t/\tau_2) \qquad (6.6)$$

where a_1 and a_2 are the relative areas of the two components, and $a_1 + a_2 = 1$. Here the time constants τ_1 and τ_2 do not measure mean conformational state lifetimes, which cannot therefore be determined directly from the closed dwell time histogram. The reason for this is that when the channel moves from C_1 to C_2 it is still closed, so no element of the closed dwell time histogram corresponds exactly to the dwell time in C_1. If there are further closed states then equation 6.6 will need to have extra terms, as in

$$f(t_{closed}) = a_1 \tau_1^{-1} \exp(-t/\tau_1) + a_2 \tau_2^{-1} \exp(-t/\tau_2) + a_3 \tau_3^{-1} \exp(-t/\tau_3) + \ldots \qquad (6.7)$$

There may also be more than one open state, in which case an equation similar to 6.6 or 6.7 will be needed to describe the open time distribution. Finding suitable values of the appropriate parameters (the as and τs) to fit equations like these involves various statistical procedures, some of them demanding appreciable amounts of computer time (Horn & Lange, 1983; Horn, 1987; Colquhoun & Sigworth, 1995).

In general if we need n exponentials to fit a closed dwell time histogram, then there must be at least n different closed states. The same applies to open states, of course. To take an example, the open time histogram in fig. 3.19 can be described by a function like equation 6.3a, suggesting that there could be just one open state. The closed time histogram in fig. 3.19, however, cannot

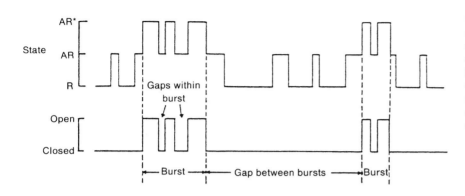

Fig. 6.1. How state transitions in a three-state scheme such as 6.5 or 6.12 give rise to bursts of channel openings separated by longer gaps. (From Colquhoun & Hawkes, 1983. Reproduced with permission, copyright Plenum Publishing Corporation.)

be fitted by a single exponential similar to equation 6.3b, but requires a function like equation 6.6 with at least two exponentials, hence there must be at least two closed states.

Bursts and clusters

A phenomenon commonly seen in single channel records is that channel openings are grouped together. They tend to occur in 'bursts' of a small number of openings, separated by longer closed periods. This is a consequence of multiple closed states such as in scheme 6.5, provided that the mean lifetime of C_1 is longer than that of C_2. After a change from O to C_2, the channel is quite likely to revert back to O instead of changing to C_1. When it eventually returns to C_1, however, there is no chance of opening until it has switched to C_2 again, and it may be some time before this happens (Colquhoun & Hawkes, 1981, 1995). Figure 6.1 illustrates this idea. For scheme 6.5 the number of closures per burst is equal to k_{+2}/k_{-1}.

If there is another closed state that interchanges with C_1 or with O with much slower rate constants than in the rest of the scheme, then bursts may themselves be grouped into 'clusters' separated by relatively long quiescent periods. Sometimes channels are described as being in different 'modes' when this happens. A further use of the modes concept occurs when the channel behaves as if most or all of the rate constants change to different values. Thus, a calcium channel may switch spontaneously between different modes, but is stabilized in one particular mode by the action of certain drugs (Hess *et al.*, 1984).

Openings after a jump

So far we have considered the kinetics of channels in a steady state situation, where conditions are constant. It is also useful to find out what happens when the conditions are changed abruptly in a stepwise fashion. With a voltage-gated channel we could depolarize the membrane suddenly in a voltage jump experiment. With a neurotrasmitter-gated channel, we could suddenly introduce the neurotransmitter in a concentration jump experiment.

In such cases it is useful to measure the time to first opening, commonly called the 'first latency'. The mean first latency will probably be longer than the mean closed time. Let us assume that scheme 6.5 applies. Before the jump the channel will be in state C_1 so it will have to pass through C_2 before it can open. In later closed periods this may not be so, since the channel may spend the whole of the time in state C_2. In terms of equation 6.6, the ratio of a_1 to a_2 will be higher in the first latency condition.

Transition times

How long does it take for the channel to make the final jump from the closed to the open state? Most kinetic schemes simply assume that the transition is effectively instantaneous, but we could imagine a situation in which the channel opens gradually so that the ionic current takes some time to reach its maximum value. The question has been examined by Maconochie and his colleagues (1995) in mouse muscle nicotinic acetylcholine receptor (nAChR) channels expressed in fibroblasts. They recorded single channel currents using an unusually high bandwidth, aligned their onset so that the times at which the current crossed 50% of its final value were identical, and used signal averaging to determine the time course of the current. They found that the change from closed to open took no more than 3 μs, the limit on the precision of their alignment procedure, so it may well have been much less. In comparison with a lower time resolution limit of 25 μs or more for most single channel records, it is clear that the change really is effectively instantaneous.

Non-Markovian models

So far we have assumed that the behaviour of channels is stochastic and best described by Markov models. The basic assumptions are that there are a small number of distinct channel states, and that the rate constants for transition between these states are independent of time. This is the approach used by the majority of investigators in the field and which is adopted in this book. There are, however, other possibilities. There could be a large number of interconvertible states, and their dynamics could be described in terms of diffusion or fractals.

In diffusion models a closed channel is viewed as diffusing away from a gateway state (from which it opens) through a large number of equivalent closed states (Millhauser *et al.*, 1988; Oswald *et al.*, 1991). Fractal models start from the observation that the pattern of openings and closings at one temporal resolution is similar to that viewed at other temporal resolutions. At low time resolution, it is said, bursts look like single openings and clusters look like bursts. Modelling here may be based on deterministic chaos rather than stochastic events (Liebovitch & Toth, 1991; Bassingthwaighte *et al.*, 1994).

Sansom and his colleagues (1989) have compared some particular sets of experimental data with the predictions of the various models. They used single channel records from delayed rectifier potassium channels and from locust muscle glutamate receptor channels. The results showed that the Markov model provided better descriptions of channel open and closed time distributions than did the alternatives.

Ligand–receptor interactions

The trigger for opening in many channels is the binding of a substance called the ligand; we can call the binding site on the channel the receptor. If we have a ligand that combines with a receptor to produce a response of some kind, then a very obvious experiment is to measure the size of the response at different concentrations of the ligand. We obtain a dose–response curve, as in fig. 6.2. A simple theory to explain the form of the dose–response curve was developed by A.J. Clark in 1926, based on the law of mass action (see Clark, 1933). It assumes that the amount of ligand taken up by the receptors is negligible relative to the total amount available, that the receptors are identical and do not interact with each other, and that the response is proportional to the number of the receptors that are occupied by the ligand. (This last assumption may be appropriate for ligand-gated ion channels, but it probably does not apply to most receptors coupled to second-messenger systems.) What follows is a brief account of elementary ligand–receptor theory; more extensive accounts can be found elsewhere (e.g. Triggle, 1979; Kenakin, 1984; Williams & Sills, 1990; Gibb, 1993).

The ligand L combines with the receptor R to form a complex LR:

$$L + R \underset{k_{-1}}{\overset{k_{+1}}{\rightleftharpoons}} LR \tag{6.8}$$

The equilibrium dissociation constant K_d, equal to k_{-1}/k_{+1}, is given by

$$K_d = \frac{[L][R]}{[LR]} \tag{6.9}$$

The proportion of receptors that are occupied by the ligand (the *occupancy* of the receptors) is

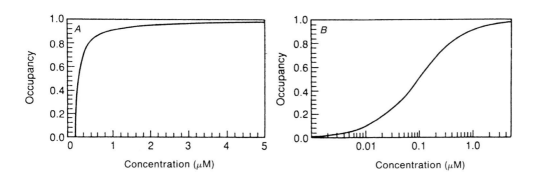

Fig. 6.2. Receptor occupancy curves for a hypothetical drug–receptor interaction with a dissociation constant $K_d = 10^{-7}$ M. The curve is a rectangular hyperbola in accordance with equation 6.10. The left hand curve (A), on a linear scale, shows the rapid rise in occupancy at low concentrations and the asymptotic approach to saturation at high concentrations. On the right (B) is the same curve plotted on a logarithmic scale of concentration, showing the sigmoid form. (From Gibb, 1993.)

$$p_o = \frac{[LR]}{[R] + [LR]}$$

and this can be combined with equation 6.9 to give

$$p_o = \frac{[L]}{K_d + [L]}$$

or
$$p_o = \frac{1}{1 + (K_d/[L])} \qquad (6.10)$$

Equation 6.10 is sometimes known as the Langmuir isotherm or the Hill-Langmuir equation. It was first used by A.V. Hill to describe the binding of nicotine in muscle, and later by I. Langmuir to describe the adsorption of gases onto metal surfaces (Hill, 1909; Langmuir, 1918). Notice that K_d is the concentration at which half the receptors are occupied, since when $K_d = [L]$, $p_o = 0.5$. The relationship between p_o and [L] is hyperbolic in form as in fig. 6.2A, but it is customary to use a logarithmic scale for [L], which produces a sigmoid curve as is shown in fig. 6.2B. K_d in fig. 6.2 is 10^{-7} M, and at this concentration p_o is 0.5. It is easy to calculate from equation 6.10 that p_o is 0.091 at 10^{-8} M and 0.909 at 10^{-6} M. If K_d were 10^{-5} M then p_o would be 0.091 at 10^{-6} M and 0.909 at 10^{-4} M. In other words, the shape of the binding curve when plotted with a logarithmic concentration scale is constant; different values of K_d simply move it sideways.

Since we usually cannot measure p_o directly, some other property has to be used as a measure of ligand–receptor action, such as the contraction of a strip of muscle, the intensity of a whole-cell current under voltage clamp, or the proportion of time that a single channel is open. The relation between the concentration of the ligand and the size of the response is then called the dose–response curve. The concentration at which the response is 50% of its maximum value is called ED_{50} or EC_{50} (effective dose or effective concentration). Often the dose–response curve is fitted quite well by equation 6.10, in which case we can assume that K_d is equal to the ED_{50}.

K_d is an inverse measure of the *affinity* of the ligand for the receptor: low values of K_d indicate a high affinity and vice versa. The reciprocal of the dissociation constant K_d is sometimes used: it is called the association constant, K_a.

Sometimes the dose–response curve is steeper or less steep than is predicted from equation 6.10. This effect might be produced by interactions of some sort between the receptors. A useful modification of equation 6.10 is the Hill equation:

$$p_o = \frac{[L]^n}{K_d^{\ n} + [L]^n} \tag{6.11a}$$

or

$$p_o = \frac{1}{1 + (K_d/[L])^n} \tag{6.11b}$$

where n is known as the Hill coefficient. The value of n can be determined as follows. Equation 6.11a can be rearranged to give

$$\log \frac{p_o}{1 - p_o} = n\log[L] - n\log K_d$$

so a graph of $p_o/(1 - p_o)$ against $[L]$ on logarithmic scales (a Hill plot) will be linear with a slope of n. If n is greater than 1 then there is cooperativity between the ligand molecules, i.e. binding of one molecule promotes the binding of another. If n is less than one there may be negative cooperativity or multiple receptor types, or desensitization, a process described later in this chapter. For ligand-gated channels, a value of n greater than 1 suggests that two or more ligand molecules need to be bound to a receptor before it becomes fully active.

Gating of the nicotinic acetylcholine receptor channel

The nAChR channels at the neuromuscular junction are in nature gated by combination with the acetylcholine molecules released from the motor nerve ending. In the laboratory they can also be gated by some other compounds known as agonists, such as carbachol or suberyldicholine.

We have seen in chapter 4 that the muscle or electric organ nAChR is a pentameric complex formed from four different subunits with the stoichiometry $\alpha_2\beta\gamma\delta$ and that acetylcholine binds only to the α subunits. Thus there are two acetylcholine binding sites in each receptor. Further evidence is provided by plotting the dose–response curve for mass responses to acetylcholine on logarithmic scales. The slope n of this graph (the Hill coefficient) gives a measure of the degree of cooperativity between different binding sites. Values of 1.5

to 2 for n have been obtained, suggesting that there are at least two binding sites per receptor, consistent with the two sites known from the molecular structure, and that binding of acetylcholine to one of them promotes binding to the other (Dionne et al., 1978).

Kinetics

Del Castillo & Katz (1957) suggested that activation of the nicotinic acetylcholine receptor channel is a two-stage process in which firstly the acetylcholine A combines with the receptor R to form a complex AR, and secondly this complex undergoes a conformational change so as to open the ion channel. This view can be represented by the reaction scheme

$$A + R \underset{k_{-1}}{\overset{k_{+1}}{\rightleftharpoons}} AR \underset{\alpha}{\overset{\beta}{\rightleftharpoons}} AR^* \qquad (6.12)$$

$$\text{closed} \qquad\qquad \text{closed} \qquad\qquad \text{open}$$

Here AR* represents the channel-open state and k_{+1}, k_{-1}, α and β are rate constants.

With the realization that two molecules of acetylcholine are bound to each channel complex, scheme 6.12 was modified to become:

$$2A + R \underset{k_{-1}}{\overset{2k_{+1}}{\rightleftharpoons}} A + AR \underset{2k_{-2}}{\overset{k_{+1}}{\rightleftharpoons}} A_2R \underset{\alpha}{\overset{\beta}{\rightleftharpoons}} A_2R^* \qquad (6.13)$$

$$\text{closed} \qquad\qquad \text{closed} \qquad\qquad \text{closed} \qquad\qquad \text{open}$$

The factor of 2 for the rate constants k_{+1} and k_{-2} arises because there are two possible forms of AR according to which of its two binding sites is occupied.

Patch clamp records of nAChR activity can be analysed to show the open and closed dwell time characteristics. These can be related to the rate constants of scheme 6.13 as follows. An open channel A_2R^* can close only by reverting to A_2R, and the rate constant for this is α. Hence

$$\text{mean duration of channel opening} = 1/\alpha \qquad (6.14)$$

The rate constant for the departure from the A_2R state will be the sum of the two rate constants leading away from it, $\beta + k_{-2}$. So, for gaps within a burst,

$$\text{mean gap duration} = 1/(\beta + 2k_{-2}) \qquad (6.15)$$

and $$\text{mean number of gaps per burst} = \beta/2k_{-2} \qquad (6.16)$$

By applying equations 6.14 to 6.16 to their data from patch clamp records of frog muscle end-plate nAChR channels, Colquhoun & Sakmann (1985)

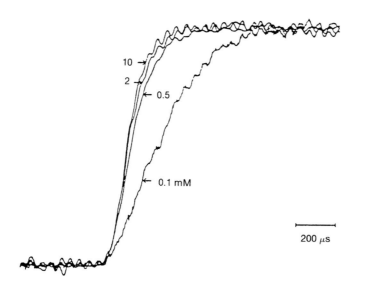

10 →

2 →

← 0.5

← 0.1 mM

200 μs

Fig. 6.3. Concentration-jump experiments with acetylcholine. An outside-out patch of mouse cell line (BC3H-1) membrane containing about 100 channels was rapidly perfused with acetylcholine. Currents are averages at the different concentrations, scaled to the same maximum. Notice that the rate of rise reaches a maximum at the higher concentrations, suggesting that β, the rate constant for the change $A_2R{\rightarrow}A_2R^*$ in scheme 6.13, is limiting. (From Liu & Dilger, 1991.)

were able to calculate the values of the rate constants involved. They found that for acetylcholine $\alpha = 714\ \mathrm{s}^{-1}$, $\beta = 30\,600\ \mathrm{s}^{-1}$ and $k_{-2} = 8150\ \mathrm{s}^{-1}$; different values were obtained for other agonists. Since β is much greater than α and k_{-2}, these values imply that the conformational change from A_2R to A_2R^* is energetically favoured, so that a channel with two molecules of acetylcholine bound to it will open rapidly and spend most of its time in the open condition. Similar conclusions have been reached from experiments on *Torpedo* nAChR channels expressed in mouse fibroblasts (Sine *et al.*, 1990).

Confirmation of the view that β is sufficiently high to lead to rapid channel opening comes from an ingenious experiment by Liu & Dilger (1991). They prepared an outside-out patch containing many acetylcholine receptors from cultured BC3H1 cells, a clonal mouse cell line that expresses muscle nicotinic acetylcholine receptors. The tip of the patch electrode projected into a stream of fast-moving saline solution whose source could be switched rapidly from one solution to another. This provided a means of changing the acetylcholine concentration in contact with the receptors very rapidly, within a matter of microseconds. The sudden jump in acetylcholine concentration produced a rapid change in current flow through the patch, as is shown in fig. 6.3.

Liu & Dilger found that the onset time (the time taken for the response to go from 20% to 80% of completion) fell as the acetylcholine concentration was raised, reaching a minimum of about 110 μs at 5 mM and above. This suggests that β, the rate constant for channel opening, is limiting at these levels, which allows it to be calculated: a value of 12 000 s^{-1} was obtained, which is in reasonable agreement with the value obtained from the frog neuromuscular junction channels by single channel analysis.

Although most channel openings occur when two acetylcholine molecules are bound to the nAChR, some recordings suggest that occasionally the channel will open either spontaneously with no acetylcholine binding or when only one molecule is bound (Colquhoun & Sakmann, 1985; Jackson 1986, 1988). Jackson (1988) has formalized this situation as follows, with a scheme in which there are three open and three closed states, corresponding to binding of zero, one or two molecules of acetylcholine or other agonist:

$$
\begin{array}{ccccc}
\text{C} & \underset{\alpha_0 \Updownarrow \beta_0}{\overset{K_1}{\rightleftharpoons}} & A_1C & \underset{\alpha_1 \Updownarrow \beta_1}{\overset{K_2}{\rightleftharpoons}} & A_2C \quad \text{closed} \\[2mm]
\text{O} & \underset{J_1}{\rightleftharpoons} & A_1O & \underset{J_2}{\rightleftharpoons} & A_2O \quad \text{open}
\end{array}
\qquad (6.17)
$$

Here the αs and βs are rate constants, the Ks and Js are equilibrium constants. Estimates for the rate constants for opening, β_0, β_1 and β_2, were respectively 0.0028, 1.1 and 2800 s^{-1}, using carbachol as the agonist, in accordance with the low chances of a receptor opening with no or one molecule of agonist bound.

Jackson's estimates of the equilibrium constants in scheme 6.17 suggested that the two binding sites are not precisely equivalent, so that the first acetylcholine molecule is bound more tightly to the closed channel than the second, and also that both molecules are bound more tightly when the channel is in the open configuration. He argues that these are essential features of the functioning of the channel, allowing rapid opening in the presence of high acetylcholine concentrations (as when acetylcholine is released from the nerve terminal) and rapid termination of the response afterwards. He also suggests that the conformational change involved in opening of the channel requires an amount of energy that is normally released only by binding two molecules of acetylcholine. So the allosteric properties of the nAChR channel make it well adapted to its function as a rapidly activated neurotransmitter-gated channel (Jackson, 1989, 1994).

The situation gets more complex at higher acetylcholine concentrations. When acetylcholine is continuously present at the neuromuscular junction, it becomes insensitive to further application; the nAChR channels will no longer open. This phenomenon is called *desensitization*. It is evident in single channel records as long periods of inactivity, interrupted by clusters of channel openings. At high concentrations a channel may spend most of its time in the desensitized state, with the binding sites occupied by acetylcholine but the channel closed. Under these circumstances the characteristics of the 'normal' state may be investigated by restricting measurements to clusters only.

The other phenomenon that emerges at higher acetylcholine concentrations is an increase in the number of very short channel closures. These appear to be caused by temporary blockage of the channel by acetylcholine

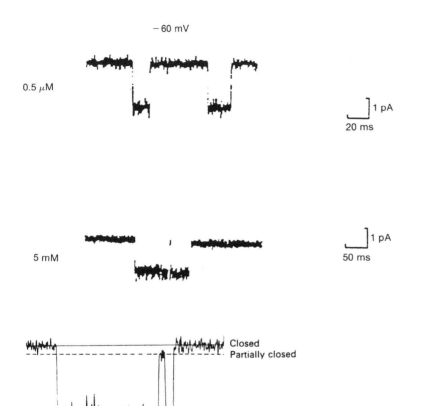

Fig. 6.4. Single channel currents in a mouse cell line (BC3H-1) in response to acetylcholine at two different concentrations, showing agonist block. At the higher concentration the currents are noisier and reduced in amplitude. The probable explanation for this is that acetylcholine molecules briefly occupy the open pore, so that the channel flickers between the open and the blocked state. The frequency of this flicker is too high to be detected by the recording system. (From Sine & Steinbach, 1984.)

Fig. 6.5. A subconductance state in the nAChR channel, seen in a cell-attached patch clamp record from the end-plate of frog muscle fibre. Current through the fully open channel was −3.71 pA. During the partial closure it fell to −0.52 pA, i.e. 14% of the full value. In other cases the subconductance levels were commonly 18% and 71%, respectively, of the full value. (From Colquhoun & Sakmann, 1985.)

molecules themselves. They may be evident as 'flickering' of the channel rapidly between the open and closed state. High frequencies of flicker may not be resolvable by the recording equipment, in which case we shall see a decrease in the mean conductance of the apparently open channel and an increase in its noise level. This effect is shown in fig. 6.4.

A further complexity in nAChR channel kinetics is the occasional occurrence of subconductance states, in which the channel conductance is lower than normal. Figure 6.5 shows an example. Subconductance states have been seen in a variety of different channels. A satisfactory explanation for their existence has, in most cases, yet to be produced.

All this serves to show that the kinetic analysis of nAChR channel behaviour is a complicated business. Further progress may come from more experiments with the fast flow concentration jump technique, since this can provide estimates for some reaction rates independent of particular kinetic schemes (Liu & Dilger, 1991; Lingle *et al.*, 1992).

The acetylcholine binding site

There has been much interest in just how the binding sites on the two α subunits of the nAChR are made up. One way of approaching this problem is to use some chemical that will attach to the binding site and then to find which amino acid residues it has become connected to. Changeux and his colleagues at the Institut Pasteur in Paris have used a radioactive photoaffinity probe called DDF for this purpose. DDF is an aryldiazonium compound that will bind reversibly to the nAChR in the dark. Ultraviolet light makes it highly reactive, so that it forms irreversible links with adjacent amino acid residues in the protein chain. For some of the amino acid residues, the amount of DDF binding was reduced in the presence of carbamylcholine, an acetylcholine agonist, so these residues would appear to be the ones associated specifically with the acetylcholine binding site. They were Tyr-93, Trp-149, Tyr-190, Cys-192 and Cys-193, plus weakly labelled Trp-86, Tyr-151 and Tyr-198. These particular residues are conserved in the α subunits from various different species from *Torpedo* to the rat, but are mostly not found at the corresponding positions in the other subunits (Dennis *et al.*, 1988; Galzi *et al.*, 1990; Changeux *et al.*, 1992).

The tyrosine residue at 190 seems to be a particularly important component of the binding site. It can be converted to phenylalanine by site-directed mutagenesis and the mutants can then be studied by expression in *Xenopus* oocytes. The mutants have a much reduced affinity for acetylcholine and are much less responsive to it (Tomaselli *et al.*, 1991). Similar experiments have been done with the α7 neuronal nAChR; mutation of the residues corresponding to those labelled by DDF in the *Torpedo* α subunit reduced the binding of acetylcholine and nicotine (Galzi *et al.*, 1991).

All these residues are in the long N-terminal part of the nAChR chain on the external synaptic side of the membrane. Their positions suggest that at least three loops of the protein chain are involved in forming the binding site. There is some evidence that the γ and δ subunits may also be involved in binding, since both are labelled by [³H]d-tubocurarine and γ is labelled by DDF. Figure 6.6 shows a model of the binding site derived from these labelling studies.

Study of the quaternary structure of the nAChR by high resolution electron image analysis showed three dense rods halfway along the synaptic part of each subunit, about 30 Å above the lipid bilayer (Unwin, 1993a). The rods are presumably α-helices, and they seem to bound a cavity near the middle of each subunit (this can be seen as a small circular contour half up the right hand side of the nAChR section in fig. 4.8). These cavities are particularly pronounced in the α subunits, so it seems very likely that they are the acetylcholine binding sites. The three α-helices adjacent to them correspond well with the three loops found by photoaffinity labelling.

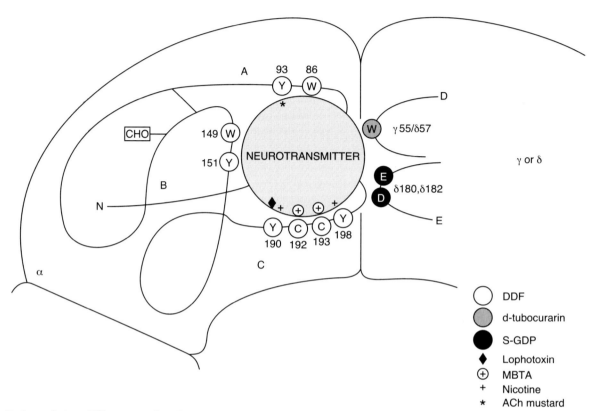

Roles of the different subunits

The importance of the different subunits in gating has been investigated using *Xenopus* oocytes. Messenger RNAs for the different subunits, in different combinations or from different species or after modification by site-directed mutagenesis, are injected into them and then the activity of the receptors that are produced is determined. Full responses are found only when the full complement of four subunits is used, but effective receptors can also be produced in the absence of γ or δ subunits, suggesting that these two can substitute for each other to some extent. Such substitutions have some effect on gating kinetics: the mouse δ-less receptor (which presumably has the composition $\alpha_2\beta\gamma_2$) has an average burst duration only half that of the normal receptor (Kullberg *et al.*, 1990), whereas the cow γ-less receptor has longer burst lengths and also many spontaneous openings in the absence of acetylcholine (Jackson *et al.*, 1990).

Acetylcholine receptors from different species show quantitative differences in kinetics when expressed in oocytes. Thus *Torpedo* electric organ receptors have shorter open times and shorter burst lengths than do those from mammalian muscle. Chimeric proteins have been made in oocytes, with the

Fig. 6.6. (left) Model of the acetylcholine binding site in the nAChR channel. The plasma membrane is below the plane of the paper and we are looking at the receptor from the extracellular (synaptic cleft) side. The line from the N terminus to Y198 represents part of the α subunit amino acid chain, and loops A, B and C are involved in the acetylcholine binding site. Also shown are residues involved in the binding site from the D loop in the γ and δ subunits and the E loop of the δ subunit. The sphere represents the photoactivatable binding agent DDF C p-((dimethylamino)benzenediazonium fluoroborate) in all possible orientations, and shows where acetylcholine (the neurotransmitter) is bound. Circles show amino acid residues labelled by DDF. Residues labelled by other compounds, and affected by site-directed mutagenesis, are also shown. ACh, acetylcholine; MBTA, 4-(N-maleimido)benzyltrimethylammonium iodide. (Diagram kindly supplied by Professor J. P. Changeux; see also Changeaux et al. 1992.)

different subunits derived from different species. The δ subunit seems to be particularly important in gating, since chimeras with αβγ from *Torpedo* and δ from the cow showed kinetics more like pure cow receptors than pure *Torpedo* ones (Sakmann *et al.*, 1985).

Mammalian muscles show developmental differences in their acetylcholine receptors. Foetal muscle receptors and the extrajunctional receptors of denervated muscles show channel openings that are longer and of lower conductance than those shown by receptors from adult muscles. The difference is due to the replacement of the foetal γ subunit by an alternative called the ε (epsilon) subunit in the adult end-plate, as is demonstrated in fig. 6.7 (Mishina *et al.*, 1986).

How does the channel open?

We can regard the nAChR as an allosteric protein, one that changes its shape in response to combination of acetylcholine molecules with its α subunits. The distance between the acetylcholine binding sites and the narrowest part of the permeant pore is about 50 Å. In view of the effects of different non-α subunits on gating, we might expect the shape change to be widely spread through the whole molecule rather than restricted to some particular part of it (Lingle *et al.*, 1992). Confirmation of this view, and some remarkable detail on the changes in the transmembrane pore, has been provided by Unwin (1995), using high resolution analysis of electron microscope images.

Tubular vesicles of postsynaptic membrane from *Torpedo* electric organ contain arrays of nAChRs that can be analysed by the image processing method, as described in chapter 4. Simply applying acetylcholine to such tubes would result in rapid desensitization of the receptors, so that their channels would be closed. Unwin met this problem by applying acetylcholine by means of an atomizer spray just 5 ms before freezing the tubes in liquid ethane at −178 °C. This timing ensured that all the receptors would be frozen with their channels open. They could then be compared with receptors treated similarly but without meeting the acetylcholine spray.

Unwin found that binding of acetylcholine produces little shape change at the extreme outer mouth of the receptor, but that some movement and twisting of the subunits is evident in most of the rest of the molecule. Much larger changes, not investigated in detail, were seen when the receptors had been in contact with acetylcholine for some seconds before freezing, by which time they would have reached the desensitized state.

At the level of the binding sites activation produces movement of the dense rods surrounding them in the two α subunits. There is also a displacement of the subunit between the two α subunits (probably the γ subunit) clockwise as seen from the outer side. Sections through the shaft between the

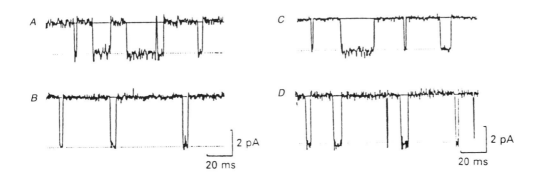

Fig. 6.7. Single channel currents from foetal and adult nAChR channels in bovine (cow) muscle. The traces on the left show patch clamp records from foetal (*A*) and adult (*B*) muscle. Those on the right show similar records from *Xenopus* oocytes which had previously been injected with mRNA coding for (*C*) the α, β, γ and δ subunits, or (*D*) the α, β, δ and ε subunits. The change from the γ to the ε subunit leads to channels with shorter opening times and higher conductances. (From Mishina *et al.*, 1986. Reprinted with permission from *Nature* **321**, pp. 408–9, Copyright 1986 Macmillan Magazines Limited.)

binding sites and the membrane show some clockwise twisting of both the α subunits by about 4°, suggesting that this is the means whereby changes at the binding sites produce effects 30 Å or more away at the membrane level.

The transmembrane pore is lined with the five M2 α-helices, each with a bend or kink in the middle, as we have seen in chapter 4. At rest in the closed state the kink forms the narrowest part of the pore, and it seems likely that the conserved leucine residues (L251 in the α subunits) at the kink form a hydrophobic ring which acts as a block to any ion movement through the pore. Activation leads to a marked change in the position of these helices: the kink is withdrawn from the axis of the pore and the 'lower' halves (the halves in the inner leaflet of the bilayer) of the helices swing round to become much more tangential to the pore, as is shown in fig. 6.8. This removes the large hydrophobic leucine residues from near the axis of the pore and replaces them with a line of smaller polar residues. The narrowest part of the pore is now a ring of threonine residues (T244 in the α subunits) at the lower ends of the M2 helices, with a diameter of 9 to 10 Å.

The results of this beautiful investigation have provided us with some of our most detailed information about the shape changes associated with gating in any channel molecule. It seems very likely that similar changes occur in other neurotransmitter-gated channels. Some general features of the changes (the slight twisting of subunits, the withdrawal of large hydrophobic residues from near the pore axis and their replacement by small polar ones, for example) may be worth looking for in other types of channel as well.

Voltage-gated channel gating

As was superbly demonstrated by Hodgkin & Huxley in 1952, the nerve action potential depends upon changes in the permeability of the axon membrane to sodium and potassium ions, and these changes are triggered by changes in membrane potential. The sodium and potassium channels that mediate these

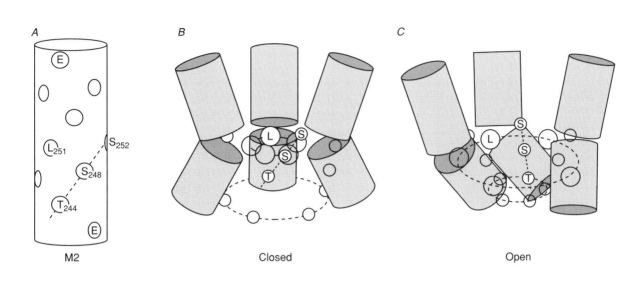

A

M2

B

Closed

C

Open

Fig. 6.8. Movement of the pore-lining M2 segment of the nAChR during gating, as deduced from image processing of arrays of receptors from *Torpedo* electric organ. *A* shows pore-facing amino acid residues on an α subunit M2 α-helix; corresponding residues are found on the other subunits. In the closed condition (*B*) the five M2 segments are kinked so that their leucine residues (L251 and corresponding positions) line the pore at its narrowest part. On opening (*C*), the M2 segments move so as to withdraw the leucine residues and line the 'lower' (inner) half of the pore with a row of smaller polar residues; the lowest of these (T244 and corresponding ones) now form the narrowest part of the open pore. (From Unwin, 1995. Reprinted with permission from *Nature* **373**, p. 42, Copyright 1995 Macmillan Magazines Limited.)

events are closed at rest but open on depolarization of the membrane; that is to say they are voltage gated. Much effort has gone into investigations of the mechanism of voltage gating since 1952.

Gating currents

We can tell whether or not a particle has a charge on it by seeing whether it will move in an electric field: uncharged particles do not move. So any detector for an electric field must have some electric charge in it somewhere, and its response to change in the field must involve some degree of movement of the charge. In the protein structure of a voltage-gated channel, therefore, we might expect to see some charges that are capable of moving when the potential across the membrane changes, and these charges would be intimately associated with the gating process.

This argument was clearly stated by Hodgkin & Huxley (1952b), and they predicted the existence of *gating currents*, i.e. currents produced in nerve membranes by the synchronous movements of the gating charges just prior to the onset of ionic flow. Gating currents are much smaller than ionic currents and it was some years before electronic techniques were good enough for them to be measured. They were finally demonstrated in 1973 by Armstrong & Bezanilla and by Keynes & Rojas.

In order to demonstrate gating currents it is necessary to block the ionic currents with suitable agents, such as external tetrodotoxin for the sodium channels and internal caesium fluoride (or, better, trimethylammonium fluoride) for the potassium channels. This leaves just the gating currents and the current associated with the membrane capacitance. These two can be

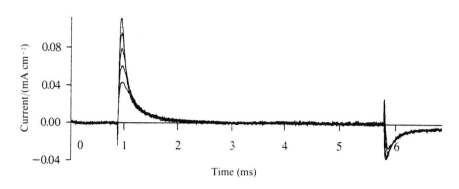

Fig. 6.9. Sodium channel gating currents from squid giant axon. Sodium ionic currents were eliminated by using a sodium-free external solution containing tetrodotoxin; potassium ionic currents were eliminated by using an internal tetramethylammonium fluoride solution. The currents were produced by clamped depolarizations lasting 5 ms, from −70 mV to a range of membrane potentials from −3 mV (for the smallest current) to +74 mV (for the largest). Symmetrical capacity transients were eliminated by adding appropriately scaled responses to hyperpolarizations from −150 to −180 mV (so, to get the outward gating current for the −70 to −3 mV depolarization, for example, the inward capacity current for the −150 to −180 mV hyperpolarization has to be multiplied by 67/30 before using it to cancel out the outward capacity current). (From Keynes *et al.*, 1992.)

distinguished by differences in their linearity. Capacity currents are essentially linear and independent of membrane potential, whereas gating currents are not. A common way of utilizing this property is to use the $P/4$ method. Here we depolarize the membrane by P mV from a holding potential of −70 mV. Then we take the membrane potential down to −180 mV and depolarize by $P/4$ mV four times. There will be almost no gating current component in the $P/4$ pulses, so we can add them up to get the capacity current produced during P pulses. Subtraction of this sum from the total current produced during P pulses then gives the gating current. Analogous methods, such as the use four negative-going reference pulses (the $P/-4$ method), a symmetrical negative-going pulse (the $P\pm$ method), or scaling a negative-going pulse have also been used.

Figure 6.9 shows the sodium channel gating current obtained from a squid giant axon during a clamped depolarization. The current is a brief outward movement of charge that largely precedes the onset of the inward ionic current. At the end of the depolarizing pulse there is an *off* current whose peak value is less than the *on* current but which lasts longer; the total charge flow is eventually equal and opposite to that in the *on* current. (The *off* current in fig. 6.9 demonstrates the phenomenon called 'charge immobilization' whereby some of the gating charge has still not returned to its original position by the end of the record; we shall return to this point later.) The maximum total amount of charge moved is about 30 nC cm^{-2}, or about 1900 electronic charges per square micrometre of membrane (Keynes & Rojas, 1974).

The gating charge per channel

How much gating charge is associated with each channel? An answer to this question is very relevant to the sensitivity of the channels to membrane potential changes: other things being equal, the relation between membrane potential and the probability of opening is steeper the greater the gating charge.

Discussions on the subject have produced varying answers over the years, and have not yet concluded (see Hodgkin & Huxley, 1952b; Ehrenstein & Lecar, 1977; Almers, 1978; Labarca *et al.*, 1980; Bezanilla & Stefani, 1994; Keynes, 1994; Sigworth, 1994); our treatment here is much simplified, but it may help one to understand what it is that researchers are grappling with and why certain measurements are made.

We consider first a simple two-state system in which each channel is opened by the movement of a single gating particle which carries a charge z. At any moment the particle is in one of two positions, 1 and 2, and these are associated respectively with the closed and open states, C and O in scheme 6.1. In Eyring rate theory terms, positions 1 and 2 correspond to two wells in an energy profile, and there is a single energy barrier between them. In a population of N similar channels, n_1 will have their gating particles in position 1 and n_2 in position 2, so that $n_1 + n_2 = N$.

The Boltzmann distribution tells us how the thermal energy is distributed in a population of molecules:

$$n = n_0 \exp(-\epsilon/kT) \qquad (6.18)$$

Here ϵ (epsilon) is the energy per molecule in joules, n_0 is the number of molecules in any particular state with energy ϵ_0, n is the number with energy ϵ greater than ϵ_0, T is the absolute temperature in K, and k is the Boltzmann constant, 1.38×10^{-23} J K^{-1}.

Applying equation 6.18 to our simple model, with n_1 gating particles with energy ϵ_1 in position 1 and n_2 particles with energy ϵ_2 in position 2, we get

$$n_2/n_1 = \exp[-(\epsilon_2 - \epsilon_1)/kT]$$

We can rewrite this as

$$n_2/n_1 = \exp(w/kT)$$

where w is the work done in moving a gating particle from position 1 to position 2, i.e. the energy difference between the two positions, in the absence of a membrane potential. To include the effect of the membrane potential, we have to add the electrical energy possessed by each particle to its positional energy. Electrical energy is given by the product of the charge and the potential difference, as in equation 2.1. Here this will be ze_0V, where z is the number of charges on each particle, e_0 is the elementary electronic charge and V is the potential difference between the two positions. When we include this electrical energy component we get

$$n_2/n_1 = \exp[(w + ze_0V)/kT] \qquad (6.19)$$

In our simple model the channels are open when the gating particle is at 2 and closed when it is at 1 and there are no other possible states to consider.

The proportion P_O of the channels that are open is therefore given by

$$P_O = n_2/(n_1 + n_2).$$
(6.20)

By combining equations 6.19 and 6.20 we get

$$P_O = \frac{\exp[(w + ze_0V)/kT]}{1 + \exp[(w + ze_0V)/kT]}$$
(6.21)

Another way of writing equation 6.21 is

$$P_O = \frac{1}{1 + \exp[-(w + ze_0V)/kT]}$$
(6.22)

We can rewrite this as

$$P_O = \frac{1}{1 + \exp[-(ze_0(V - V_{1/2})/kT]}$$
(6.23a)

where $V_{1/2}$ (equal to $-w/ze_0$) is the voltage at which half the channels are open. An alternative form of this, using F/RT in place of e_0/kT (see chapter 2), is

$$P_O = \frac{1}{1 + \exp[-zF(V - V_{1/2})/RT]}$$
(6.23b)

Equation 6.22, or its modification 6.23, is sometimes known as the Boltzmann relation. It gives a sigmoid curve on a linear plot of P_O against V, with the steepness of the relation being greater with higher values of z. It can be very useful in describing the activation of voltage-gated channels. In the conductance–voltage curves from pronase-treated squid axon sodium channels shown in fig. 6.10, for example, the right hand curve is described by the parameters $V_{1/2}$ (the potential at which half the channels are open) equal to -14 mV and kT/ze_0 equal to 8.5 mV. Since $kT/ze_0 = 25/z$ mV, the value of z for this curve is 2.9.

When V is large and negative (i.e. when the depolarizing pulse from a negative membrane potential is very small so that few channels are opened) then w will be much less than ze_0V, and $\exp[(w + ze_0V)/kT]$ will be much less than 1, so that equation 6.21 simplifies to

$$P_O = \exp(ze_0V/kT)$$
(6.24a)

Taking logarithms of this we get

$$\ln P_O = ze_0V/kT$$
(6.24b)

Since kT/e_0 is 25 mV at room temperature, we would expect a plot of P_O (or conductances g_{Na} or g_K etc) on a logarithmic scale against V to have a slope of $25/z$ mV per e-fold change in P_O at low values of P_O. Figure 6.11 shows such plots for the sodium and potassium conductances of squid axons, as

Fig. 6.10. Conductance–voltage curves for squid axon sodium channels. Points show conductances determined from steady-state currents obtained by voltage-clamp depolarizations from a negative holding potential, expressed as a proportion of the maximum conductance. Inactivation was removed by internal perfusion with pronase. White circles show the relation without further treatment, and the curve is drawn according to equation 6.23 (the Boltzmann relation) with $V_{1/2} = -14$ mV and kT/ze_0 (or RT/zF) = 8.5 mV. Black circles show the response after internal perfusion with batrachotoxin and conditioning at +40 mV for 10 min; the curve is drawn with $V_{1/2} = -74$ mV and kT/ze_0 = 9.4 mV. (From Tanguy & Yeh, 1991. Reproduced from *The Journal of General Physiology* 1991, **97**, pp. 499–519, by copyright permission of The Rockefeller University Press.)

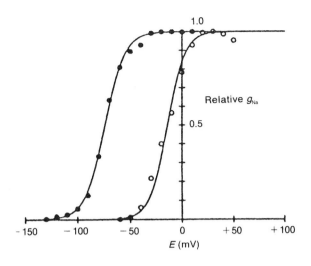

determined by Hodgkin & Huxley. At low conductance values, they show slopes of 3.9 mV per e-fold change in g_{Na} and 4.8 mV per e-fold change in g_K, suggesting values for z of at least 6 and 5, respectively, for the sodium and potassium channels.

Application of the Boltzmann principle in this way gives us a lower limit to the amount of gating charge per channel. Because of the simplification involved in deriving equation 6.24a from 6.21, the slope of the log conductance–voltage curves is only really equal to $25/z$ at very low conductance values, when very few channels are open. An alternative approach is to make long single channel records at negative membrane potentials and measure the open probability in the absence of inactivation. Such measurements on a non-inactivating mutant sodium channel expressed in *Xenopus* oocytes have revealed a steep relationship between $\log p_O$ and membrane potential, suggesting (from equation 6.24b) a value of at least 10 to 11 for z (Patlak *et al.*, 1995).

The mathematics of multi-state kinetics is more complicated than that for two states which we have considered, and the values of z may be correspondingly difficult to extract. Equation 6.24 still applies, but simulations show that for some kinetic models with many closed states the correct value of z emerges only at conductances so low that they would be difficult or impossible to measure experimentally (Bezanilla & Stefani, 1994).

What does a value for z mean? The simplest interpretation, implicit in our analysis so far, is that it is simply the amount of charge moved from one side of the membrane to the other during the gating process. There are, however, other possibilities. The charge may not move across the whole of the membrane (i.e. it may not traverse the whole of the electric field), in which case the exponent $ze_0 V$ should be $z\delta e_0 V$, where δ represents the fraction of the field that the gating charge moves across. Thus, 12 charges moving across 0.25 of the field may give an apparent value of 3 for z.

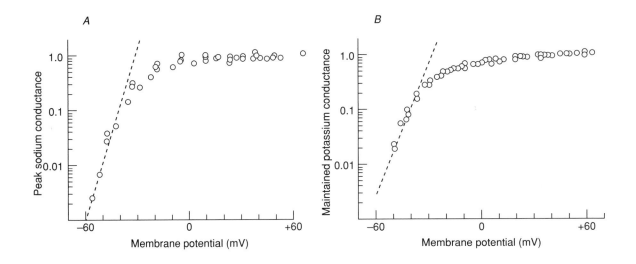

Another possibility (a very strong one in view of the molecular nature of voltage-gated channels, as we shall see shortly) is that there is more than one gating particle per channel. If there were four particles whose movements were completely independent of each other, then the values of z obtained by application of equation 6.23 would apply to the individual particles, and the gating charge for the whole channel would be $4z$. If their movements were not independent, then we would get an apparently higher value for z, but the total charge for the whole channel would be less than $4z$.

A different way of determining the gating charge per channel is to compare the gating charge density in the membrane, as determined by measurement of the gating currents, with the density of the channels. In squid axon, noise analysis suggests that there are about 180 sodium channels μm^{-2} (Bekkers *et al.*, 1986); saxitoxin binding assays suggest a somewhat larger figure at 290 channels μm^{-2}, but this will include the channels in the Schwann cell membranes as well as in the axon (Keynes & Ritchie, 1984). With a figure of 2500 charges μm^{-2} for the total gating charge movement (Keynes, 1994) we get about 14 gating charges per channel.

The number of charges per voltage-gated potassium channel has recently been estimated in *Shaker* channels expressed in *Xenopus* oocytes (Schoppa *et al.*, 1992). A plot of $\log g_K$ against membrane potential had a slope of 2.4 mV per e-fold change at low values of g_K, suggesting a value for z of 10.4. Comparison of gating currents with channel density by noise analysis suggested a similarly high value of 12.4.

Overall, it seems reasonable to conclude that there are about 12 gating charges per channel in voltage-gated sodium and potassium channels. There are clearly still some uncertainties in this area, but the figure is perhaps unlikely to be less than 10 or more than 15.

Fig. 6.11. Determination of the gating charge per channel from the relation between conductance and voltage in the squid giant axon. The graphs show peak sodium conductance (*A*) and maintained potassium conductance (*B*) during clamped depolarizations to different membrane potentials. Conductances are plotted as proportions of the maximum, on a logarithmic scale. The slope of the initial part of the curve gives the gating charge per channel, as z in equation 6.24. (Redrawn after Hodgkin & Huxley, 1952a.)

S4$_I$

215	A	L	**R**	T	F	**R**	V	L	**R**	A	L	**K**	T	I	S	V	I	P	G	L	**K**	Rat brain
205	G	L	**R**	T	F	**R**	V	L	**R**	A	L	**K**	T	V	S	I	M	P	G	L	**K**	Fly
208	A	L	**R**	T	F	**R**	V	L	**R**	A	L	**K**	T	I	T	I	F	P	G	L	**K**	Eel
144	G	L	**R**	T	F	**R**	V	L	**R**	A	L	**R**	T	L	S	I	I	P	G	L	**K**	*L. bleekeri*
216	A	L	**R**	T	F	**R**	V	L	**R**	A	L	**K**	T	V	A	V	I	P	G	L	**K**	*L. opalescens*
217	A	L	**R**	T	F	**R**	V	L	**R**	A	L	**K**	T	I	T	V	I	P	G	L	**K**	SCN4A gene
			+			+			+			+									+	

S4$_{II}$

848	V	L	**R**	S	F	**R**	L	L	**R**	V	F	**K**	L	A	**K**	S	W	P	T	L	N	Rat brain
631	V	L	**R**	G	L	**R**	L	L	**R**	A	L	**K**	L	A	**K**	S	W	T	T	M	**K**	Fly
655	V	L	**R**	S	L	**R**	L	L	**R**	I	F	**K**	L	A	**K**	S	W	P	T	L	N	Eel
499	V	F	**R**	S	F	**R**	L	L	**R**	V	F	**H**	L	A	Q	S	W	T	T	M	**R**	*L. bleekeri*
680	V	L	**R**	S	F	**R**	L	L	**R**	V	F	**K**	L	A	**K**	S	W	P	T	L	N	*L. opalescens*
667	V	L	**R**	S	F	**R**	L	L	**R**	V	F	**K**	L	A	**K**	S	W	P	T	L	N	SCN4A gene
			+			+			+			+			(+)						(+)	

S4$_{III}$

1299	L	G	A	I	**K**	S	L	**R**	T	L	**R**	A	L	**R**	P	L	**R**	A	L	S	**R**	F	Rat brain
1090	L	**K**	V	L	**R**	S	L	**R**	T	L	**R**	A	L	**R**	P	L	**R**	A	I	S	**R**	W	Fly
1091	L	G	A	I	**K**	N	L	**R**	T	I	**R**	A	L	**R**	P	L	**R**	A	L	S	**R**	F	Eel
876	F	**R**	S	L	**R**	T	L	**R**	A	L	**R**	P	L	**R**	A	A	V	S	**R**	W	Q	G	*L. bleekeri*
1114	L	T	A	F	**R**	S	M	**R**	T	L	**R**	A	L	**R**	P	L	**R**	A	V	S	**R**	S	*L. opalescens*
1125	I	**K**	S	L	**R**	T	L	**R**	A	L	**R**	P	L	**R**	A	L	S	**R**	F	E	G	M	SCN4A gene
		(+)			+			+			+			+		(+)	(+)(+)		(+)			

S4$_{IV}$

1626	**R**	V	I	**R**	L	A	**R**	I	G	**R**	I	L	**R**	L	I	**K**	G	A	**K**	G	I	**R**	Rat brain
1413	**R**	V	V	**R**	V	F	**R**	I	G	**R**	I	L	**R**	L	I	**K**	A	A	**K**	G	I	**R**	Fly
1417	**R**	V	I	**R**	L	A	**R**	I	A	**R**	V	L	**R**	L	I	**R**	A	A	**K**	G	I	**R**	Eel
1197	**R**	V	A	**R**	M	F	**R**	I	G	**R**	I	I	**R**	L	I	**K**	W	A	**K**	G	M	**R**	*L. bleekeri*
1439	**R**	V	V	**R**	V	F	**R**	V	G	**R**	V	L	**R**	L	V	**K**	S	A	**K**	G	I	**R**	*L. opalescens*
1448	**R**	V	I	**R**	L	A	**R**	I	G	**R**	V	L	**R**	L	I	**R**	G	A	**K**	G	I	**R**	SCN4A gene
	+			+			+			+			+			+			+			+	

Fig. 6.12. Amino acid sequences in the four S4 segments of various voltage-gated sodium channels. The positively charged arginine (R) and lysine (K) residues are shown in bold type. The sequences are from rat brain type II, the fruit fly *Drosophila*, the electric eel *Electrophorus*, two squids *Loligo*, and the human muscle sodium channel gene *SNC4A*. (From Keynes, 1994.)

The molecular basis of gating

When the primary structure of the electric eel sodium channel was first determined by the Kyoto University group (see chapter 4), one of its striking features was the nature of the S4 segments. In each of the four domains there are stretches of this segment where every third residue is either an arginine or a lysine, both of which are positively charged. There are five such residues in S4$_I$ (the S4 segment of domain I) and S4$_{II}$, six in S4$_{III}$ and eight in S4$_{IV}$. The intervening pairs of residues are mostly non-polar. This suggested to Numa and his colleagues that the S4 segments together make up the voltage sensor, and that some partial movement of them across the membrane will give rise to the gating current. Patterns very similar to that of the electric eel channel are found in other voltage-gated sodium channels, as is shown in fig. 6.12.

Site-directed mutagenesis, as we have seen, can be a very useful way of testing ideas about the operation of particular parts of channel molecules. It was used by Stühmer and his colleagues (1989) to replace some of the arginine and lysine residues from the S4 segments of domains I and II of rat

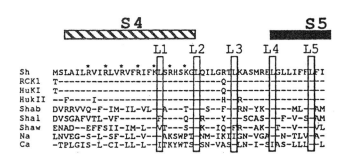

Fig. 6.13. Amino acid sequences in the S4 and leucine-heptad regions of various voltage-gated channels. The top seven sequences are from potassium channels, the sodium channel sequence (Na) is the second domain of the rat brain IIa sodium channel, and the calcium channel sequence (Ca) is from that of the skeletal muscle dihydropyridine receptor. Amino acids identical to *Shaker* (Sh) are shown by dashes. Asterisks show positively charged R and K residues in S4. Boxes show leucine residues in the heptad repeat. (From McCormack *et al.*, 1991.)

sodium channels. They injected mRNAs made from the altered cDNAs into *Xenopus* oocytes, so that the mutant sodium channels would be expressed in the oocyte membrane and could be investigated by voltage clamping large patches of membrane. They found that the steepness of the relation between channel opening and the membrane potential was progressively reduced as the positively charged residues of the $S4_I$ segment were replaced by neutral or negatively charged residues. A few similar changes in the $S4_{II}$ segment produced a similar but less marked change. The voltage for half-activation ($V_{1/2}$, as in equations 6.23) was also altered from its wild-type value of -32 mV; mutations towards the C-terminal (cytoplasmic) end of S4 moved $V_{1/2}$ to less negative values whereas those towards the N-terminal (extracellular) end moved it to more negative values.

The importance of the S4 segment in the gating of voltage-gated channels is further shown in calcium and potassium channels. They have the same characteristic arrangement of positively charged residues as do sodium channels, as is shown in fig. 6.13. Site-directed mutagenesis experiments have also been carried out on the S4 segments of potassium channels, and again it is found that replacement of the charged residues alters the relationship between channel opening and membrane potential (Papazian *et al.*, 1991).

So how does the S4 segment move in response to membrane potential changes? One way in which it might happen has been suggested by Catterall (1986) and by Guy & Seetharamalu (1986), following an earlier idea by Armstrong (1981, 1992). The model (fig. 6.14) assumes that each positively charged residue in the S4 segment is paired with some negative charge on the adjacent S1 to S6 segments. Depolarization would cause an S4 segment to move outwards by one step on this array, i.e. by 4.5 Å, corresponding to three residues on the α-helix. This would expose a negative charge at the inside of the membrane and a positive charge at the outside, which is equivalent to the movement of one charge across the membrane.

There seems to be little doubt that the positive charges in the S4 segments must be stabilized by the formation of ion pairs with negatively charged residues on other transmembrane segments if they are to remain in the

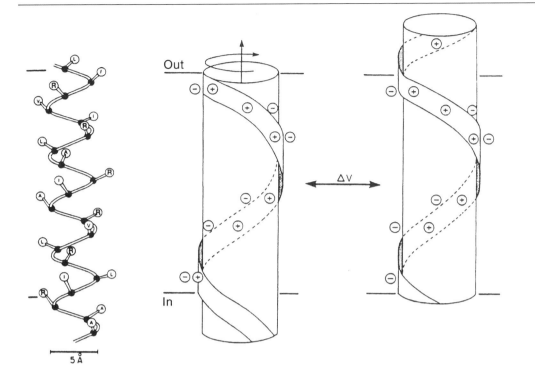

Fig. 6.14. Catterall's sliding helix model of gating charge movement in the voltage-gated sodium channel. Segment S4 in each domain forms an α-helix crossing the membrane; the sequence for the electric eel S4$_{IV}$ segment is shown here. Its arginine (R) or lysine (K) residues in every third position form a helix of positive charges. The model proposes that these form ion pairs with an array of negative charges on adjacent membrane-crossing segments, and that depolarization allows an outward movement of the S4 segment by one or perhaps two steps along this array. A movement of one step, exposing a positive charge at the outer side of the membrane and a negative charge at the inner side, is equivalent to the movement of one charge across the membrane. (From Catterall, 1992.)

hydrophobic environment of the lipid membrane. Evidence for the existence of such ion pairs comes from some mutagenesis experiments in which removal of positive charges from S4 in *Shaker* potassium channels was balanced by the removal of negative charges on the S2 and S3 segments (Papazian *et al.*, 1995). In these circumstances functional channels are formed in *Xenopus* oocytes, whereas with the S4 mutations alone this is not so. It seems that the capability of ion pair formation has to be present in order for the protein to fold properly into its place in the membrane.

The gating charge in *Shaker* potassium channels seems to be about 12 e_0, and this would demand an exposure of three charges per S4 segment, which implies a total movement of at least 13.5 Å. To do this the S4 segment would have to stick out into the hydrophilic environment on the outside of the membrane. Sigworth (1994) suggests that this is inherently improbable, and proposes instead that the S4 segment does not remain as a rigid α-helix during gating, but undergoes some other secondary structure change. Nevertheless, work with the sodium channel mutants R1448H and R1448C, both of which occur in myotonia congenita (see chapter 8) and have slowed rates of inactivation, suggests that the outer end of S4 does have an extracellular location, since increase in extracellular pH makes the rate of inactivation of R1448H similar to that of R1448C (Chahine *et al.*, 1994).

The S4 segment contains a number of leucine residues, and the stretch from the second half of S4 to the first half of S5 contains a well-conserved section in which every seventh residue is leucine (fig. 6.13). This heptad repeat is reminiscent of that in the 'leucine zipper' group of DNA binding proteins, where adjacent α-helices are held together by rows of leucine residues (O'Shea *et al.*, 1991). Mutation of the leucine residues causes large changes in the voltage–conductance and voltage–inactivation curves. For example, substitution of valine for the leucine at the inner end of S4 moved the voltage at which half the channels are opened from about zero mV to about $+70$ mV, and moved the voltage at which half of them were inactivated from -34 mV to $+26$ mV (McCormack *et al.*, 1991). Substitution of other leucine residues in the S4 segment has comparable effects (Lopez *et al.*, 1991). Perhaps the leucines serve to stabilize the different conformational states of the open and closed channel by means of hydrophobic interactions between the S4 segment and the rest of the molecule.

There is good evidence that gating leads to movement of water molecules into the channel. The potassium conductance increase produced by depolarization in squid axons is reduced under osmotic stress produced by increasing the sucrose or sorbitol content of the internal and external solutions, suggesting that some compartment of water that the sugar molecules cannot enter is added to the channel when it opens. Calculations suggest that the volume of this compartment is about 1350 Å^3, equivalent to about 45 water molecules (Zimmerberg *et al.*, 1990). This figure represents about 0.4% of the total volume of the channel molecule.

Osmotic experiments on sodium channels in crayfish axons have led to similar conclusions, with water entering a volume of 700 Å^3 on activation (Rayner *et al.*, 1992). Furthermore the gating currents are not affected by increases in osmotic pressure, suggesting that movement of the gating charge finishes before the channel opens.

Overall the studies on water movement imply that opening the pore involves fairly extensive structural adjustments in the channel molecule. How the postulated movement of the S4 segments produces them is not yet clear. Theoretical models, such as that produced by Durell & Guy (1992) for potassium channels, are interesting and instructive, but they are not as yet based on independent structural evidence.

Inactivation

Voltage clamp records of the response to maintained depolarization of the nerve cell membrane show sodium currents rising to a peak and then falling back to zero or a low level (figs. 3.4. and 3.5). Hodgkin & Huxley called the initial opening of the sodium channels activation and their later closing

Fig. 6.15. The 'ball and chain' model for the inactivation of voltage-gated potassium channels, such as the fruit fly *Shaker* A2 channels and some mammalian channels. Only three of the four channel subunits are shown. It is assumed that inactivation occurs once any one inactivating particle (or 'ball') docks into its receptor site near the mouth of the pore. (From Rehm, 1991.)

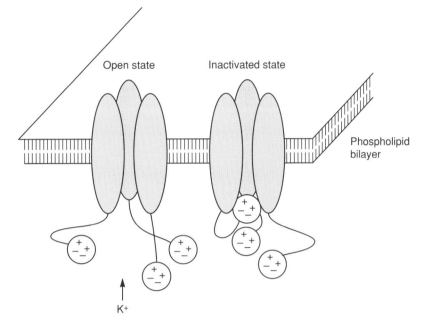

inactivation. The delayed rectifier potassium channels of squid giant axons show little inactivation in most experiments; currents are maintained for the duration of depolarization if this is limited to a few milliseconds. They do show, however, some inactivation on a time scale of seconds (Chabala, 1984). Other potassium channels may show relatively rapid inactivation when the depolarization is sustained.

An important clue as to the mechanism of inactivation came from experiments in which the proteolytic enzyme pronase was injected into squid giant axons. The sodium currents then did not inactivate, suggesting that inactivation was dependent upon some part of the channel molecule that was readily accessible from the inside of the membrane (Armstrong *et al.*, 1973). This led to the 'ball and chain' model: inactivation is produced by a mobile part of the channel protein that swings into the open channel pore so as to block it, as is shown in fig. 6.15 (Armstrong & Bezanilla, 1977). Experiments with block by internal quaternary ammonium ions, including triethylnonylammonium, which has a hydrophobic 'chain' attached to its pore-blocking head, were also suggestive in this respect (Armstrong, 1969, 1971).

Good evidence for the ball and chain model comes from site-directed mutagenesis investigations on *Shaker* B potassium channels by Hoshi and his colleagues (1990). They prepared deletion mutants in which various sections of the N-terminal cytoplasmic region of the molecule had been removed, and expressed them in oocytes. They found that deletions in the first 22 residues slowed or removed inactivation. Deletions of sufficient length in the sequence

A

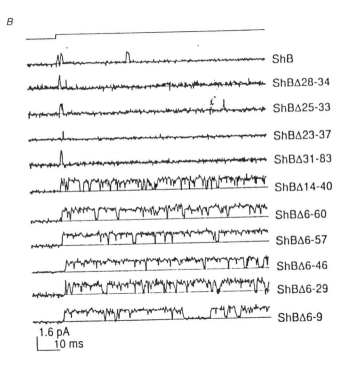

between residues 23 and 83 tended to speed up inactivation. Examples of these effects are shown in fig. 6.16.

These results fit the ball and chain model very well: deletions from the ball alone (residues 1 to 19), or from the ball and part of the chain, disrupt inactivation. Deletions from the chain alone tend to speed it up, as if a longer chain gives more freedom of movement to the ball so that it takes longer to find the open channel. The ball itself contains a concentration of positively charged residues, which we might suppose to be important in holding it in contact with the negatively charged pore of a cation-selective channel.

In further experiments a synthetic peptide with the same amino acid

Fig. 6.16. (left) Effects of removing parts of the N-terminal 'ball and chain' on inactivation of *Shaker* B potassium channels (these are *Shaker* A2 channels in the terminology of fig. 4.19) expressed in *Xenopus* oocytes. Part A shows the first 90 amino acid residue of a subunit. The bars show the sections removed in different deletion mutants: black bar deletions removed inactivation; white bar ones did not. Sample single channel records from the *Shaker* B channel and the deletion mutants (arranged in the same order as in *A* from the top downwards) are shown in *B*. Notice the presence of inactivation in the first five records and its absence in the last six. The stimulus in *B* was a voltage step from −100 to +50 mV. (From Hoshi *et al.*, 1990. Reprinted with permission from: Biophysical and molecular mechanisms of *Shaker* potassium channel inactivation. *Science* **250**, pp. 533–8. Copyright 1990 American Association for the Advancement of Science.)

sequence as the first 20 residues of the normal channel was prepared (Zagotta *et al.*, 1990). Mutant *Shaker B* channels with part of the ball sequence removed were expressed in oocytes and their activity was measured using the inside-out patch clamp method; the channels opened on depolarization and there was little or no inactivation. Then a solution of the synthetic peptide was brought into contact with the cytoplasmic side of the patch, and this restored the inactivation process. So the ball is sufficient to block the channel! There are probably up to four balls with chains in each potassium channel, since they are tetramers; just one of them seems to be sufficient to block the channel.

Is there a receptor for the ball, a particular section of the channel molecule that the ball can lock onto when it blocks the pore? The S4–S5 cytoplasmic loop contains a number of residues that are conserved in a wide variety of *Drosophila* and mammalian voltage-gated potassium channels. Mutations to some of these in the *Shaker* B channel reduced the degree of inactivation, suggesting that this loop is near to the internal channel mouth and forms part of the receptor for the inactivation ball (Isacoff *et al.*, 1991).

The inactivating ball peptide from *Shaker* B channels will also block other voltage-gated potassium channels, and mammalian calcium-activated channels (Toro *et al.*, 1992), all of which are in the same superfamily. It will not, however, block the ATP-dependent channels of mammalian muscle, which are sufficently different in molecular structure to lack the receptor for the peptide (Beirão *et al.*, 1994).

The ball and chain inactivation system in potassium channels has become known as N-type inactivation because it is identified with part of the N-terminal region of the molecule. A second type of inactivation has been associated with the C-terminal region, and so is called C-type. Thus when N-type inactivation is absent in deletion mutants such as ShAΔ6–46, a slower inactivation system remains (fig. 6.17). The time course of C-type inactivation is much slower in the ShB splicing variant than in ShA. These two variants have different carboxyl terminals. One of the differences between them is in the S6 membrane-crossing segment, where the valine residues at 463 and 464 in ShA are replaced by alanine and isoleucine, respectively, in ShB. Point mutations at 463 can change the time course of the slow inactivation. Thus V463A makes the inactivation of ShAΔ6–46 as slow as that of ShBΔ6–46, while A463V in ShBΔ6–46 makes it as fast as it is in ShAΔ6–46 (Hoshi *et al.*, 1991).

C-type inactivation is also affected by mutations in the H5 region, especially replacement of threonine at 449. Thus, mutations T449E and T449K have much increased inactivation rates, whereas T449V does not inactivate. These mutations are at the outer mouth of the channel pore and also have effects on ion permeation. Inactivation is slowed if the external potassium ion concentration is raised. It seems likely that C-type inactivation is associated with some change in the permeability characteristics of the pore (Lopez-Barneo *et al.*,

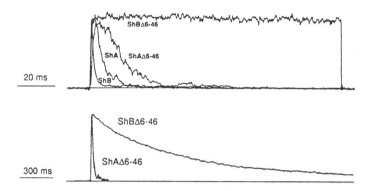

Fig. 6.17. N- and C-type
inactivation in *Shaker*
potassium channels. Records
show potassium currents from
Xenopus oocyte membranes
expressing two different
products of the *Shaker* gene,
ShA and ShB (these are named
Shaker A1 and A2 in the
terminology of fig. 4.19), and
from deletion mutants of these
two with residues 6 to 46
removed so that they have no
inactivating 'ball'. The
currents were produced in response to
clamped depolarizations from
−100 to +50 mV and have
been scaled to approximately
the same peak level; they show
the activity of some hundreds
of potassium channels. Intact
ShA and ShB channels show
rapid N-type inactivation.
Removal of the 'ball' leaves
slower C-type inactivation
(seen easily on a slower time
scale in the lower record),
which is much slower in
ShBΔ6–46 than in ShAΔ6–46.
(From Hoshi *et al.*, 1991.
Reproduced with permission
from *Neuron* **7**, copyright Cell
Press.)

1993). The very slow inactivation in squid axon potassium channels is probably C-type.

A further method of inactivation has been discovered in mammalian voltage-gated potassium channels. Such channels can be isolated from brain tissue as α-dendrotoxin (DTX) receptors, which consist of two types of subunit, α and β. The α subunits are of the familiar *Shaker*-related Kv1 (RCK) family, but the β subunits are quite different in structure, with no obvious hydrophobic membrane-crossing segments. It looks as though they are attached to the cytoplasmic end of the channel, probably with $\alpha_4\beta_4$ stoichiometry. Rettig and his colleagues (1994) found that their β subunits (called Kvβ1) would not form channels when expressed in *Xenopus* oocytes, but they greatly enhanced the inactivation of channels formed from Kv1 α subunits. Deletion of part of the N-terminal region from the β subunit removed its inactivation capability, and a peptide with the same sequence as its first 24 amino acid residues would produce inactivation in the absence of the rest of the β subunit. All this suggests strongly that the β subunit carries a ball and chain section that can block the channel in the same way as the N-terminal ball and chain of the pore-forming subunits does in the *Shaker* B channels of *Drosophila*. This idea is illustrated in fig. 6.18.

The molecular basis of inactivation in voltage-gated sodium channels is somewhat different. Site-directed mutagenesis experiments on rat sodium channels showed that changes to the cytoplasmic section of the molecule between domains III and IV greatly reduced the amount of inactivation (Stühmer *et al.*, 1989). This region contains a cluster of conserved positively charged residues, and also three adjacent hydrophobic residues, Ile-1488, Phe-1489 and Met-1490. Substitution of glutamine for any of these three slows inactivation, and substitution for all three removes it (West *et al.*, 1992). The loop structure for this region suggests that it may be less flexible than the ball and chain found in potassium channels, so a 'hinged lid' model has been suggested. Perhaps the three hydrophobic residues act as a latch that stabilizes the

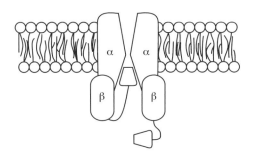

Fig. 6.18. Inactivation in some mammalian voltage-gated potassium channels, where the inactivating ball and chain is part of the auxiliary β subunit. The channels have $\alpha_4\beta_4$ structure, where the α subunits form the membrane-spanning pore and the β subunits are attached to them on the cytoplasmic side. (From Rettig *et al.*, 1994. Reprinted with permission from *Nature* **369**, p. 291, Copyright 1994 Macmillan Magazines Limited.)

lid when it folds over the open channel. Mutation of three adjacent hydrophobic residues (Val-Ile-Leu) at the inner end of $S6_{IV}$ greatly reduced inactivation, suggesting that these residues may act as the hydrophobic receptor site for the fast inactivation gate (McPhee *et al.*, 1994). The III–IV linker acted as a ball and chain inactivating system when attached to a non-inactivating potassium channel, even though there is little sequence similarity between it and the ball and chain sections of potassium channels (Patton *et al.*, 1993).

There is evidence that $S4_{IV}$ is involved in the regulation of inactivation in sodium channels. In the hereditary disease paramyotonia congenita (chapter 8), where muscles become stiff during exercise, sodium channel inactivation is slowed and less dependent on voltage, and recovery from inactivation is more rapid. In one form of this disease the arginine (R) at position 1448, at the outer end of $S4_{IV}$, is mutated to cysteine (C) or histidine (H). The R1448H mutant shows inactivation properties that are dependent on external pH, implying as we have seen that the residue is at least sometimes exposed to the external medium. At high pH, when the histidine will be uncharged, inactivation is slowed so that the channels behave like those of the R1448C mutant. At low pH, when the histidine will be charged like the arginine it has replaced, inactivation is more or less normal. This suggests that inactivation and activation are linked processes, and that $S4_{IV}$ plays an important role in coupling them (Chahine *et al.*, 1994).

The presence of auxiliary sodium channel β1 subunits in mammalian cells increases the rate of inactivation of the sodium channel principal α subunit (Isom *et al.*, 1994).

Inactivation rates in calcium channels are much slower than in sodium channels. They are affected by mutations in and about the S6 segment of the first domain, and perhaps have something in common with the C-type inactivation of sodium and potassium channels (J.F. Zhang *et al.*, 1994). Different inactivation rates in the α_1 principal subunit are produced by different auxiliary β subunits (Isom *et al.*, 1994).

Inactivation is evident in the gating currents of squid axons. As it proceeds the *off* gating current is reduced in size, as if some of the charges cannot immediately return to their initial position. This phenomenon is called charge immobilization. Since the charge movement recovers with the same time course as the recovery of the ionic currents from inactivation, it is assumed that the immobilized charge returns gradually to the normal resting position and that the small current generated by this is lost in the baseline noise (Armstrong & Bezanilla, 1977). Charge immobilization can also be demonstrated in membrane patches of oocytes expressing *Shaker* channels, but mutant channels with no inactivation do not show it (Bezanilla *et al.*, 1991). It looks as though the inactivation particle is preventing the return of the S4 segments to their original position.

Kinetics

Kinetic analyses for the nAChR channel have often been made under stationary or equilibrium conditions. We might look at the records of single channel activity, for example, when the acetylcholine concentration has been constant for some time. This is not, however, appropriate for many voltage-gated channels: inactivation is common, and it is easy to change the voltage suddenly to some new value and then follow the time course of the channel response. Changes of activation and inactivation with time are major features of interest in the study of these channels, and so have to be taken account of in kinetic analyses of them.

In constructing models for voltage-gated channels, we make the familiar assumptions of stochastic changes from one state to another by Markov processes. For a first attempt at a model we could use the simple two-state system of scheme 6.1:

$$C \underset{k_{OC}}{\overset{k_{CO}}{\rightleftharpoons}} O \qquad (6.25)$$

We assume that the rate constants k_{CO} and k_{OC} are dependent upon the membrane potential. If we suddenly depolarize the membrane by a suitable amount from its resting potential, then we would expect k_{CO} to jump from near zero to some appreciable value, and so the probability of any particular channel opening would rise accordingly. The number of open channels in a patch of membrane would then increase exponentially from zero at $t = 0$ to reach an equilibrium level:

$$N_t = N_\infty[1 - \exp(-t/\tau)] \qquad (6.26)$$

Here N_t is the number of channels open at time t, N_∞ is the number open at the new equilibrium level, and τ is the observed time constant of the change. τ is equal to $1/(k_{CO} + k_{OC})$.

A simple exponential curve like this does not in fact agree with experimental determinations of the activation (opening) of voltage-gated channels. Hodgkin & Huxley (1952b) found that the time courses of activation of the sodium and potassium channels in squid axon were initially slower than would be predicted by equation 6.26; this is evident in the S-shaped start to the traces *b* and *c* in fig. 3.5. The potassium conductance changes could be described by a fourth-power relation, involving $[1 - \exp(-t/\tau)]^4$ for the change on depolarization and $\exp(-4t/\tau)$ for that on repolarization.

Hodgkin & Huxley's model for potassium channel opening was that four charged particles moved independently to a certain part of the membrane under the influence of the electric field. If *n* is the probability that one of these particles is in the right place, then the probability that they are all in the right place is n^4, and so the conductance g_K (which is proportional to the number of channels open) of a patch of membrane is related to the maximum possible conductance $\bar{g}_K$ by the relation

$$g_K = \bar{g}_K n^4 \tag{6.27}$$

The parameter *n* follows relatively simple kinetics similar to scheme 6.25, such that

$$dn/dt = \alpha_n(1 - n) - \beta_n n \tag{6.28}$$

where α_n and β_n are rate constants which depend upon the membrane potential. α_n is analogous to k_{CO} in scheme 6.25 and β_n is analogous to k_{OC}. α_n increases with depolarization and β_n decreases.

The Hodgkin–Huxley model for the sodium conductance had to take account of inactivation. It assumed that a channel was opened if three charged particles (each with a probability *m* of being in the right place) came together and was inactivated by a fourth particle with a probability $(1 - h)$ of being in the blocking position. Thus, the equation for the sodium conductance is

$$g_{Na} = \bar{g}_{Na} m^3 h \tag{6.29}$$

Here *m* and *h* are described by equations similar to 6.28 with their own rate constants α_m and β_m, α_h and β_h, all of them dependent upon the membrane potential.

The Hodgkin–Huxley equations were derived entirely from measurements on macroscopic ionic currents. They were enormously successful in predicting the form of the nerve action potential and of many other properties of nerve axon membranes. More recently, however, measurements on gating currents and on single channel activity have displayed phenomena that cannot be accounted for within the Hodgkin–Huxley kinetic framework. The gating currents are more complex than the framework predicts (they do not rise instantaneously to their peak value and their falling phase has several components) and the *off* gating current is not slower than the corresponding ionic current (as it should be if the third or fourth power law applies). Single channel measurements are also at variance with the Hodgkin–Huxley framework: in the absence of inactivation they show longer open times than would be expected (Patlak, 1991).

Kinetic schemes since that of Hodgkin & Huxley have introduced various new elements. One is that the channel may have to pass sequentially through a number of different closed states before it can open. We may write, for example, for a system with five closed states:

$$C_1 \underset{k_{-1}}{\overset{k_{+1}}{\rightleftharpoons}} C_2 \underset{k_{-2}}{\overset{k_{+2}}{\rightleftharpoons}} C_3 \underset{k_{-3}}{\overset{k_{+3}}{\rightleftharpoons}} C_4 \underset{k_{-4}}{\overset{k_{+4}}{\rightleftharpoons}} C_5 \underset{k_{-5}}{\overset{k_{+5}}{\rightleftharpoons}} O \qquad (6.30)$$

Here the population of channels in the open state O is not fed from a population that is falling exponentially; indeed conversions to state O may well occur while the number of channels in state C_5 is rising. Hence the number of channels in state O will not follow an exponential time course like that in equation 6.26. The gating charge movement needs to be apportioned between the various steps in this sequence; in one scheme, similar in part to 6.30, the first four transitions were each associated with a charge movement of 1 e_0 and the last with a movement of 2 e_0 (Armstrong & Gilly, 1979).

Inactivation needs to be included in kinetic schemes for the sodium channel and for some potassium channels. Questions arise as to whether the inactive state can be entered directly from one or more of the closed states, and whether the channel has to pass through the open state in returning from the inactive state to the closed state. Thus, in the following scheme the channel can enter and leave the inactive state I only via the open state:

$$C_1 \rightleftharpoons C_2 \rightleftharpoons O \rightleftharpoons I \qquad (6.31)$$

Experiments with cloned rat brain Raw3 (Kv3.4) potassium channels, which inactivate rapidly, suggest that some rigidly coupled scheme like 6.31 is operative, and that the channels reopen during recovery from inactivation (Ruppersberg *et al.*, 1991). Similar conclusions have obtained from experiments with cloned *Shaker* channels (Demo & Yellen, 1991). One explanation for this is the 'foot in the door' model whereby the inactivation particle blocking the channel pore prevents the activation gate from closing (Armstrong, 1981). The return of the S4 segments may be physically prevented by the inactivation particle; this might well account for the charge immobilization seen in gating current measurements.

A further general feature of some recent schemes, as indeed in the Hodgkin–Huxley scheme, is that changes may occur in parallel. This idea

Fig. 6.19. The Keynes kinetic model of the voltage-gated sodium channel. S4 segments in the four homologous transmembrane domains undergo independent transitions from the resting state R, via the primed state P to the activated state A. When all four S4 segments have reached A the channel takes up 20 to 30 water molecules (Rayner *et al.*, 1992) to move to the CI state, and from there to the open OI state. Inactivation is associated with a third transition of the $S4_{IV}$ segment to the B state, and this converts the channel to the mode II condition, in which opening is much less probable. Charge movements are associated with the R→P, P→A, A→B and C→O transitions (and their reversals), but not with the hydration step. (From Keynes, 1994.)

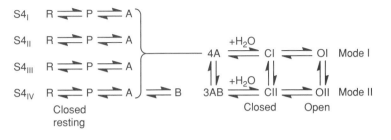

reflects the tetrameric structure of the channels, and supposes that changes may occur simultaneously in the different subunits. Keynes (1990, 1992, 1994) was one of the originators of this idea. He has produced an evolving scheme for the sodium channel that embodies parallel voltage-controlled transitions in the four S4 segments, based on detailed measurements of squid axon gating currents.

The 1994 version of Keynes's model is shown in fig. 6.19. Activation is brought about by parallel changes in the four S4 voltage sensors, from a resting state R via a primed state P to an activated state A. Charge movements of about 0.9 e_0 are associated with each transition; these correspond to movements of the S4 segment from one semi-stable position (i.e. when its positively charged arginine residues are paired with putative negatively charged residues in the rest of the molecule) to another. The activated channel state 4A then takes in water to reach the still closed state CI which then moves to the open state OI with a further charge transfer of about 0.8 e_0. The total charge transfer for activation is thus about 8 e_0. Inactivation in this model is represented as a mode change dependent upon a third transition of the S4 segment in the fourth transmembrane domain, associated with a further 0.9 e_0 of gating charge movement. This $S4_{IV}$ segment is then in state B and the whole channel is in the inactivated state 3AB, which can then change to the closed state CII and the open state OII. The rate constants for the C ⇌ O transition are such that the probability of the channel being open is high in mode I but much lower in mode II.

The existence of a second open state in mode II neatly explains the residual non-inactivating sodium current seen in a clamped depolarization after the main transient sodium current has died away. Further evidence for two open states comes from the observation that the membrane appears to be less permeable to calcium ions during the late non-inactivating period than during the initial sodium current (Keynes *et al.*, 1992).

A similar parallel-activation scheme has been produced by Aldrich and his colleagues to describe the behaviour of potassium channels (Hoshi *et al.*, 1994; Zagotta *et al.*, 1994a,b). They used a mutant *Shaker* channel (ShΔ6–46) that had no N-type inactivation, expressed it in *Xenopus* oocytes, and measured the

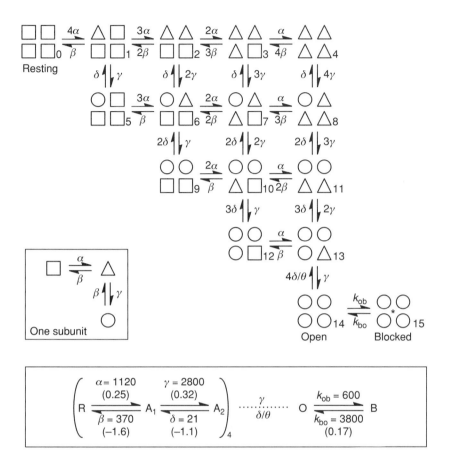

Fig. 6.20. The Zagotta–Hoshi–Aldrich kinetic model of *Shaker* potassium channel activation. Subunits undergo independent changes from the resting state R (squares) through an intermediate state A_1 (triangles) to a permissive state A_2 (circles), with rate constants α to γ as shown in the upper inset panel. The channel opens when all four subunits are in the A2 state. The sequences whereby a channel can move from the resting to the open state are shown in the main diagram. The lower panel gives model values for the various rate constants at $E = 0$; the figures in brackets are the effective valences z for the transitions. The rate constants are voltage dependent; for example $\alpha = 1120 \exp(0.25\, e_0 E/kT)\ \mathrm{s}^{-1}$ and $\beta = 370 \exp(-1.6\, e_0 E/kT)\ \mathrm{s}^{-1}$. The factor θ, representing cooperativity between the subunits, which keeps the channel longer in the open state, has a value of 9.7. (From Sigworth, 1994, based on Zagotta *et al.*, 1994b).

ionic currents, the gating currents, and single channel currents produced by depolarization. They produced a detailed kinetic scheme to explain their results, which is summarized in fig. 6.20. Subunits undergo independent changes from the resting state R through an intermediate state A_1 to a permissive state A_2. When all four subunits are in the permissive state the channel opens. Beyond the open state an unstable 'blocked' state is included to account for brief interruptions in the single channel currents. Charge movements are 1.85 e_0 for R$\rightarrow$A$_1$ and 1.42 e_0 for A$_1\rightarrow$A$_2$, so each subunit contributes a gating charge movement of 3.3 e_0 to the total of 13.4 e_0.

The rate constants in fig 6.20 are worth a look. For an individual subunit the rate constant for the transition from R to A_1 is α and that from A_1 to A_2 is γ, and those for the corresponding reverse transitions are β and δ respectively. All these are affected by the membrane potential, such that the αs and γs increase with depolarization and the βs and δs decrease. The rate constants for the transitions between the different states of the tetramer have to take account of the different ways in which this may be done. Thus the transition from RRRR to A_1RRR at the top left in the diagram can be done by change in any one of the four subunits, so its rate constant is 4α, whereas the reverse change can only involve one of the subunits so its rate constant is β. Again, change from A_1RRR to A_1A_1RR could involve any of three subunits so its rate constant is 3α, whereas the rate constant for the reverse transition is 2β.

The change from the open state $A_2A_2A_2A_2$ to the closed state $A_1A_2A_2A_2$ is much slower than would follow from complete independence of the subunits, so the rate constant is set at $4\delta/\theta$, where the factor θ is 9.7. This represents a positive cooperativity between the four subunits so that the channel tends to stay open when it has reached the open state.

It is clear from all this that much has been discovered about the gating process in voltage-gated channels since Hodgkin & Huxley's seminal analysis in 1952. It is also clear that there are still many ambiguities left in our understanding. They are likely to remain at least until we have better information about the structures of the channel molecules in their open and closed states.

Gating of some other channels

We know more about the gating processes in voltage-gated channels and in the nAChR than in most other channels because these are the ones involved in nervous conduction and neuromuscular transmission, and so they have been classical subjects for investigation by physiologists and were among the first to have their molecular structures determined. In this section we look briefly at some examples of gating in other channels.

Cyclic-nucleotide-gated channels

Many ion channels are gated by internal ligands. Cyclic-nucleotide-gated channels, such as those involved in phototransduction in the retinal rods or in olfactory transduction in the cilia of olfactory sensory cells, provide an example. Goulding and his colleagues (1994) have investigated the gating of these channels by comparing the responses of retinal and olfactory channels, and of molecular chimeras made from them, to cGMP and cAMP.

Cyclic-nucleotide-gated channels, as we have seen in chapter 4, are similar in structure to voltage-gated potassium channels. They are probably tetramers, with each subunit sequence having a cytoplasmic N-terminal, a membrane-crossing domain with S1 to S6 segments and a P or H5 pore-lining loop, and a C-terminal section containing a cyclic nucleotide binding site. The retinal channel is much less responsive to cAMP than to cGMP, whereas the olfactory channel is equally responsive to both. Chimeric channels made with the nucleotide binding site from one channel and the rest of the sequence

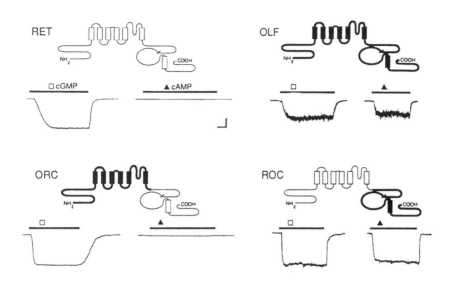

Fig. 6.21. Experiments on agonist selectivity in natural and chimeric cyclic-nucleotide-gated channels, expressed in *Xenopus* oocytes. Records show currents through inside-out patches in response to application of 30 μM cGMP or cAMP to their cytoplasmic surfaces. Channels are: RET, bovine retinal photoreceptor channel; OLF, catfish olfactory neuron channel; ORC, olfactory channel with retinal channel C-terminal; ROC, retinal channel with olfactory channel C-terminal. The currents (various scales) are larger for RET and ORC than for OLF and ROC. The sequence diagrams show the S1 to S6 transmembrane helices and the α-helix in the cyclic nucleotide binding site as rectangles, with thin lines for retinal and thick lines for olfactory sequences. Notice that the C-terminal region, containing the cyclic nucleotide binding site, largely determines the relative size of the responses to cGMP and cAMP. (From Goulding *et al.*, 1994. Reprinted with permission from *Nature* **372**, p. 370, Copyright 1994 Macmillan Magazines Limited.)

from the other show the nucleotide selectivity characteristic of the channel from which their binding site comes (fig. 6.21). The crucial part of the binding site seems to be the third of three α-helices that it contains. Inclusion of just this 24-residue section from the retinal channel into the olfactory channel sequence gives it the increased sensitivity to cGMP and reduced sensitivity to cAMP that is characteristic of the retinal channel, and in the reverse experiment the corresponding section of the olfactory channel makes the retinal channel equally responsive to the two cyclic nucleotides.

A second part of the sequence seems to determine how easy it is for the channel to open; this runs from the middle of the cytoplasmic N-terminal region to the end of the S2 transmembrane segment. It seems likely that binding of the cyclic nucleotide to the C-terminal region produces a conformational change in this part of the subunit, leading eventually to opening of the channel.

Goulding and his colleagues produced a kinetic scheme to model the gating process in these channels, based on the original Monod–Wyman–Changeux (Monod *et al.*, 1965) model for allosteric actions, and shown in fig. 6.22. The channel can exist in two conductive states, T (closed) and R (open), with $L_o = ([T]/[R])$ as the equilibrium constant when no cyclic nucleotide is bound. Equilibrium constants K_T and K_R apply to the binding reaction of the agonist A, a cyclic nucleotide; A binds more easily to the R state than to the T state so K_R is smaller than K_T. Each successive binding of an A molecule to the channel causes a progressive shift in the equilibrium towards the open state, by the factor C equal to K_R/K_T. Plausible values for the variables in the scheme gave good fits to the data. For example, for the retinal channel binding cGMP, L_o was 100 000, K_T was 33 μM and K_R was 1 μM; this means that with no cGMP

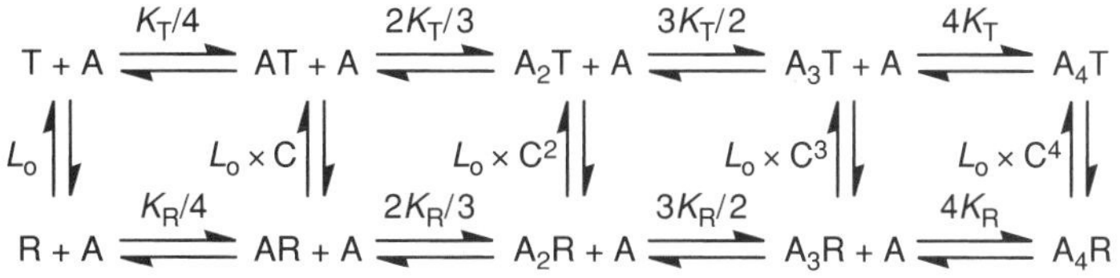

Fig. 6.22. Kinetic scheme to model the behaviour of cyclic nucleotide-gated channels using an allosteric framework. The channels are assumed to be tetrameric with four independent binding sites. T and R represent closed and open states respectively, and A represents the agonist (cGMP or cAMP). K_T and K_R are dissociation constants for binding for the two states. The equilibrium constant for the T⇌R transition is L_o, equal to [T]/[R]; for each bound A molecule this is multiplied by the factor C, equal to K_R/K_T. Some values fitted to data for the retinal channel are given in the text. (From Goulding *et al.*, 1994. Reprinted with permission from *Nature* **372**, p. 371, Copyright 1994 Macmillan Magazines Limited.)

$$T + A \; \underset{}{\overset{K_T/4}{\rightleftharpoons}} \; AT + A \; \underset{}{\overset{2K_T/3}{\rightleftharpoons}} \; A_2T + A \; \underset{}{\overset{3K_T/2}{\rightleftharpoons}} \; A_3T + A \; \underset{}{\overset{4K_T}{\rightleftharpoons}} \; A_4T$$

$$L_o \updownarrow \qquad L_o \times C \updownarrow \qquad L_o \times C^2 \updownarrow \qquad L_o \times C^3 \updownarrow \qquad L_o \times C^4 \updownarrow$$

$$R + A \; \underset{}{\overset{K_R/4}{\rightleftharpoons}} \; AR + A \; \underset{}{\overset{2K_R/3}{\rightleftharpoons}} \; A_2R + A \; \underset{}{\overset{3K_R/2}{\rightleftharpoons}} \; A_3R + A \; \underset{}{\overset{4K_R}{\rightleftharpoons}} \; A_4R$$

bound only 1 in 10^5 channels would be open whereas 92% of the channels with four molecules bound would be open.

One implication of this scheme is that there is two-way interaction between the gating process apparently located in the N-S2 region and the binding that takes place in the C terminus. Perhaps gating involves a change in subunit orientation (as apparently in gap junctions and the nAChR) and this also produces a change in the characteristics of the binding site. Does the S4 segment have any gating role in cyclic-nucleotide-gated channels? We shall see.

Calcium-activated potassium channels

Calcium-activated potassium channels (K(Ca) channels) open in response to increases in cytoplasmic calcium ion concentrations; they are also sensitive to simultaneous membrane depolarization. Consequently they are ideally suited to restore the membrane potential when depolarizations are accompanied by a rise in internal calcium (Barrett *et al.*, 1982; Latorre, 1994).

Injection of calcium chloride into snail neurons causes a hyperpolarization attributable to K(Ca) channels, and the form of the dose–response curve suggests that each channel is opened by the binding of at least three calcium ions (Meech & Thomas, 1980). The interplay of membrane potential and calcium ion concentration as gating agents is illustrated in the heroic experiment shown in fig. 6.23 (Barrett *et al.*, 1982). At −50 mV, channel opening does not begin until the internal calcium ion concentration rises above 1 μM, and p_O is still only about 0.4 at 100 μM. But at +50 mV, some channel opening can already be recorded at 0.01 μM calcium ions and p_O is close to 1.0 at 100 μM. Similar properties are found in channels expressed in oocytes after cloning from the *slo* gene locus of *Drosophila* and from the *mSlo* locus in mice (Adelman *et al.*, 1992; Butler *et al.*, 1993).

The voltage sensitivity of K(Ca) channels is much less than that of voltage-gated potassium channels. The initial rise of conductance with depolarization at constant calcium ion concentrations is exponential, with an e-fold change for every 25 mV change in membrane potential (Gorman & Thomas, 1980). This suggests a value for $z\delta$ (as in equation 5.18) of 1; if the binding of calcium ions is involved here, then z is 2 and δ, the electric distance through the membrane, is 0.5. This implies that the calcium binding sites are set within the electric field of the membrane, but it is not evident from the molecular

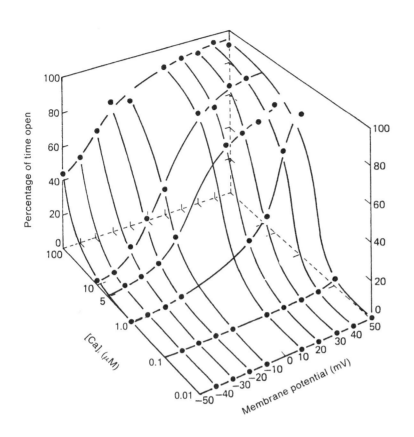

Fig. 6.23. Gating of a calcium-activated potassium channel. The vertical scale shows the open probability (proportion of time open) of a K(Ca) channel from rat muscle recorded using an inside-out patch with various membrane potentials and (internal) calcium ion concentrations. (From Barrett et al., 1982.)

structure of the channels that this is so. Perhaps the S4 segments provide the voltage sensitivity and the binding of calcium ions modulates this in some way.

The kinetics of large-conductance K(Ca) channels have been investigated in single channel patches from cultured rat muscle; they turn out to be quite complicated, with many different kinetic states and at least four different modes (McManus & Magleby, 1988). For most of the time the channel behaviour suggests three or four open states and six to eight shut states; this is known as the normal mode. Some of these states will presumably reflect the binding of no, one, two or three calcium ions. On occasion the channel switches briefly to one of three other modes in which the open times are briefer; these are known as intermediate open, brief open and 'buzz' modes.

Inward rectifier potassium channels

Inward rectifier potassium channels readily allow potassium ions to flow through them into the cell when the membrane potential E is more negative than the potassium equilibrium potential E_K. They also permit some outward flow when E is a little more positive than E_K, but at larger depolarizations they shut completely. This property tends to stabilize the resting potential near to E_K, but it also has some other useful effects (Stanfield, 1988). In heart muscle cells and egg cells it aids in the production of action potentials with very long plateaus. In muscle cells it aids inward spread of excitation through the transverse tubular system. In glial cells it promotes the buffering and disposal of potassium ions released during neuronal activity.

Patch clamp studies on isolated single heart muscle cells showed that the rectification disappeared if the patch was removed from the cell and its cytoplasmic surface exposed to a solution with potassium as the only cation (Matsuda et al., 1987; Vandenberg, 1987). It could be restored by the addition of magnesium ions (up to 1 mM) to the solution bathing the cytoplasmic side. This suggested that magnesium ions might act as a blocking particle in the cytoplasmic mouth of the channel, being held there by the potential gradient on depolarization, but being swept out of the way by inwardly moving potassium ions when E is more negative than E_K. Magnesium also revealed the presence of three substates of equal conductance (Matsuda, 1988).

Block by magnesium ions is not the whole story, however, since rectification can occur in their absence. After repolarization from potentials less negative than E_K the whole-cell current takes some milliseconds to reach its final value, and similarly on depolarization there is an initial flow of current that takes some milliseconds to inactivate. These effects still occur with magnesium-free solutions (Ishihara et al., 1989; Stanfield et al., 1994a). This means that it takes some time for the channels to open and shut when the membrane potential changes; the phenomenon became known as 'intrinsic' rectification.

The cloning of inward rectifier channels and their expression in oocytes or other cells has allowed their properties to be studied in some detail. IRK1 channels show strong rectification (they do not permit outward currents) and high sensitivity to magnesium ions, whereas ROMK1 channels, which have a similar structure and 40% amino acid sequence identity, show less rectification and are less sensitive to magnesium ions. The magnesium sensitivity and strong intrinsic rectification of IRK1 were both reduced if the negatively charged aspartate residue D172 on the M2 segment was converted to an uncharged residue (Stanfield et al., 1994b; Wible et al., 1994). The mutant IRK1 channel had properties much more like ROMK1, as is shown in fig. 6.24. Conversely, conversion of the homologous residue in the M2 segment of ROMK1 from asparagine to aspartate gives the mutant channel the magnesium sensitivity and strong rectification characteristic of the IRK1 channel (Lu & MacKinnon, 1994).

The problem of 'intrinsic' rectification was solved by the discovery that certain polyamines, particularly spermine and to a lesser extent spermidine, will block the outward currents in the strongly rectifying channels IRK1 and HRK1 expressed in *Xenopus* oocytes (Ficker et al., 1994; Lopatin et al., 1994;

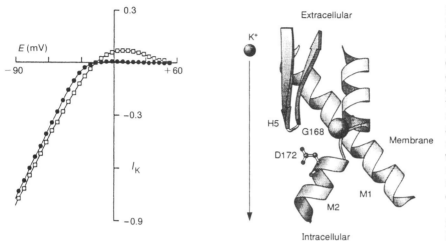

Fig. 6.24. Inward rectification in the IRK1 channel. The graph (left) shows the current–voltage relations of normal IRK1 (alternatively named Kir2.1) channels expressed in mouse erytholeukaemia cells (wild-type, black circles) and of mutant channels with the aspartate residue at 172 replaced by glutamine (D172Q, white squares). The weaker rectification shown by the mutant (seen as outward currents between −15 and +50 mV) is similar to that seen in ROMK1 (alternatively Kir1.1a) channels. Currents scaled to −1 at −107 mV. At the right is a possible model of subunit structure in the vicinity of the channel pore showing how the aspartate residue D172 may form a ring of four positive charges (if there are four subunits) inside the inner mouth of the pore. The M2 segment may bend at the glycine residue G168. (From Stanfield *et al.*, 1994b.)

Fakler *et al.*, 1995). Rectification is lost from a patch of membrane when it is excised from the oocyte as an inside-out patch and bathed on both sides with a magnesium-free solution. It is restored by addition of a soluble fraction of the cytoplasm to the cytoplasmic surface and this action is mimicked by spermine, as is shown in fig. 6.25.

Spermine is a polyamine with the formula $H_2N(CH_2)_3NH(CH_2)_4$ $NH(CH_2)_3NH_2$. It has four ionizable amino groups, and so one can imagine that this highly positively charged molecule is attracted by the negative charges on the Asp-172 residues mentioned above and so is drawn into the inner mouth of the channel. In the BIR10 channel the affinity for spermine falls dramatically when the equivalent negatively charged residue (glutamate in this case) is replaced by the neutral asparagine (Fakler *et al.*, 1994). On hyperpolarization, the charged spermine would move down the electrical gradient out of the pore, thereby unblocking the channel.

Spermidine is a substance similar to spermine but with only three amino groups. The rectification it produces is less pronounced than that mediated by spermine and the concentrations required are about a hundred times higher. Both compounds are present in the cytoplasm of most cells. Most of the spermine in cells is bound to ATP and other nucleotides, but the 1% to 5% that is not bound gives a free concentration of 8 to 75 µM. Since the blocking effect is 50% complete at 31 nM, this concentration is easily sufficient to account for spermine's role in inward rectification (Fakler *et al.*, 1994, 1995).

One particular inward rectifier is gated by combination with a G protein. This is the potassium channel in the atrial muscle of the heart that is activated by acetylcholine and has been cloned as GIRK1. When acetylcholine combines with atrial muscarinic acetylcholine receptors, this activates the G protein so that it takes up GTP in place of its bound GDP and splits so as to

Fig. 6.25. The role of spermine in strong inward rectification of IRK1 inward rectifier potassium channels expressed in *Xenopus* oocytes. Curves show the current–voltage relations for a large excised membrane patch in the inside-out configuration, obtained by measuring the current during 5 s clamped voltage ramps from −80 to +120 mV; hence they show the summed activity of a large number of channels. Cell-attached patches show strong rectification that is not mimicked in the inside-out patch when 1 mM Mg²⁺ is present (*A* and, on a larger scale, *B*). The strong rectification can be mimicked by a cell lysate (a soluble fraction of the oocyte cytoplasm), even if EDTA is present to remove any magnesium ions (*C*). Notice the absence of rectification in the control curve with neither lysate nor magnesium present. Spermine (SPM) in low concentration (10 μM) produces strong rectification similar to that of the cell lysate and the cell-attached patch (*D*). Solutions on both sides of the membrane patch contained equal concentrations of potassium ions, so that the membrane potential was zero in the absence of applied current. (From Fakler *et al.*, 1995. Reproduced with permission from *Cell* **80**, copyright Cell Press.)

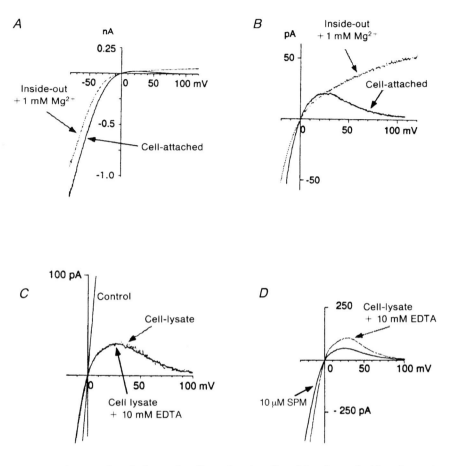

separate the α subunit from the βγ subunits. Combination of either its activated α subunit or its βγ subunit with the potassium channel then opens it (Logothetis *et al.*, 1987; Brown, 1990; Reuveny *et al.*, 1994; Wickman *et al.*, 1994). This is the means whereby acetylcholine, released on stimulation of the vagus nerve as Loewi showed in his classic experiments in the 1920s, produces inhibition of the heart beat. We shall take a closer look at G proteins later in this chapter.

ATP-sensitive potassium channels

ATP-sensitive potassium channels are widely distributed in vertebrate tissues. As we have seen in chapter 4, they probably belong to the same gene family as inward rectifier potassium channels, although they allow greater potassium ion efflux on depolarization. They show complex kinetic behaviour in the absence of intracellular ATP, with multiple open and closed states (Spruce *et al.*, 1987a). The channels shut when they bind ATP and will open again when it is released (Noma, 1983). ATP hydrolysis is not required for shutting or opening the channel, since non-hydrolysable analogues of ATP have the same effect as ATP itself. In heart muscle channels ATP binding is cooperative, since the Hill coefficient (equation 6.11) is greater than 1 (Nichols *et al.*, 1991).

If the intracellular ATP concentration is stepped up, K_{ATP} channels close rapidly and p_O falls exponentially. But reopening after a stepwise reduction in ATP follows after a delay (Qin *et al.*, 1989; Nichols *et al.*, 1991). These features can be explained if there are ATP binding sites on more than one of the sub-units that form the channel, but only one of them has to be occupied in order to shut the channel. The rate constant for binding of ATP was estimated to be 600 mM^{-1} s^{-1} in frog muscle, whereas that for unbinding was only 10 s^{-1} (Davies *et al.*, 1992). This relatively low unbinding rate, together with the need for all sites to be free from ATP before the channel will open, results in long shut intervals in the presence of low concentrations of ATP. At higher, saturating concentrations the channels are always shut.

The very high affinity for ATP raises the question of how the channels are regulated in the cell under normal physiological conditions (Ashcroft & Ashcroft, 1990; Davies *et al.*, 1991; Standen, 1992). In muscle, for example, K_d for the ATP binding is in the range 20 to 140 μM, yet the internal ATP concentration is in the millimolar range, so the great majority of the channels will be blocked for most of the time. Under extreme metabolic exhaustion, the channels do open and there is a loss of potassium from the muscle (Castle & Haylett, 1987). In frog muscle a rise in intracellular hydrogen ion concentration, such as occurs during exercise, reduces the affinity of the binding sites for ATP and so can lead to some channel opening (Davies *et al.*, 1992). In pancreatic β cells an increase in glucose metabolism raises the intracellular ATP concentration and this does result in K_{ATP} channel closure, leading ultimately to insulin secretion (fig. 1.2) (Ashcroft & Rorsman, 1989).

The CFTR channel

The structure of the CFTR channel, as deduced from its amino acid sequence and shown in fig. 4.24, has possible implications for its gating mechanism; one would expect the two nucleotide-binding domains NBD1 and NBD2 to be involved. Such considerations led Anderson and his colleagues (1991a) to examine the effect of ATP on channel activity. They found that the channels had to be phosphorylated by cAMP-dependent protein kinase A before any opening could occur, after which ATP was the necessary trigger for opening. Other hydrolysable nucleotides such as GTP were also effective, although less so than ATP. Non-hydrolysable analogues of ATP, however, were ineffective.

Evidence that ATP is actually broken down during the gating process has been provided by experiments using inorganic phosphate analogues such as orthovanadate, VO_4^{3-}. ATP breakdown in cells usually occurs after it has been bound to a protein, so that the products ADP and inorganic phosphate continue to be bound for a short time; they are then released and the binding site can react with another ATP molecule. Orthovanadate displaces phosphate from the binding site and sticks to it, so that the cycle of ATP breakdown is stopped in its tracks. CFTR channels opened by ATP are stabilized in the open state in the presence of orthovanadate, so that their mean open times are increased from a few seconds to several minutes (Baukrowitz *et al.*, 1994).

Gating by ATP is quite closely dependent upon phosphorylation of the R domain of the CFTR channel by protein kinase A. In the absence of such phosphorylation, the channels remain closed and unresponsive to ATP. There is some evidence that two degrees of phosphorylation exist, and that these are related to separate activity of the two NBDs. The non-hydrolysable ATP analogue 5'-adenosine (β,γ-imino)triphosphate (AMP-PNP) will not open CFTR channels in the absence of ATP but it does lead to long open times when ATP is present and the channel open probability is high (Hwang *et al.*, 1994). This has been interpreted as suggesting the regulatory scheme shown in fig. 6.26. In this scheme channels are phosphorylated in two stages. The first stage allows ATP to bind and be hydrolysed at one of the NBDs; this leads to brief openings and an overall low open probability. A further stage of phosphorylation enables the second NBD to bind and hydrolyse ATP (or to bind AMP-PNP); when this second ATP molecule is bound the channel remains open until the ATP is hydrolysed.

The need for ATP hydrolysis in the CFTR before the channel will function is a most unusual gating requirement. It is quite different from the need for ATP in transmembrane pumps such as the sodium pump. Many thousands of chloride ions flow across the membrane down their electrochemical gradient for each ATP molecule split by the CFTR, whereas the corresponding figures for the sodium pump are just three sodium ions and two potassium ions. Perhaps the free energy change in the CFTR molecule in its transition from closed to open is larger than could be met simply by binding a ligand molecule.

Mechanosensitive channels

Mechanosensitive channels are channels whose probability of opening is altered by mechanical stresses applied to the cell membrane (see Morris, 1990; Sackin, 1995). They have long been postulated as the basis for the activity of sense organs such as Pacinian corpuscles and muscle stretch receptors. More recently the patch clamp technique has demonstrated their presence in a large number of non-sensory cells. The channels become evident when suction is applied to the patch pipette so as to stretch the membrane containing them, hence they are also known as stretch-activated channels.

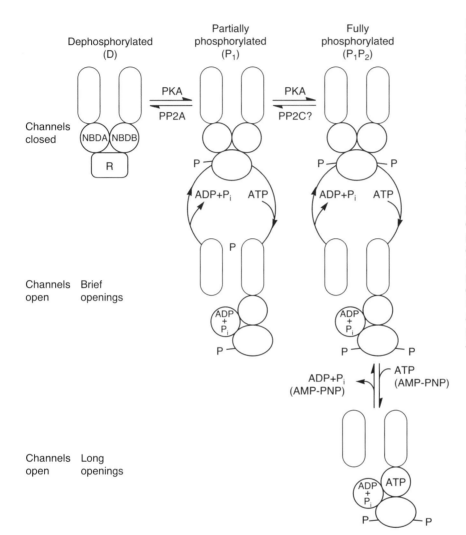

Fig. 6.26. Schematic model for regulation of the CFTR channel by two-stage phosphorylation. In the partially phosphorylated state (P_1), ATP binding and hydrolysis opens the channel and loss of the products (ADP and inorganic phosphate) closes it. In the fully phosphorylated state (P_1P_2), binding of ATP (or AMP-PNP) to the second nucleotide binding site keeps the channel open until the ATP is hydrolysed. Dephosphorylation from P_1 is only by okadaic-acid-sensitive phosphatase (PP2A), whereas dephosphorylation from P_1P_2 to P_1 can involve another phosphatase (perhaps PP2C). (From Gadsby & Nairn, 1994. Reproduced from *Trends in Biochemical Sciences*, with permission from Elsevier Trends Journals.)

Many stretch-activated channels are permeable to cations in general. Their conductances are in the range 25 to 35 pS in sodium or potassium salt solutions, and they show significant permeability to divalent cations, which means that calcium may enter the cell during stretch. These properties may allow them to act as important components of a cell volume-regulating system. Thus, in cells of the choroid plexus in the ventricles of the brain, hypotonic shock makes the cell swell as a consequence of osmotic uptake of water. Then stretch-activated channels open, calcium enters the cell so that calcium-activated potassium channels open and potassium ions leave the cell, hence some water leaves as well and so the cell volume is stabilized (Christensen, 1987). Other stretch-activated channels are selectively permeable to potassium ions or to anions, the latter with relatively high conductances.

Fig. 6.27. Activity of a stretch-activated channel in a patch of membrane from embryonic chick muscle. Downward deflections show inward currents as the channel opens, thin lines show the duration of bursts. The figures at the right give the suction pressure in the patch pipette. Notice how increased suction increases the rate of bursting.
(1 cmHg ≈ 1330 Pa.) (From Guharay & Sachs, 1984.)

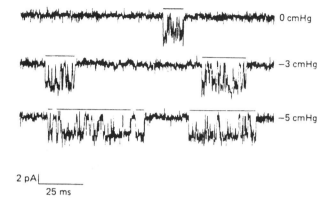

Stretch-activated channels often show bursts of activity separated by periods of inactivity, and membrane stretch then reduces the lengths of the inactive periods, as in fig. 6.27. This suggests a kinetic scheme in which there are a number of closed states in linear sequence, such as C_1 to C_3, with only the rate constant for the transition from the state furthest from the open state, k_{12}, being affected by stretch:

$$C_1 \underset{k_{21}}{\overset{k_{12}}{\rightleftharpoons}} C_2 \underset{k_{32}}{\overset{k_{23}}{\rightleftharpoons}} C_3 \underset{kO_3}{\overset{k_3O}{\rightleftharpoons}} O \tag{6.32}$$

The cation-selective stretch-activated channels of embryonic chick muscle are quite well described by this scheme (Guharay & Sachs, 1984).

The means whereby membrane stretch leads to channel opening is in most cases not too clear. One suggestion is that membrane fatty acids act as gating agents, possibly via activation of membrane-bound phospholipases (Ordway *et al.*, 1991). Another, which must surely be operative in many cases, is that the cytoskeleton just below the membrane is involved (Guharay & Sachs, 1984).

A rather different type of mechanosensitive channel is found in the hair cells of the vertebrate ear. The channels seem to be located on the stereocilia of the hair bundles and are probably opened by the fine filaments seen in electron micrographs, each one running from the tip of one hair to near the tip of its taller neighbour. They are opened by bending of the hair bundle in one direction but not in the other, and it seems likely that tension in the tip link opens the channel (Howard *et al.*, 1988; Hudspeth, 1989).

Gap junctions

The main function of gap junctions seems to be to maintain pathways of communication for ions and small molecules between two cells that are

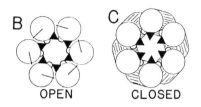

Fig. 6.28. A model for gating in gap junctions. The six M3 segments surrounding the channel pore are shown. In the open state the pore is lined with a number of small polar residues (white semicircles). When the M3 segments tilt, their large hydrophobic phenylalanine residues (black triangles) are brought into place to close the channel. (From Bennett *et al.*, 1991. Reproduced with permission from *Neuron* **6**, copyright Cell Press.)

usually of similar types. The channels may close if the membrane potential between the two cells changes or there is rise in cytoplasmic calcium or hydrogen ion concentration (see Bennett *et al.*, 1991). Such changes may be brought about by injury to one or both of the cells, so it seems likely that channel closure is largely a safety measure whereby cells can isolate themselves from damaged neighbours (Peracchia *et al.*, 1994).

Diffraction analysis of electron microscopic images of arrays of gap junctions shows structural differences between channels prepared in the presence or absence of calcium ions. With calcium present during the preparation process the axes of the six connexin subunits in each connexon are parallel to the central axis of the channel; in its absence they are tilted by 7° to 8° tangential to the axis (Unwin & Ennis, 1984). It is tempting to see these differences as the basis for channel gating. According to one model, shown in fig. 6.28, the open channel is lined by a series of small polar residues on the inner surface of each M3 α-helix when it is aligned parallel to the channel axis. The channel closes when the M3 segments tilt and so bring a larger phenylalanine residue into a blocking position in the pore (Unwin, 1989; Bennett *et al.*, 1991).

Subconductance states have been observed a number of times in gap junction channels. A modification of the tilting-helix model suggests that individual helices can tilt independently of each other, and that subconductance states are produced by the partial obstruction of the channel pore by fewer than six phenylalanine residues (Chen & DeHaan, 1993).

Modulation by phosphorylation

Modulation is a term usually used to describe some agent or process that modifies the gating of a channel. The probability of a channel opening may be altered by factors other than the primary gating trigger. For example, channels may be affected by attachment of phosphate groups to their cytoplasmic surfaces, by changes in the cytoplasmic or external calcium ion concentration, by the attachment of specific ligands, and so on. For the rest of this chapter we take a look at some of these processes.

The attachment of phosphate groups to proteins or their detachment from them is of major importance in cell biochemistry. This process is called *phosphorylation*. It is brought about by the action of a protein kinase, an enzyme that transfers a phosphate group from ATP to the protein:

$$\text{protein} + \text{ATP} \xrightarrow{\text{protein kinase}} \text{protein-P} + \text{ADP}$$

Dephosphorylation is the removal of the phosphate group by hydrolysis to leave the protein and inorganic phosphate. It is catalysed by a different enzyme called a protein phosphatase:

$$\text{protein-P} + \text{H}_2\text{O} \xrightarrow{\text{phosphatase}} \text{protein} + \text{P}_i$$

The net effect of these two reactions is the hydrolysis of ATP, so both reactions are energetically favourable and will take place readily and rapidly provided the protein kinase or phosphatase is active. The phosphorylated protein usually has properties different from its non-phosphorylated form, hence regulation of the protein's activity can take place by control of the protein kinase or phosphatase activity (Cohen, 1988; Barritt, 1992; Hunter, 1995).

There are a number of different types of protein kinase, distinguished by the different ways in which their activity is controlled. Cyclic AMP-dependent protein kinase (often called protein kinase A or PKA) is one of the best known (Taylor, 1989). The enzyme molecule is a complex of two regulatory subunits, each binding two molecules of cAMP, and two catalytic subunits which catalyse the phosphorylation of the substrate protein. The complex is inactive in the absence of cAMP, but it dissociates to release the active catalytic subunits when the cAMP is bound.

In other protein kinases the regulatory and catalytic parts of the molecule are in different domains of the same protein chain rather than in separate subunits. Cyclic GMP-dependent protein kinases are activated by binding cGMP. The (Ca^{2+} + calmodulin)-dependent protein kinases are activated by calmodulin to which calcium ions have bound. The various forms of protein kinase C (there are at least eight of them) are activated by diacylglycerol and calcium ions. Diacylglycerol is a membrane-soluble second-messenger molecule formed by the hydrolysis of phosphoinositol 4,5-bisphosphate. Protein-tyrosine kinases phosphorylate tyrosine residues; they appear to be largely concerned with growth regulation in cells.

The phosphate groups are usually attached to the protein chain at serine or threonine residues, less often at a tyrosine residue, and they replace the hydrogen atom of the $-$OH group. Cyclic AMP-dependent protein kinase usually phosphorylates the serine residue in the consensus sequence -Arg-Arg-$(Xaa)_n$-Ser-, where n is usually 1 but can be 0 or 2. Consensus sequences for other serine/threonine kinases are similar (Creighton, 1993).

How does phosphorylation produce its effects? After phosphorylation the particular residue is larger than it was and carries two negative charges, so it is

not too surprising that the properties of the protein may be changed somewhat. X-ray diffraction studies on the enzyme glycogen phosphorylase have
shown how phosphorylation of the Ser-14 residue produces small but significant alterations in the conformation of the molecule, involving interaction
with the positive charges of nearby arginine residues and resulting in the alteration of the binding sites for allosteric effectors and substrates (Sprang *et al.*,
1988; Johnson & Barford, 1990).

The initial work in this field was concerned with the regulation of enzyme
activity, but later it became clear that many different types of protein, including nuclear proteins, ribosomal proteins, contractile proteins, cytoskeletal proteins, membrane proteins and others, could also have their properties
modified by phosphorylation (Krebs, 1983). It was not surprising, then, to
find that ion channels are also modulated in this way (Levitan, 1985, 1994).
Let us have a look at some examples.

Voltage-gated sodium channels

The α subunit of rat brain sodium channels is readily phosphorylated by
cAMP-dependent protein kinase (Rossie *et al.*, 1987; Rossie & Catterall, 1989;
Catterall, 1992). The particular residues at which phosphate groups are
attached are all in the cytoplasmic loop between the transmembrane domains
I and II, as is shown in fig. 4.18. Phosphorylation reduces the amplitude of the
macroscopic sodium currents produced by depolarization. Patch clamp
experiments (fig. 6.29) showed that this is caused by a reduction in the probability of opening; the individual channel currents are not affected by phosphorylation (Li *et al.*, 1992b).

A separate site in the cytoplasmic loop between domains III and IV,
Ser-1506, is phosphorylated by protein kinase C (Numann *et al.*, 1991; West *et
al.*, 1991). This again produces a fall in the probability of the channel opening,
but there is also a marked decrease in inactivation rate, so that channel open
lifetimes are longer. We have seen that this cytoplasmic loop appears to be
directly involved in the inactivation process, and we can imagine that the addition of a negatively charged phosphate group interferes with the operation of
the 'hinged lid' formed by the loop.

The existence of two separate ways of phosphorylating the sodium
channel, with somewhat different effects on its function, implies that there are
complex possibilities for control of its functions in the intact cell.
Phosphorylation by cAMP-dependent protein kinase will be dependent on
the presence of cAMP in the cell. This is produced from ATP by the action
of the enzyme adenylyl cyclase, which is itself activated by a number of membrane receptors for various hormones and neurotransmitters, and inhibited by
others. Phosphorylation by protein kinase C is dependent upon the enzyme

Fig. 6.29. Effect of phosphorylation by cAMP-dependent protein kinase (PKA) on single channel currents of rat brain voltage-gated sodium channels. Records show the response of an inside-out patch from a cultured rat brain neuron to a depolarization from −120 to −30 mV. Those on the right were obtained after exposure to the protein kinase and 1 mM ATP. There were several channels in the patch, and sometimes two or more were open at once. The lower record in each set shows the current averaged from 500 such records. Notice that fewer channels are open after phosphorylation, but that single channel currents are not affected; we can conclude from this that the probability of channel opening is reduced. (From Li *et al.*, 1992b. Reproduced with permission from *Neuron* **8**, copyright Cell Press.)

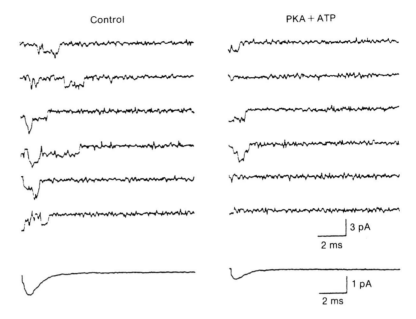

Control   PKA + ATP

3 pA
2 ms

1 pA
2 ms

being activated by diacylglycerol, itself produced from the membrane phospholipid phosphatidylinositol by the action of a different set of receptors for hormones and neurotransmitters. In this way it is clear that agents external to the cell could produce changes in the activity of the sodium channels. Such changes might not be of much importance in the all-or-nothing propagation of nerve action potentials along the axon, but they could have marked effects on thresholds and frequencies of firing at impulse initiation sites or on transmitter release at nerve terminals (Catterall, 1992).

Voltage-gated calcium channels

Action potentials in vertebrate heart muscle last much longer than those in skeletal muscle; after an initial peak depolarization there is a plateau lasting tens or hundreds of milliseconds before the membrane potential returns to its original level. During this time there an inward flow of calcium ions through L-type voltage-gated calcium channels. The rise in internal calcium ion concentration triggers further calcium ion release from the sarcoplasmic reticulum, and this results in contraction of the heart muscle. The heart beat can be modified by the action of the sympathetic nervous system, since activation of β-adrenergic receptors increases the contraction at each beat.

This increased contraction is brought about via an increased calcium influx through voltage-gated calcium channels as a result of their phosphorylation. Protein kinase A phosphorylates the accessory β subunit of the calcium channel and leaves the pore-containing α_1 subunit unaffected (Haase *et al.*, 1993). In the next section we look at how noradrenaline brings this about.

In the skeletal muscle calcium channel α_1 subunit there is a phosphorylation site at Ser-687 in the cytoplasmic section linking domains II and III, but this is absent from cardiac and brain channels (Röhrkasten *et al.*, 1988; see Mori, 1994). It may be relevant that it is this section that endows the skeletal channel with its characteristic properties for excitation–contraction coupling in linking to the sarcoplasmic reticulum ryanodine channel (Tanabe *et al.*, 1990).

The nicotinic acetylcholine receptor

Phosphorylation of the γ and δ subunits of the nAChR by cAMP-dependent protein kinase increases the rate of desensitization (Huganir *et al.*, 1986). The sites of phosphorylation appear to be Ser-353 and 354 on the γ subunit and Ser-361 and 362 on the δ subunit. In mutants that had these serine residues removed by site-directed mutagenesis, the rate of desensitization was slower than in wild-type phosphorylated channels. But if the serines were replaced by glutamate residues, which are negatively charged like the phosphate groups, then the rates of desensitization were similar to those of the phosphorylated wild-type (Hoffman *et al.*, 1994).

Modulation by neurotransmitters and G proteins

A number of channels are modulated by the action of neurotransmitters. In modulation the neurotransmitter is not the prime gating agent (as it is in neurotransmitter-gated channels such as the nAChR or the NMDA receptor) but it somehow alters the probability of channel opening. One of the first examples of this was discovered in chick sensory neurons of the dorsal root ganglion, where action potentials are shortened by the action of a number of different neurotransmitters, including noradrenaline, GABA, serotonin, and the neuropeptides enkephalin and somatostatin. These agents all serve to reduce the inward calcium currents associated with the action potential, by modulating the voltage-gated calcium channels involved (Dunlap & Fischbach, 1981). Such modulatory effects are usually mediated by second-messenger systems involving G proteins (Hille, 1992b, 1994).

Second-messenger systems

We have already touched on the concept of the second messenger in considering how protein kinases are activated. The idea was introduced by

Fig. 6.30. Signal transduction by G proteins. *A* shows the seven-transmembrane-segment receptor R combining with its specific ligand L to form an activated receptor R*L. This induces the α subunit of the αβγ inactive form of the G protein to bind GTP in place of GDP, and the G protein to dissociate then into its α and βγ subunits. (*B*) shows the G protein cycle; the activated α and βγ subunits interact with a variety of different effectors. The α subunit is a GTPase, so that after a short time αGTP becomes αGDP, which then reassociates with the βγ subunit so that the G protein becomes inactive. Pertussis toxin (PTX) blocks the catalysis of GTP exchange in G proteins of the G_i group, and cholera toxin (CTX) blocks the GTPase activity of some G proteins of the G_s and G_i groups. The cycle repeats until R*L becomes inactive by densensitization or is removed by dissociation. (From Simon *et al.*, 1991. Reprinted with permission from: Diversity of G proteins in signal transduction. *Science* **252**, pp. 802–8, Copyright 1991 American Association for the Advancement of Science.)

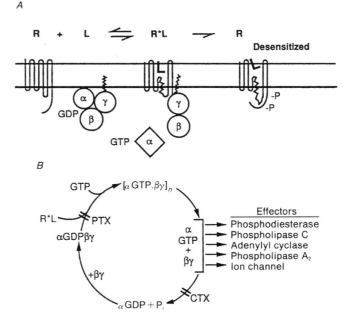

Sutherland and his colleagues following their discovery of the roles of cAMP in cells (Sutherland & Rall, 1960; Sutherland, 1971). A hormone or neurotransmitter (the 'first messenger') may bind to a receptor in the plasma membrane so as to induce a series of changes that result in a diffusible substance (the 'second messenger') that can then affect particular target molecules, often protein kinases, in other parts of the cell. G proteins act as links between receptor activation and the later links in the chain. They usually serve to activate a membrane-associated enzyme, which then produces the second messenger.

G proteins are so called because they bind guanosine diphosphate and triphosphate, GDP and GTP. They are activated by receptors of the seven-transmembrane-segment group such as the β-adrenergic receptors, the muscarinic acetylcholine receptors and many others. They consist of three protein chains, α, β and γ. At rest, the whole heterotrimeric protein is attached to the cytoplasmic side of the plasma membrane and the α subunit binds GDP. Activation of the receptor induces a series of conformational changes in the G protein so that its α subunit releases its GDP molecule and replaces it by GTP, and the α and βγ subunits are then released from the complex and can activate their target molecules (fig. 6.30). The cycle comes to an end when the α subunit, which is a GTPase, splits its GTP molecule: the βγ subunit now reassociates with the α-GDP subunit and the whole complex returns to its inactive state (see Gilman, 1987; Linder & Gilman, 1992; Neer, 1995).

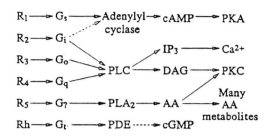

Fig. 6.31. Second-messenger pathways involving G proteins. Full arrows mean 'promotes the activity of' or 'increases the concentration of'; dashed arrows indicate the reverse of this. R_1, R_2 etc. are groups of seven-transmembrane-segment receptors; Rh is rhodopsin. G_s, G_i etc. are families of related G proteins. The enzymes activated by G proteins are adenylyl cyclase, phospholipase C (PLC), phospholipase A_2 (PLA$_2$) and phosphodiesterase (PDE). The products of activity by these enzymes are the second messengers cAMP, inositol trisphosphate (IP$_3$), diacylglycerol (DAG) and arachidonic acid (AA), and the non-messenger GMP, produced from cGMP. These second messengers may activate cAMP-dependent protein kinase (PKA) and protein kinase C (PKC), among others. (From Hille, 1992b. Reproduced with permission from *Neuron* 9, copyright Cell Press.)

The major targets for G proteins are enzymes that produce second messengers. Thus adenylyl cyclase is activated by G_s to produce cAMP, which then activates protein kinase A, which in turn leads to phosphorylation of various target proteins that may include ion channels. Such indirect actions may lead to considerable biochemical amplification, since one receptor molecule can activate a number of G protein molecules, each of which will activate its target enzyme, and the activated enzyme may produce relatively large quantities of the second messenger.

There are a number of different G proteins, each specific for their own sets of receptors and targets. Sequence analysis suggests that the α subunits fall into four main groups, G_s, G_i (including G_o), G_q (including G_{11}) and G_{12} (Simon *et al.*, 1991). The main G-protein-coupled signalling pathways involving second messengers are summarized in fig. 6.31. Ion channels are frequently implicated as the ultimate targets of second-messenger cascades of this type. Let us look at some examples:

(1)   We have seen in the previous section that adrenergic stimulation of the heart beat is brought about by phosphorylation of the L-type voltage-gated channels in the atrium. The chain of events begins with the binding of the neurotransmitter noradrenaline or the hormone adrenaline to the β_1-adrenergic receptor. This activates a G protein (G_s), so that the Gα subunit binds GTP and is released from the receptor and the βγ subunit, and the α subunit in turn acts as an activator for the membrane-bound enzyme adenylyl cyclase, which converts ATP to cAMP. Cyclic AMP activates cAMP-dependent protein kinase and this then phosphorylates the calcium channels. When the cell is next depolarized, the open probability for the calcium channels is increased, so more calcium ions enter the cell (Trautwein & Hescheler, 1990; Hartzell *et al.*, 1991).

(2)   A low level voltage-dependent potassium current in sympathetic ganglion neurons, known as the M current, is suppressed by a muscarinic action of acetylcholine. This suppression of the M current, as it is called, is the basis of the slow excitatory synaptic potential in these neurons (Brown & Adams,1980; Adams & Brown, 1982). Injection of antibodies specific for a sequence common to G_q and G_{11} reduces muscarinic

M current inhibition. Activation of $G_{q/11}$ by the muscarinic receptor would activate phospholipase C, leading to the production of diacylglycerol as a second messenger. The usual action of diacylglycerol is to activate protein kinase C, so it seems likely that this phosphorylates the potassium channels that carry the M current and so closes them (Brown, 1990; Caulfield *et al.*, 1994).

(3)   Substance P is a neuropeptide transmitter substance that produces slow excitation in a variety of different neurons, including the cholinergic neurons of the nucleus basalis whose degeneration seems to be one of the causes of dementia in Alzheimer's disease. This action is brought about by a closure of inward rectifier potassium channels, so that the membrane potential can more easily be depolarized (Stanfield *et al.*, 1985; Yamaguchi *et al.*, 1990).

Substance P receptors have the seven-transmembrane-segment structure found in many cell membrane receptors, and, like other members of that group, act via activation of a G protein; in this case it is probably G_q or G_{11}. Inhibitors of protein kinase C suppressed the action of substance P on inward rectifier channels, and okadaic acid, an inhibitor of serine/threonine protein phosphatases, suppressed the recovery of the potassium current afterwards. Hence it seems probable that protein kinase C acts directly on the channels (Takano *et al.*, 1995). The most likely sequence of events in the second-messenger cascade is therefore as follows: activation of the substance P receptors, activation of phospholipase C by $G_{q/11}$, hydrolysis of inositol phospholipids to produce diacylglycerol and IP_3, and activation of protein kinase C by diacylglycerol, and finally phosphorylation of the potassium channels leading to their closure. The channels would open again as a result of dephosphorylation by the okadaic-acid-sensitive phosphatase.

(4)   In some cases there is evidence for different G proteins acting in opposite ways in channel modulation. Somatostatin is a small neuropeptide that is widely distributed in the brain. In rat pituitary tumour cells it activates a large conductance calcium-activated potassium channel. This effect can be blocked by pertussis toxin (which blocks the action of certain G proteins) and by okadaic acid (which inhibits serine/threonine protein phophatases), so it probably depends on dephosphorylation of the channel as a result of G protein action, perhaps G_o (White *et al.*, 1991). Somatostatin has a similar pertussis-sensitive action on the inward rectifier potassium channels of some rat brain neurons. Substance P, however, has quite the opposite effect: it activates a separate G protein (perhaps G_q or G_{11}) that is not sensitive to pertussis toxin and leads to closing of the channels (Inoue *et al.*, 1988; Velimirovic *et al.*, 1995). The experiments on a different group of neurons, referred to in

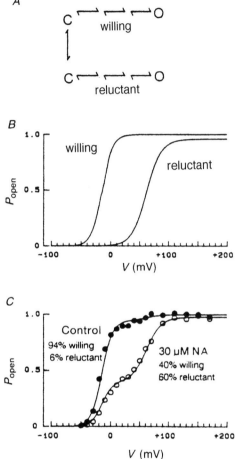

Fig. 6.32. Modulation of neuronal voltage-gated calcium channels by neurotransmitters that activate G proteins. *A* shows the kinetic model whereby each channel can exist in two modes in equilibrium, 'willing' and 'reluctant'. In the absence of activated G protein most channels would be in the willing mode. *B* models the activation curves for the two modes. Each curve is described by the Boltzmann equation $P_{open} = P_{max} / [1 + \exp -(V - V_{1/2})/k_s]$ where P_{max} is 1 for willing channels, 0.94 for reluctants; the midpoint $V_{1/2}$ is -15 mV for willing channels, $+62$ mV for reluctants, and the slope factor k_s is 9 mV for willing channels, 13 mV for reluctants. *C* shows actual activation data (circles) obtained in the presence and absence of 30 μM noradrenaline (NA); P_{open} is calculated from currents during clamped depolarizations. The data are fitted by curves drawn as the sum of two Boltzmann curves, assuming that some channels are in the willing mode and others are reluctant. (From Bean, 1989. Reprinted with permission from *Nature* **340**, p. 155, Copyright 1989 Macmillan Magazines Limited.)

the previous example, suggest that the substance P effect would be a phosphorylation by protein kinase C, mediated by the second messenger diacylglycerol (Takano *et al.*, 1995)

Direct action of G proteins on channels

In addition their role in the production of second messengers, G proteins may also act directly on ion channels. We have already seen the direct action on the G-protein-gated inward rectifier (GIRK) of the heart, where the G protein βγ subunit appears to be the gating agent. Here we look at modulatory effects of the G protein, acting directly on the channel and altering the probability of opening in response to the primary gating agent.

In heart muscle G_s appears combine directly with voltage-gated calcium channels so as to increase their open probability P_O, and with voltage-gated sodium channels so as to decrease P_O (Mattera *et al.*, 1989; Schubert *et al.*, 1989). Since G_s also stimulates adenylyl cyclase so as to raise the concentration of the cAMP, we have here an example of a single G protein affecting three separate targets.

The effects of G protein modulation on N-type calcium channels in sympathetic ganglia have been investigated by Bean, Elmslie and others (Bean, 1989; Elmslie *et al.*, 1990; Elmslie & Jones, 1994). They concluded that the channels can exist in two modes, 'willing' and 'reluctant' and that direct combination with a G protein converts willing channels to reluctant ones. The difference between the two is in their readiness to be opened by depolarization; in reluctant channels the voltage-P_{open} curve is moved to the right (more positive membrane potentials) by 77 mV, as is shown in fig. 6.32. Injection of the antibodies specific to particular Gα protein sequences shows that G_o is the G protein involved (Caulfield *et al.*, 1994).

Direct action of neurotransmitter on a channel

A direct modulatory action of a neurotransmitter or hormone on a membrane channel is much less common than an action via G proteins. The NMDA receptor channel, however, is remarkable in that its response to glutamate or NMDA is considerably potentiated by glycine (Johnson & Ascher, 1987). Single channel currents are unchanged in amplitude, but the probability of a channel opening in response to NMDA is greatly increased. The effect is evident in *Xenopus* oocytes expressing the cloned NMDA receptor, showing that there is no separate glycine receptor molecule (Moriyoshi *et al.*, 1991). Thus, glycine is a positive allosteric modulator of glutamate action at NMDA receptors.

Some other modulators

Calcium ions

The free calcium ion concentration in resting cells is usually in the range 10^{-8} to 10^{-7} M. This is a very low level in comparison with external concentrations of the order of 10^{-3} M. It is maintained by the sequestration of calcium in organelles such as the endoplasmic reticulum and mitochondria and by its binding to a variety of different proteins. There are also mechanisms for extruding calcium from the cell.

Entry of calcium ions into the cell via ion channels, or release from the endoplasmic reticulum via the IP_3 or ryanodine receptor channels, may raise the internal ionic calcium level to 10^{-5} M or locally even higher. Because of the many calcium-binding proteins present in the cytoplasm, free calcium ions are unable to diffuse far or for long before they are bound, so their effective diffusion rate is low and there can be considerable differences in free calcium ion concentrations in different locations in the cell. The possibility of changing internal calcium ion concentrations by 100-fold or more very rapidly and very locally makes calcium particularly suitable for use as an intracellular messenger (Clapham, 1995).

We have already looked at the gating of calcium-activated potassium channels, which open in response to internal calcium ions and a simultaneous depolarization. It is a matter of definition as to whether we regard calcium ions as a joint gating agent or as a modulatory agent in this case.

L-type voltage-gated calcium channels show inactivation during maintained depolarizations. This inactivation is enhanced if calcium ions are injected into the cell, so the inactivation is brought about by the very ions that enter through the channel. Such calcium-induced inactivation was first seen in the protozoan *Paramecium*, was soon found in snail neurons and insect muscle, and seems to be a very general occurrence (Brehm & Eckert, 1978; Tillotson, 1979; Ashcroft & Stanfield, 1982). Further experiments have utilized 'caged' calcium ions: a substance that releases calcium ions when illuminated by a flash of light is injected into the cell, so the calcium ion concentration can be suddenly brought to a new level. In heart muscle cells the inactivation that this produces is prevented by β-adrenergic stimulation, which, as we have seen, promotes phosphorylation of calcium channels. Two possibilities as to how the calcium works have been suggested: either it binds directly to the channel but only if the channel is not phosphorylated by cAMP-dependent protein kinase, or it activates a phosphatase which then dephosphorylates the channel and so closes it (Chad & Eckert, 1986; Hadley & Lederer, 1991).

Gap junction channels in most cells are sensitive to cytoplasmic calcium; they close when it rises (Bennett *et al.*, 1991; Peracchia *et al.*, 1994). A cell with a high internal calcium ion concentration will thus become isolated from its neighbours, which may be a useful characteristic if it is damaged. There is some evidence that calcium may act via calmodulin, a calcium-binding protein that interacts with a wide variety of different intracellular targets.

Fatty acids

Fatty acids have widespread effects on ion channels (Ordway *et al.*, 1991). Figure 6.33 shows an example: potassium channels in smooth muscle cells opened in response to a low concentration of myristic acid, a C_{14} saturated fatty acid. Similar responses were obtained when the patch was flushed with a number of other fatty acids, including palmitoleic acid (16:1, i.e. an unsaturated fatty acid with 16 carbon atoms and one double bond), linoelaidic acid

Fig. 6.33. Activation of potassium channels by a fatty acid in an excised outside-out patch from a toad stomach smooth muscle cell. Myristic acid (40 μM, MA) was applied to the patch at the time shown by the arrow, causing channel opening. The channels closed when the fatty acid stream was washed away for a few seconds. The lower trace shows part of the upper trace (at the asterisk) on an expanded time scale. (From Ordway *et al.*, 1989. Reprinted with permission from: Arachidonic acid and other fatty acids directly activate potassium channels in smooth muscle cells. *Science* **244**, pp. 1176–9. Copyright 1989 American Association for the Advancement of Science.)

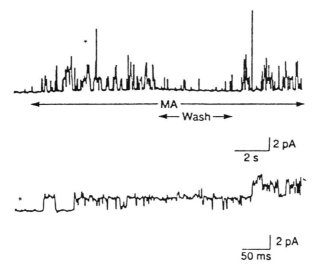

(18:2 *trans*), linolenic acid (18:3) and arachidonic acid (18:4) (Ordway *et al.*, 1989). These effects are dependent upon a medium- or long-chain lipid chain plus a negatively charged group at one end. Similar compounds that are neutral have no effect and positively charged ones inhibit it. The results suggest that the fatty acids bind to the channel in order to produce their effect (Petrou *et al.*, 1994).

Arachidonic acid differs from other fatty acids in that it is the starting point for a whole series of biochemical conversions, producing prostaglandins, leukotrienes and other eicosanoids (C_{20} compounds). Many of these substances are highly bioactive, often producing their effects via second-messenger systems and cAMP. Hence it is not surprising to find that arachidonic acid may produce effects on ion channels indirectly as a result of enzyme action as well as directly by binding to the channel (Attwell *et al.*, 1993; Meves, 1994).

Nitric oxide and cGMP

The discovery in 1987 that the gas nitric oxide is formed in endothelial cells and that it induces relaxation of vascular smooth muscle led to a great surge of experimental work on the biology of this surprising cellular messenger (see Moncada *et al.*, 1991; Nathan, 1992; Bredt & Snyder, 1994; Garthwaite & Boulton, 1995). It is produced when arginine is oxidized by the action of the enzyme nitric oxide synthase. It forms in brain neurons as a result of calcium influx into the cell, either when glutamate opens NMDA receptor channels or when voltage-gated calcium channels open; the calcium binds to calmodulin, which then activates nitric oxide synthase. Nitric oxide is a small molecule so it diffuses readily within the cell and across cell membranes, but it does not get far because it is highly reactive.

The usual action of nitric oxide is to activate soluble guanylyl cyclase, leading to production of cGMP. This acts as a trigger for cGMP-gated channels in retinal photoreceptors and bipolar cells, and also activates phosphodiesterases, a protein kinase, and other enzymes, all of which may themselves have effects on ion channels. Nitric oxide may act directly to inhibit NMDA receptor channels, perhaps by modifying free sulphydryl groups on the receptor molecules (Lei *et al.*, 1992). Calcium-activated potassium channels in excised membrane patches from vascular smooth muscle can be opened by nitric oxide in the absence of cGMP (Bolotina *et al.*, 1994).

External calcium ions and surface charges

It is well known that nerve cells become more excitable if the extracellular calcium ion concentration is reduced. Frankenhaeuser & Hodgkin (1957) concluded that these effects were caused by changes in the voltage-conductance relations for both sodium and potassium channels. A five-fold decrease in external calcium ion concentration was equivalent to a depolarization of 10 to 15 mV. One way in which these results may be interpreted is in terms of fixed charges at the surface of the membrane.

Hille and his colleagues (1975) carried out a study of surface charges in frog nerve fibres. They found that the effects of increased external calcium ion concentration could to some extent be mimicked by increases in the external concentrations of other divalent cations, monovalent cations or hydrogen ions. They suggested that all these agents produce their effects by altering the negative surface charge on the external face of the plasma membrane. The surface charge arises from acidic groups in the membrane phospholipids and proteins, including the sialic acid carbohydrate groups attached to the external face of many membrane proteins.

The existence of the surface charge will alter the potential distribution across the membrane, as is indicated in fig. 6.34. The membrane potential gradient, instead of being contained within the membrane as it would be in a simple constant field model, extends some way into the external solution, by the Debye length (see chapter 2) of about 8 Å. This means that the potential gradient within the membrane, which is what we might assume would be sensed by the gating mechanism, is less than it would be without the surface charges. Binding of calcium ions to the external surface would neutralize some of the surface charges and so would make the field within the membrane steeper. The effect of this would be to reduce the gating effect of depolarization, and so make the nerve fibre less excitable.

Fig. 6.34. Schematic diagrams to show how fixed surface charges may affect the potential distribution in the membrane and its immediate vicinity (heavy lines). Dashed lines show the potential distribution assumed by a constant field model and disregarding fixed charges. The diagrams show low and high calcium concentrations, with membrane potentials at the resting level and partly depolarized. Notice how calcium ions alter the potential gradient within the membrane. (From Aidley, 1990.)

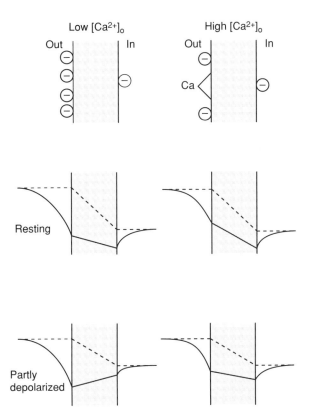

These effects of external calcium and hydrogen ions are of some clinical importance, since hyperexcitability is associated with hypoparathyroidism (which produces low serum calcium ion concentrations) and with the alkalosis occurring in hyperventilation.

7 Drugs and toxins

In this chapter we look at some of the ways in which various pharmacological agents interact with some particular channels. Pharmacology is the study of the responses of living systems and their component cells to drugs, toxins and other chemical agents. The molecules of these various agents exert their actions on their targets by binding to them, and different agents may produce different effects by binding to different sites on the target molecule.

Following Hille (1992a), we can distinguish *blocking agents* from *modifiers of gating*. Block occurs when some substance prevents the passage of ions through the channel, sometimes by binding wholly or partly within the pore and thereby occluding it, sometimes by an allosteric action in which the agent binding elsewhere in the molecule prevents the channel from opening. Modifiers of gating are substances or agents that affect the gating of the channel in some way, such as by changing the probability of the channel being open under a particular set of conditions. These definitions are not always precisely applicable to particular pharmacological actions, and they may overlap.

Many drugs and toxins are of natural origin and were originally developed by the organisms that produce them as mechanisms of defence or offence in the struggle for life. Ion channels are complex proteins with varied roles in cell functions, and they are mostly accessible at the outer surface of cells, so it is not too surprising that they are targets for the action of a wide variety of different natural products. Their crucial role in cell physiology also makes them prime targets for synthetic drugs of medical importance.

Natural neurotoxins, in particular, are usually highly selective in their action and will bind to channels with a high affinity. Such agents have been crucially useful over the years in understanding how channels work. Tetrodotoxin, for example, which binds tightly to voltage-gated sodium channels and blocks them, has been used to count the density of the channels in nerves, to act as a marker in the biochemical extraction of electric organ sodium channels, to eliminate sodium currents in the study of potassium currents and gating currents in nerve axons, and for many other operations: it is an essential part of the sodium channel investigator's toolkit.

Simple models for block

We can set up a simple model for block based on the two-state channel considered in chapter 6 as scheme 6.1. We add to this a blocking agent B, one molecule of which combines with the open channel to form the complex OB, as in scheme 7.1:

$$C \underset{k_{-1}}{\overset{k_{+1}}{\rightleftharpoons}} O + B \underset{k_{-b}}{\overset{k_{+b}}{\rightleftharpoons}} OB \qquad (7.1)$$

When the channel is in the blocked state OB it is not permeable to ions. In this model the channel cannot close without first releasing the blocking molecule B. The right hand side of this scheme is similar to the ligand–receptor model in scheme 6.8. Units for the rate constants are s^{-1} for k_{-b} and $M^{-1} s^{-1}$ for k_{+b}. The rate of the blocking reaction is proportional to the concentration of the blocking agent, so the actual transition rate for the forward reaction is equal to $[B]k_{+b}$, with dimensions of s^{-1} (see Palotta, 1991; Moczydlowski, 1992; Gibb, 1993; Colquhoun & Hawkes, 1994, 1995).

The dissociation constant K_d for the blocking reaction, given by k_{-b}/k_{+b}, is inversely proportional to the affinity of the blocking agent for the channel. Thus, tetrodotoxin blocks voltage-gated sodium channels and has a high affinity for them ($K_d = 3$ nM), whereas the tetraethylammonium (TEA) ion has a much lower affinity for voltage-gated potassium channels ($K_d = 1$ mM). Sometimes the term K_i is used instead of K_d to emphasize the inhibitory effect of the blocking action.

The effect of the blocking agent on the time course of single channel currents depends markedly on the values of the rate constants k_{+b} and k_{-b}. If these are much lower than k_{+1} and k_{-1}, so that the blocking agent binds to, and then leaves, the channel much more slowly than the channel switches between the open and closed states, then the single channel record shows periods of complete inactivity separated by periods of 'normal' activity similar to that which occurs in the absence of the blocking agent. This behaviour is sometimes known as 'slow' block. A simulated example is shown in fig. 7.1B. If, however, k_{+b} and k_{-b} are much higher than k_{+1} and k_{-1}, then channel openings will be frequently interrupted as the blocking agent binds and then unbinds from the channel (fig. 7.1C). This phenomenon is known as 'flickery

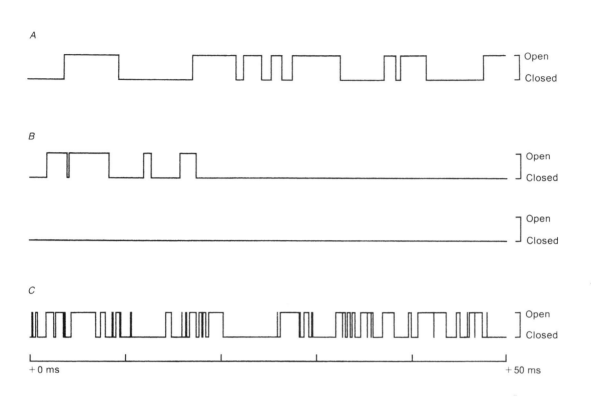

Fig. 7.1. Simulated single channel currents in 'slow' and 'intermediate' block. The simulations are based on scheme 7.1 with $k_{+1} = 200$ s^{-1} and $k_{-1} = 400$ s^{-1}. In trace A there is no block so the channels are always either in state C or state O. B shows two consecutive traces in the presence of a slow blocking agent, with $k_{+b} = k_{-b} = 10$ s^{-1}. In C the rate constants for blocking action are higher, with $k_{+b} = k_{-b} = 1500$ s^{-1}, giving an example of 'intermediate' block. A record for 'fast' block, with $k_{+b} = k_{-b} > 10000$ s^{-1}, would look like that in A but with the current pulse amplitude reduced by half and the average pulse duration doubled. Simulations were done with W.J. Heitler's *NeuroSim* program (Heitler, 1992).

block' or 'intermediate' block, and sometimes the term 'slow block' is extended to cover it.

It may be that k_{+b} and k_{-b} are so high that the rate at which the channel switches back and forth between the blocked and unblocked condition is too rapid for the recording system to resolve. This means that the apparent 'openings' of the channel are actually comparable to the bursts seen in flickery block. The measured conductance during these single channel 'openings' will be lower than in the absence of the blocking agent (as in fig. 6.4). The duration of the 'openings' will be longer than the single channel openings in the absence of block, since they include the time spent in state OB as well as in state O, or, to put it another way, the blocking agent holds the channel away from state O and therefore away from state C. This type of behaviour occurs in the block of node of Ranvier sodium channels by hydrogen ions (Woodhull, 1973; Sigworth, 1980b) and the block of ATP-sensitive potassium channels in frog muscle by external TEA ions (Davies *et al.*, 1989). It is known as 'fast' block.

The dissociation constant K_d for the blocked state can be deduced from single channel records. In the case of intermediate block the record consists of a series of bursts during which the channel oscillates between the open and blocked states, separated by periods when the channel is in the closed state.

We can measure the total times spent in the open and blocked states, t_O and t_B, during all the bursts. The ratio of these two times is in inverse ratio to the rate constants leading away from the two states (see chapter 6), so

$$t_O/t_B = k_{-b}/[B]k_{+b}$$

Therefore
$$K_d = [B]\, t_O/t_B$$

For fast block the situation is somewhat different since the recorded 'open' event actually masks a large number of very rapid openings and closings, so we cannot measure t_O and t_B directly. But the size of the single channel current, measured as the fraction f of the current in the absence of the blocking agent, is equal to the fraction of time that the channel is actually in state O during the apparent 'open' event:

$$f = \frac{t_O}{t_O + t_B}$$

So
$$\frac{t_O}{t_B} = \frac{1}{(1/f) - 1}$$

Hence
$$K_d = \frac{[B]}{(1/f) - 1}$$

Scheme 7.1 implies that the blocked state can be entered only from the open state and that the channel cannot enter the closed state while the blocking agent is still bound to it. This is not always the case, and we can imagine a situation in which the blocking agent can bind to or dissociate from the closed state as well:

$$
\begin{array}{ccc}
 & \overset{k_{+1}}{\underset{k_{-1}}{\rightleftharpoons}} & \\
C & & O \\
k_{-b} \Big\Updownarrow k_{+b}[B] & & k_{-b} \Big\Updownarrow k_{+b}[B] \\
CB & \overset{k_{+1}}{\underset{k_{-1}}{\rightleftharpoons}} & OB
\end{array}
\qquad (7.2)
$$

In this case the presence of the blocking agent does not hold the channel in the open state, so the durations of the apparent 'openings' are not longer than in the absence of block. This distinction can be used to distinguish between schemes 7.1 and 7.2, as, for example, in the fast block of potassium channels in frog muscle by external TEA. In voltage-gated potassium channels the apparent open time in single channel currents during block is the same as without the block, suggesting that scheme 7.2 applies (Spruce et al., 1987b). In the ATP-sensitive potassium channels, however, external TEA increases the

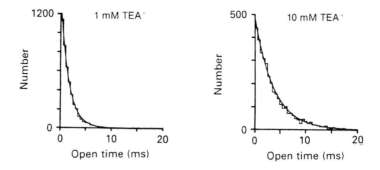

Fig. 7.2. Effect of external tetrathylammonium (TEA) on the apparent open times in single channel patch clamp records from ATP-sensitive potassium channels of frog skeletal muscle. The mean open time increases from 3.3 ms at 1 mM TEA to 7.5 ms at 10 mM. This implies that TEA binds only to the open channel as indicated in scheme 7.1. (From Davies *et al.*, 1989.)

apparent open times (fig. 7.2), showing that the TEA molecule can bind only to the open state and that scheme 7.1 applies (Davies *et al.*, 1989).

Potassium channels

We probably know more about the process of block in potassium channels than in any other group. The reason for this may be that they included some of the more accessible channels in the age before molecular cloning was possible and that, unlike voltage-gated sodium channels, many of them do not inactivate spontaneously. Thus, the voltage-gated potassium channels of squid axon plasma membranes could be investigated with the standard techniques of voltage clamp and intracellular injection, using tetrodotoxin to eliminate the sodium currents. However, the intracellular potassium channels of the sarcoplasmic reticulum could be investigated by the quite different technique of reconstitution into lipid bilayers. Some quite sophisticated experiments have been done using blocking agents on these preparations, leading to stimulating deductions about channel design (see Armstrong, 1975; Yellen, 1987; Begenisich, 1994).

Block by quaternary ammonium ions

The best known blocking agent of potassium channels is the TEA ion (Stanfield, 1983). In squid axons it blocks outward currents through the voltage-gated potassium channels when injected into the axoplasm, but has no effect when applied to the outside of the axon. In vertebrate myelinated nerve fibres, however, it is effective at both the inner and outer surfaces of the nodal membrane.

Other quaternary ammonium ions are also effective as blocking agents when injected into squid axons, and some of these have been used to probe the nature of the channel pore. Experiments by Armstrong (1971, 1975), using the compound triethylnonylammonium (abbreviated as C_9), have been

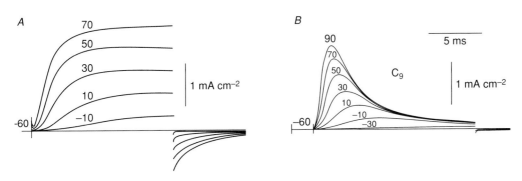

particularly informative. Figure 7.3 shows one of his experiments. In the absence of C_9 the potassium currents produced by clamped depolarizations are maintained for the duration of the depolarization; the potassium channels stay open and do not inactivate. With C_9 inside the axon, however, depolarization elicits an outward potassium current that peaks and then falls back rapidly to a low level. In other words, the block produced by C_9 is not present when the channels first open but develops progressively while they remain open. The most likely interpretation of this is that a C_9 molecule can reach its binding site only when the channel is open: at the onset of depolarization no channels are blocked, but as it proceeds more and more channels become blocked as C_9 molecules enter them. This phenomenon is known as *open channel block*.

Recovery from C_9 block takes place in two stages. The first stage takes some tens of milliseconds and corresponds to the time during which some of the potassium channels remain open after the depolarization period. Recovery is faster if there is a high external potassium ion concentration or if the membrane is held at more negative potentials. Armstrong's explanation of this is that the C_9 molecules occupy a wider inner mouth on the inner side of the transmembrane pore and that they are swept out by an inward flow of potassium ions, since both high external potassium and a more negative membrane potential will promote this flow. The second phase may take several seconds; it suggests that the C_9 molecule may be trapped within the channel when it closes. These ideas are expressed diagrammatically in fig. 7.4.

Many of the features found by Armstrong with C_9 block can also be seen to some extent with TEA block. Overall, Armstrong's experiments led him to conclude that the selectivity filter for potassium ions is towards the outside end of the channel and that the activation gate is near the inside end. The similarity of the time course of C_9 block to an inactivation process was one of the clues leading to the 'ball and chain' model of inactivation (Armstrong, 1992).

Relatively high concentrations (1 to 5 mM) of TEA have to be used to produce block. This is much higher than for most natural toxins, which have to produce their effects at very low concentrations in order to be useful to the

Fig. 7.3. Macroscopic potassium currents in squid axon showing the effect of C_9 (triethylnonylammonium) block. *A* shows a family of outward potassium currents produced by depolarizations of different extents from a holding potential of -60 mV. *B* shows potassium currents from another axon that was injected with C_9 ions. The axons were in artificial sea water; sodium currents were eliminated by tetrodotoxin. (From Armstrong, 1971. Reproduced from *The Journal of General Physiology* 1971, **58**, p. 417, by copyright permission of The Rockefeller University Press.)

Fig. 7.4. Clearing of internal quaternary ammonium (QA⁺) ion block of potassium channels by inflow of potassium ions, as proposed by Armstrong. The model assumes that the QA ions can enter or leave the internal mouth of the channel only when the activation gate is open. (From Hille, 1992a.)

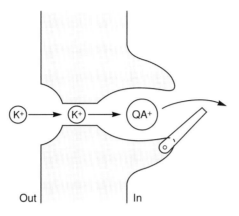

organisms that produce them. One consequence of this is that while radioactive tetrodotoxin and saxitoxin can be used to count the number of sodium channels in a tissue, since they target their receptors in a highly specific way, TEA cannot be used in this way to count potassium channels; too many of the radioactive TEA molecules would simply not be attached to their potassium channel binding sites.

The voltage-dependence of potassium channel block by TEA and other ions can be used to make further deductions about the nature of the channels. Following the approach developed by Woodhull (1973) to analyse the block of voltage-gated sodium channels by hydrogen ions (see chapter 5), one can determine the dissociation constant $K_d(E)$ at different membrane potentials E. From this one can estimate the electrical distance δ across the electric field of the membrane at which the blocking ions become bound, using the relation

$$K_d(E) = K_d(0) \exp(-z\delta FE/RT) \qquad (7.3)$$

French & Shoukimas (1981, 1985) used an approach similar to this to investigate the blocking action of various ions injected into squid axons. They found that TEA and similar larger ions such as tetrabutylammonium showed relatively low voltage dependence for block, with δ values about 0.15; this value means that the blocking ion is bound at a point 15% of the way across the electric field of the membrane, starting from the inside. Smaller ions, such as tetramethylammonium, Tris and lithium, had a higher voltage dependence, with δ values about 0.35. In other words, the smaller blocking ions penetrated further into the channel than the larger ones did. This suggests that the pore contains a narrow tunnel that only potassium and other permeant ions can enter, at the cytoplasmic end of which is a wider mouth into which TEA ions can penetrate, which itself is entered from an antechamber in which even larger blocking ions can be accommodated.

Study of the site-directed mutations that affect block of cloned potassium channels by TEA has been very useful in relating structure to function. With channels expressed in oocytes, the effects of different concentrations of TEA on the potassium currents produced by standard depolarizations can be measured, so enabling the EC_{50} for TEA block to be determined. Mutations that produce marked changes in EC_{50} probably indicate the position of the TEA binding site. Mutations at positions D431 and T449 on the *Shaker* channel (positions 1 and 19 of the P region in fig. 5.19) produce such changes for external TEA, and mutations at T441 (position 11 in the P region) do so for internal TEA (MacKinnon & Yellen, 1990; Yellen *et al.*, 1991). This provides strong evidence for the view that the P region really does form a loop in which its two ends are on the outer side of the channel and its middle is accessible from the inner side, as is shown in fig. 5.19.

Bis-quaternary ammonium block of the sarcoplasmic reticulum potassium channel

The sarcoplasmic reticulum (SR) potassium channel can be incorporated into lipid bilayers after preparation of SR vesicles by subcellular fractionation of skeletal muscle. Miller (1982b; Miller *et al.*, 1984) used this preparation for some ingenious experiments to determine the length of the electric field through the channel. He blocked the channels with quaternary ammonium compounds that consisted of a long straight alkyl chain with a trimethylammonium 'head' attached to each end, having the general formula $(CH_3)_3N^+ - (CH_2)_n - N^+(CH_3)_3$. These can be called bisQn compounds, for example bisQ6 when n is 6. Singly charged equivalents, with only one trimethylammonium head, formula $(CH_3)_3N^+ - (CH_2)_n - H$, can be called Q$n$ compounds. The SR vesicles were added to the *cis* side of the bilayer and the compounds would only block when applied to the *trans* side, which thus corresponds to the luminal side of the SR membrane.

Miller found that the degree of block varies with membrane potential. He determined the effective valency $z\delta$ of the blocking molecule from the slope of this dependence, using the relation

$$\langle \gamma \rangle = \gamma_o \left[1 + \frac{[B]}{K(0) \, \exp(z\delta FV/RT)} \right]^{-1} \tag{7.4}$$

where $\langle \gamma \rangle$ is the time-averaged channel conductance in the presence of the blocking agent, γ_o is the conductance in its absence, and the other symbols are as in equation 7.3, from which this is derived. Figure 7.5 summarizes the results of this investigation. For the singly-charged Qn compounds, the effective valency $z\delta$ is about 0.65 whatever the chain length. But for bisQn compounds there is a very interesting result: the effective valency is about

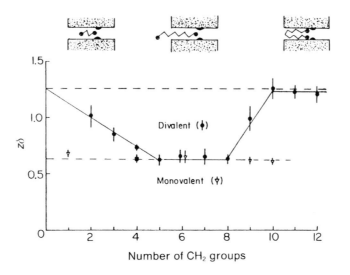

Fig. 7.5. Voltage dependence of block by bisQn compounds in sarcoplasmic reticulum potassium channels. The effective valency $z\delta$ of the blocking molecules is determined as described in the text, using equations 7.3 and 7.4. It is at a minimum when only one of the charged trimethylammonium head groups (connected by nCH$_2$ groups in the bisQn compounds) is in that part of the channel across which the electric field appears. This enables the electrical distance δ to be converted into an actual distance. (From Miller *et al.*, 1984.)

0.65 when n is in the range 5 to 8 but it rises towards 1.3 for both shorter and longer molecules. This suggests that only one charge is in the electric field or potential gradient across the channel when n is 5 to 8, but that there are two charges in it for shorter or longer lengths. Miller suggested that at shorter lengths the electric field is longer than the distance between the charges so that both charges enter that part of the pore, and at longer lengths the middle of the blocking molecule is flexible enough to allow it to bend double so as to let both charges enter the channel pore. So 0.65 of the electric field across the channel is only as long as the distance between the charges in bisQ5. This distance is in the region of 6–8 Å, implying that the field itself is 9–12 Å long.

These results lead to a model of the SR potassium channel in which there is a short, narrow tunnel whose relatively high resistance ensures that most of the transmembrane potential appears across it, leading into a wider mouth and perhaps an even wider antechamber at each end. Similar experiments with bisQn compounds on the calcium-activated potassium channels of plasma membranes again produce minimal $z\delta$ values at $n = 5$ and 6, but $z\delta$ is here only 0.26, so the postulated tunnel region of the pore must be about two and a half times as long as it is in the SR channel (Villarroel *et al.*, 1988). This fits well with the conclusion from other work that most plasma membrane potassium channels are multi-ion pores, whereas the SR potassium channel acts more like a single-ion pore (Coronado *et al.*, 1980; Yellen, 1987).

Other pharmacological agents acting on plasma membrane potassium channels

The aminopyridines, especially 4-aminopyridine (4-AP), are specifically bound to certain potassium channels. These include some but not all voltage-gated channels, the muscarinic G-protein-coupled channel, and the ATP-sensitive potassium channels.

ATP-sensitive (K_{ATP}) potassium channels are specifically blocked by sulphonylurea drugs such as glibenclamide (also called glyburide) and tolbutamide (see Davies *et al.*, 1991; Edwards & Weston, 1993). In pancreatic β cells this leads to depolarization of the cell and consequent release of insulin (see fig. 1.2), hence the use of these drugs in the treatment of diabetes. The sulphonylurea molecules bind to the sulphonylurea receptor component of the channel (Aguilar-Bryan *et al.*, 1995).

A group of drugs of much clinical interest are the potassium channel openers. The three compounds first investigated were nicorandil, pinacidil and cromakalim, and since then many similar compounds have been developed and tested (Edwards & Weston, 1990; Robertson & Steinberg, 1990). Pinacidil and cromakalim and their analogues act directly on K_{ATP} channels, and they are antagonistic to the antidiabetic drugs glibenclamide and tolbutamide. Since they open the K_{ATP} channels of vascular smooth muscle, they are effective in reducing blood pressure. Diazoxide was the first potassium channel opener to be used clinically, as an emergency treatment for high blood pressure. Some openers, including SDZ PCO 400, an analogue of cromakalim, may be useful in relaxing airway smooth muscle and thus effective in the treatment of asthma (Morley, 1994), and others, the anilide tertiary carbinols, may be specific for bladder smooth muscle (Grant *et al.*, 1994). These openers of K_{ATP} channels may act by reducing the affinity of the channel binding site for ATP (Thuringer & Escande, 1989).

A number of peptides that block potassium channels have been discovered in recent years in the venoms of various animals (Castle *et al.*, 1989; Garcia *et al.*, 1991). They include apamin from bees, charybdotoxin and noxiustoxin from scorpions, and dendrotoxin from mamba snakes (table 7.1).

Charybdotoxin is a small peptide isolated from the venom of the scorpion *Leiurus quinquestriatus*. It blocks calcium-activated and other voltage-gated potassium channels when applied to the outside of the membrane, and will bind to both open and closed channels. The toxin dissociates more readily from the channel as the internal potassium ion concentration is raised and on depolarization. Internal rubidium ions (which will pass through the channel) also displace charybdotoxin from its binding site, but non-permeant ions such as sodium and caesium do not. This suggests that charybdotoxin covers or plugs the external mouth of the channel and can be displaced when internal

Table 7.1. *Some natural neurotoxins active on potassium channels*

Toxin	Source	Action and channel affected
Apamin	Bee venom	Block of small-conductance K(Ca)
Charybdotoxin	Scorpion (*Leiurus quinquestriatus*)	Block of large-conductance K(Ca) and many voltage-gated K channels
Iberiotoxin	Scorpion (*Buthus tamulus*)	Specific block of large-conductance K(Ca)
Noxiustoxin	Scorpion (*Centruroides noxius*)	Block of delayed rectifiers, especially Kv1.2 and Kv1.3
Dendrotoxins	Green mamba snake (*Dendroaspis augusticeps*)	Block of various K channels, including Kv1.1, Kv1.2 and Kv1.6

potassium ions approach it from inside the pore (MacKinnon & Miller, 1988). Like other blocking agents, its effectiveness as a blocking agent is not the same for all potassium channels; K_d was 3 nM for Kv1.3 but over 1 μM for Kv1.1 and Kv3.1 channels (Grissmer *et al.*, 1994).

Nuclear magnetic resonance spectroscopy shows that the 37-residue chain of charybdotoxin is folded back and forth to make an ellipsoidal structure about 15 Å wide by 25 Å long. On one side of this there is a stripe of nine positively charged residues, on the other there is a group of three hydrophobic ones (Massefski *et al.*, 1990). Site-directed mutagenesis has been used to find which amino acid residues in the potassium channel molecule are important for charybdotoxin action, using a *Shaker* clone expressed in *Xenopus* oocytes (MacKinnon *et al.*, 1990). The results (table 7.2) show that residues at the outer ends of the H5 loop are critical, especially if they change the charge on the residue.

Charybdotoxin is of considerable use in investigating the biochemistry of potassium channels, since it binds only to the fully assembled tetrameric channel (Miller, 1995). Experiments on mutagenesis of both toxin and channel have been particularly productive. The toxin can be displaced from the channel if potassium ions enter the pore from the inner side, but the effect is removed if the positively charged lysine residue K27 is replaced by any other residue. This suggests that the ϵ amino group of K27 projects downwards into the pore, as though taking the place of a potassium ion (Park & Miller, 1992a,b). Since there are quite good models of charybdotoxin structure, this allows some precise positioning of the residues at which interaction takes place to be estimated. The Arg-24 and 31 residues on the related scorpion agitoxin 2 are about 25 Å apart on opposite sides of the molecule; it seems likely that they bind to the Asp-431 residues of subunits on opposite sides of the channel, suggesting that these Asp-431 residues are placed radially about 12 Å from the centre of the pore (Hidalgo & MacKinnon, 1995). Studies of this

Table 7.2. *Effect of point mutations on charybdotoxin inhibition of* Shaker *potassium channels*

Position and mutation	Position in H5	Change in charge	K_i (nM)
422 Glu→Asp	−8	0	4
Glu→Gln		+1	14
Glu→Lys		+2	48
427 Lys→Arg	−3	0	4
Lys→Asn		−1	0.55
Lys→Glu		−2	0.15
431 Asp→Glu	1	0	10
Asp→Asn		+1	>1000
Asp→Lys		+2	>1000
449 Thr→Lys	19	+1	>1000
Thr→Gln		0	>1000
Thr→Tyr		0	>500

K_i indicates the affinity of the toxin for the channel and its effectiveness in blocking it; low values indicate high blocking effectiveness. The wild-type value was 4 nM. Notice that blocking effectiveness is somewhat reduced by eliminating the negative charge 8 residues upstream from the H5 loop at Glu-422, but much greater change is produced by eliminating the negative charge at Asp-431, conventionally regarded as the beginning of the H5 loop. Block is increased by eliminating the positive charge at Lys-427. Notice also that the binding at Thr-449 does not seem to depend on the charge on the residue. (Data from MacKinnon *et al.*, 1990.)

kind lead to a model of the potassium channel vestibule as shown in fig. 7.6, in which the mouth of the narrow portion of the pore opens into a rather flat vestibule surrounded by a low wall (Miller, 1995).

The alkaloid capsaicin is the active ingredient of hot peppers; its principal action in the body is to stimulate the fine sensory nerve endings of C fibres. It will also, at somewhat higher concentrations, block some 4-AP-sensitive potassium channels, such as those expressed in Schwann cells (Baker & Ritchie, 1994).

Voltage-gated sodium channels

The voltage-gated sodium channels of nerves and muscles are crucial to effective movement in animals, since action potentials cannot occur without them. Hence they are obvious targets for toxins produced by other animals for use in predation, or by plants or animals as a defence against herbivory or predation (Strichartz *et al.*, 1987; Catterall, 1992). Different neurotoxins act at different sites on the sodium channel, and produce different effects as a result. At least six different sites of action have been distinguished, as is summarized in table 7.3. We consider first three toxins that act at 'site 1' to block the channel.

Fig. 7.6. Spatial locations of specific residues in the *Shaker* potassium channel vestibule, as deduced from comutagenesis experiments with scorpion toxins. Compare with fig. 5.19*A*. (From Miller, 1995. Reproduced with permission from *Neuron* **15**, copyright Cell Press.)

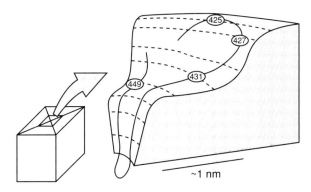

Tetrodotoxin, saxitoxin and μ-conotoxins

Tetrodotoxin is a virulent poison that blocks nervous conduction and causes death by paralysis of the respiratory muscles. It is found in the internal organs of the puffer fish and its relatives in the family Tetraodontidae. Japanese fugu restaurants specialize in removing the most poisonous organs (the ovaries, liver and intestines) from the fish before serving its flesh as a delicacy (diners get an extra frisson as their lips begin to tingle!). Nevertheless, many fatalities occur each year as a result of inexpert preparation by unlicensed cooks (Ogura, 1971; Harrison, 1991). Voltage clamp studies on lobster axons by Narahashi and his colleagues in 1964 showed that the sodium current was eliminated by external tetrodotoxin whereas the potassium current was unaffected. This was a highly significant result in that it provided conclusive evidence for the view, not yet fully accepted at that time, that the sodium and potassium channels of nerve axons were separate entities.

Saxitoxin has properties similar to those of tetrodotoxin and blocks sodium channels at the same site (Narahashi *et al.*, 1967; Hille, 1968). It is one of a group of similar compounds produced by dinoflagellates, unicellular organisms in the marine plankton. Population explosions of these may produce 'red tides' in the sea. The dinoflagellates may then be eaten by clams or similar filter-feeding shellfish, which then themselves become poisonous to eat, causing paralytic shellfish poisoning (see Viviani, 1992).

Both compounds are effective as blocking agents when applied to the outside of the channel but have no effect on the inside. This suggests that they bind to a particular array of amino acid residues at the outer mouth of the channel. Further evidence for this comes from genetic engineering of the SS2 segment, part of the H5 or P loop that is thought to line the outer mouth of the channel pore, in the first transmembrane domain. Mutation of the Glu-387 residue in the rat brain type II sodium channel to glutamine greatly reduces the sensitivity of the channel to tetrodotoxin and saxitoxin (Noda *et*

Table 7.3. *Neurotoxin binding sites in voltage-gated sodium channels*

Binding site	Toxin	Chemical type	Effect
1	Tetrodotoxin⎫ Saxitoxin ⎬	Heterocyclic	Ion channel block
	μ-conotoxins	Small peptide	Ion channel block
2	Veratridine Batrachotoxin Aconitine Grayanotoxins	Alkaloid	Persistent activation
3	Scorpion α-toxins Sea anemone toxins	Peptide	Inhibit inactivation
4	Scorpion β-toxins	Peptide	Varied effects on activation
5	Brevetoxins	Cyclic polyether	Shift voltage-dependence of activation
'6'	Pyrethrins Pyrethroids	Ester	Prolong activation

Data from Catterall, 1992; Adams & Olivera, 1994; Adams & Swanson, 1994.

al., 1989). Cardiac muscle sodium channels are less sensitive to tetrodotoxin than are those of brain and skeletal muscle: they require micromolar instead of nanomolar concentrations to cause block. When the Cys-374 residue of the cardiac channel is changed to tyrosine, which is the corresponding residue in brain and skeletal muscle sodium channels, the sensitivity to tetrodotoxin is greatly increased (Satin *et al.*, 1992).

The high affinity of tetrodotoxin and saxitoxin for most sodium channels allows them to be used as probes, usually by making them radioactive. This has proved very useful in isolating sodium channels by biochemical means: the radioactive toxin binds to the channel so that it can readily be detected at the various stages of the extraction process. Another use of radioactive toxins is to count the density of channels in particular cells such as nerve axons. By this means we have estimates of 35 channels μm^{-2} in garfish olfactory nerve and 400 to 700 μm^{-2} in the nodes of rabbit optic nerve (Ritchie *et al.*, 1976; Pellegrino & Ritchie, 1984).

Conotoxins are constituents of the venom of a remarkable group of predatory molluscs, the cone shells *Conus*. There are about 500 species in this genus, each one producing its own cocktail of venomous peptides. The venom is injected into the prey via a harpoon-like disposable tooth and rapidly causes paralysis and death. Some species feed on fish, others on molluscs or worms.

The conotoxin peptides are unusual in many ways (Olivera *et al.*, 1990, 1991; Myers *et al.*, 1993). They are only 10 to 30 amino acid residues in length, whereas most other peptide neurotoxins have 60 to 100 residues. They contain sufficient cysteine residues to form two or three disulphide links, which provide the toxins with a stable and clearly defined tertiary structure. Normally we would not expect peptides this small to adopt a defined conformation with precisely the right disulphide links. However, it seems that they are assembled from the C-terminal ends of larger propeptides and that the mature peptide is cut off after its disulphide links and particular folding pattern have been formed. The peptides often contain two unusual amino acid residues, each formed by post-translational modification: hydroxyproline and γ-carboxyglutamate. The very high specificity of many of the conotoxins has made them useful as tools for the investigation of particular types of receptors and channels and their distribution in the brain and elsewhere. The ω-conotoxins, for example, are particularly useful for investigating voltage-gated calcium channels.

The μ-conotoxins selectively block the voltage-gated sodium channels of skeletal muscle and electric organs; they are relatively ineffective on neuronal sodium channels. They probably act at the same site as tetrodotoxin and saxitoxin, since they will displace radioactive saxitoxin from eel electroplax membranes. μ-Conotoxin GIIIA is a 22 residue peptide with three disulphide bridges and three loops of the polypeptide chain between them. It has six positively charged residues, of which Arg-13 seems to be crucial. Synthetic derivatives with this replaced by glutamine showed much reduced binding, and 'blocked' channels would allow 20% to 40% of the normal current to flow through them (Becker *et al.*, 1992).

Neurotoxins that modulate sodium channel activity

A variety of different neurotoxins act on the sodium channel to modify its gating so as to keep the channel open for longer. This leads to repetitive firing of neurons so that the nervous system no longer functions properly.

Non-polar toxins are able to penetrate the lipid membrane and so can reach sodium channel sites that are embedded in it. A number of alkaloids act at site 2. They produce persistent activation by inhibiting inactivation and shifting the activation–voltage curve to more negative potentials, as is shown for batrachotoxin in fig. 6.10. Batrachotoxin, together with the similar compound homobatrachotoxin, is secreted by the skin of Columbian arrow poison frogs of the genus *Phyllobates*. The frogs themselves are insensitive to the poison they produce, probably by means of some modification of the binding site on their sodium channels (Daly *et al.*, 1980). Remarkably, homobatrachotoxin is also found in the skin and feathers of birds of the genus *Pitohui* from New Guinea, the only known example of chemical defence in birds (Dumbacher *et al.*, 1992). Batrachotoxin has been used to prevent the inactivation of sodium channels reconstituted in lipid bilayers (Khodorov, 1985).

Other alkaloids acting at site 2 are of plant origin. Veratridine is the most potent of a mixture of veratrum alkaloids isolated from the lily genus *Veratrum*. Aconitine comes from a similar mixture from the monk's hood, an ornamental flower of the buttercup family. Grayanotoxins are found in rhododendrons and other plants of the heather family. All will compete with batrachotoxin and with each other for the same binding site, and all produce persistent activation of sodium channels.

Peptide neurotoxins can usually bind only to some part of the channel exposed on the outside of the cell, since their complex and largely hydrophilic structure prevents them from passing through the plasma membrane. The scorpion α- and β-toxins must bind to different parts of the channel, since the binding of one type does not interfere with the binding of the other. Scorpion α-toxins are used by the north African genera *Androctonis*, *Buthus* and *Leirus* and the American *Centruroides*; they prevent inactivation and so produce prolonged action potentials in muscle and nerve (Strichartz *et al.*, 1987). Sea anemone toxins produce similar effects and bind at the same site (Norton, 1991). Scorpion β-toxins, produced by American scorpions, move the activation–voltage curve to more negative membrane potentials.

Brevetoxins are produced by a marine dinoflagellate *Ptychodiscus brevis*, which is responsible for Florida 'red tides' and the massive fish kills associated with them. People who eat bivalve molluscs that have been feeding on the dinoflagellates suffer from neurotoxic shellfish poisoning (Viviani, 1992). Brevetoxins are ladder-like molecules with ether linkages forming a series of adjacent ring structures. Their effects are similar to those of the alkaloid toxins, but since they do not compete with them for the binding site they must bind elsewhere. Ciguatoxin, from another dinoflagellate, binds to the same site and has similar effects.

Pyrethrins are natural insecticides produced from *Chrysanthemum cinaerofolium*, now grown in the highlands of East Africa. They are useful because they have a rapid 'knock-down' effect on insects and are largely non-toxic to mammals. Pyrethroids are synthetic analogues. In nerve axons they tend to prolong sodium channel activation and inhibit inactivation, so single channel currents last longer, macroscopic currents are maintained for some time, and the axons tend to show repetitive firing. The synthetic insecticide DDT produces similar effects.

Local anaesthetics

Local anaesthetics such as lidocaine and procaine produce their effects by blocking the sodium channels of nerve axons. They are all relatively small lipid-soluble molecules, usually with amine groups that become positively charged in acid conditions. Figure 7.7 shows the chemical formulae of some of them.

Fig. 7.7. Chemical formulae of some local anaesthetics, substances that prevent nervous conduction by blocking voltage-gated sodium channels. QX-314 is not a local anaesthetic but has proved useful in investigating local anaesthetic action.

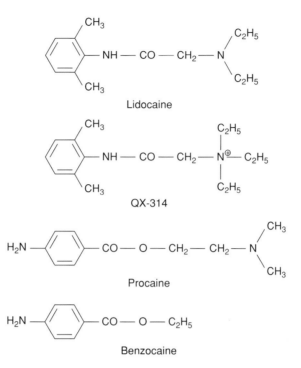

Lidocaine

QX-314

Procaine

Benzocaine

Useful information as to how local anaesthetics work came from experiments with QX-314, a quaternary analogue of lidocaine that is positively charged at all times and hence is not lipid soluble. This compound will not block sodium channels when applied from outside the cell (so it cannot be used as an anaesthetic), but it will do so if it is applied from the inside. Furthermore, the blocking action is only evident after the channel has been open in the presence of the drug, so here is another example of open-channel block. With a single voltage-clamped depolarization the sodium currents are much as normal, but they get progressively smaller with successive depolarizing pulses, as more and more sodium channels become blocked (Strichartz, 1973). This phenomenon is referred to as *use-dependent block*.

These experiments suggest that the channel is blocked when the QX-314 molecule is bound to a particular site in the channel pore. This site cannot be reached from the outside, and is only accessible from the inside when the channel is open. The cumulative effect of successive depolarizations also suggests that those channels that are blocked do not release their blocking molecules between pulses, and hence the drug must be held inside the channel when it is closed. It is as if there is gate at the inner end of the channel and a vestibule between this and the narrowest part of the channel where the drug molecule can be trapped, as is shown in fig. 7.8. The arguments are similar to

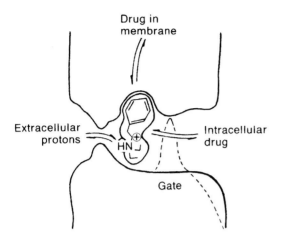

Drug in membrane

Extracellular protons

Intracellular drug

HN

Gate

Fig. 7.8. How local anaesthetics may block nerve axon sodium channels. Positively charged anaesthetic molecules can reach the receptor site only by entering the channel pore from the inside of the cell, and can leave only via the same route. This route is dependent upon the channel being open. Neutral molecules can also reach the site from the outside via the lipid cell membrane. With acidic solutions on the outside, hydrogen ions can access a neutral anaesthetic molecule and convert it to the charged form so that its escape from the channel is delayed. (From Schwarz et al., 1977.)

those used by Armstrong (1971) for the block of potassium channels by TEA ions.

Unlike QX-314, lidocaine and similar compounds with tertiary ammonium groups can reach their site of action readily from outside the cell. In solution these compounds exist as a mixture of charged and uncharged forms, with the proportion of charged molecules increasing as the solution is made more acid. Hille (1977a) found that increasing the proportion of charged forms by lowering the pH from 8.3 to 6 slowed the rate of onset of the drug action and decreased their potency. The neutral compound benzocaine, however, which has no tertiary ammonium group and therefore has no charged form, was fast in onset and was unaffected by pH. The lipid-soluble nature of the uncharged forms presumably allows them to reach the binding site by passing through the plasma membrane, whereas the charged forms can do so only via the channel pore and from the inside of the cell (Hille, 1977b). In support of this idea, the rate of onset of block is proportional to the concentration of the uncharged form when the drug is applied to the outside of sodium channels in lipid bilayers, but proportional to the concentration of the charged form when it is applied to the inside (Zamponi et al., 1993).

Local anaesthetics seem to enhance the normal sodium inactivation process. Hille (1977b, 1992a) suggested that besides blocking the channel pore they bind preferentially to the inactivated state so as to stabilize it. They also block a variety of other channels, including various potassium channels and neurotransmitter-gated channels, but their local anaesthetic action is via their action on sodium channels (Katzung, 1995).

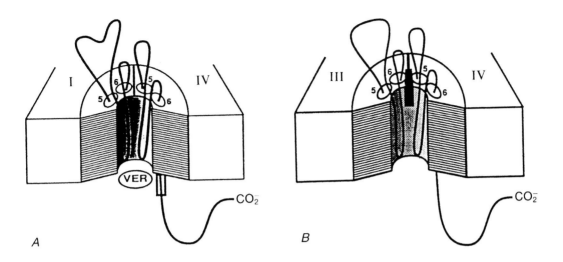

A   *B*

Fig. 7.9. Models of calcium channel block. The channel pore is shown lined by H5 segments from two of the four transmembrane domains; the S5 and S6 membrane-crossing segments are also shown. (*A*) Pore-blocking model for the action of verapamil (VER) and other phenylalkylamines on calcium channels. The box shows part of the peptide chain covalently labelled by photoactivatable verapamil derivatives. (*B*) Model for the binding of dihydropyridines to calcium channels, at the interface between domains III and IV. (From Catterall & Striessnig, 1992. Reproduced from *Trends in Pharmacological Sciences*, with permission from Elsevier Trends Journals.)

Voltage-gated calcium channels

A number of synthetic organic drugs, first discovered in the 1960s, are active on L-type voltage-gated calcium channels (Fleckenstein, 1983; Stanfield, 1986; Varadi *et al.*, 1995). The main groups are the dihydropyridines such as nifedipine and nitrendipine, the phenylalkylamines such as verapamil, and the benzothiazepines such as diltiazem. They mostly block the channel, but some of the dihydropyridines (Bay K 8644 is one) act as channel openers. The blocking agents are used therapeutically to treat ischaemic heart disease, hypertension and cardiac arrhythmias (Smith & Reynard, 1992).

Many of these drugs are lipid soluble, which makes them able to penetrate cell membranes, and so it may be difficult to locate their site of action. But this difficulty can be avoided by using quaternary analogues with a permanent positive charge at the quaternary nitrogen atom. Under these conditions it is clear that dihydropyridines are effective only at the outer side of the plasma membrane, whereas phenylalkylamines block at some site on the inside. Using photoactivatable compounds to form covalent links with the channel protein, it has been possible to locate the site of action further. Catterall & Striessnig (1992) suggested that the phenylalkylamines such as verapamil bind close to the inner end of the transmembrane pore and block it, whereas the dihydropyridines bind by slipping into the cleft between domains III and IV of the channel α subunit (fig. 7.9).

The blocking action of dihydropyridines appears to involve an allosteric mechanism rather than a simple occupancy of the channel pore. In heart muscle calcium channels they bind preferentially to particular states or modes and stabilize them. Thus nitrendipine block involves binding to the inactivated state (Bean, 1984). An alternative description of this effect is that the calcium channel can exist in three distinct 'modes' (Hess *et al.*, 1984). In mode 0 the channel remains closed, in mode 1 it opens often but only briefly, whereas in mode 2 it opens for relatively long periods of time. In the absence of drugs the channel switches between these different modes from time to time. The calcium channel antagonist nitrendipine stabilizes the channel in mode 0, whereas the agonist Bay K 8644 binds preferentially to mode 2 and stabilizes the channel in that mode so that it is open most of the time.

The ω-conotoxins are *Conus* venom peptides that bind mainly to neuronal N-type calcium channels and block them (Olivera *et al.*, 1990, 1991, 1994). They cause paralysis of the fish prey by preventing neuromuscular transmission (which is dependent upon the N-type calcium channels in the nerve terminals), and will prevent transmission at many other synapses. The peptide chain contains 24 to 27 amino acid residues and is constrained to form four loops by three disulphide bridges; this compares with 22 residues and three loops for the μ-conotoxins that bind to the sodium channel. Some indication of the binding properties of ω-conotoxin-GVIA have been obtained by measuring its binding to molecular chimeras made of parts of the N-type channel and a calcium channel that is not sensitive to the toxin. All four domains seem to be involved, but domain III, and especially the extracellular loop between $S5_{III}$ and $H5_{III}$, seems to be particularly important (Ellinor *et al.*, 1994).

Spider venom components known as ω-agatoxins are specific blockers of calcium channels (Olivera *et al.*, 1994). They have been isolated from the venoms of various spiders and particularly from the American funnel-web spider *Agelenopsis aperta*. The peptides range in length from 48 amino acid residues for ω-AgaIVA and ω-AgaIVB to 76 for ω-AgaIIIA. ω-AgaIVA seems to be specific for P-type channels (Mintz *et al.*, 1992), whereas ω-AgaIIIA has a much wider range.

Pharmacology of some neurotransmitter-gated channels

The opening of neurotransmitter-gated channels is usually affected by many different pharmacological agents. These are known as competitive agents if they attach to the neurotransmitter binding site and compete with the neurotransmitter for it; they are known as non-competitive agents if they do not. Competitive agents are known as agonists if their binding results in opening of the channel, as antagonists if it does not. Agonists can sometimes cause block by inducing desensitization. Non-competitive agents include channel blocking agents.

The nicotinic acetylcholine receptor channel

Many different compounds affect the opening of nAChR channels. Nicotine is the classical competitive agonist, curare the classical competitive antagonist.

Nicotine, as everyone knows, is found in tobacco leaves. It is an insecticide, and presumably this is its value to the tobacco plant. It acts as an agonist for the nAChR. Other agonists include the acetylcholine analogues carbachol and succinylcholine, and the alkaloid anatoxin A from the blue-green alga *Anabaena flos-aquae*.

One of the most effective nAChR antagonists is α-bungarotoxin, obtained from the venom of the banded krait *Bungarus multicinctus*. Similar α-toxin venoms are found in related snakes. They are peptides 61 to 74 amino acid residues long. α-Bungarotoxin binds firmly to the acetylcholine binding site on each of the two α subunits, and has been used as a radioactive label (after tagging with ^{125}I) in the biochemical isolation of the nAChR from electric organs and the localization and counting of nAChRs at the neuromuscular junction (Changeux *et al.*, 1970; Fertuck & Salpeter, 1974). α-Bungarotoxin does not bind to most neuronal nAChRs, but a different constituent of *Bungarus* venom, κ-bungarotoxin, does.

Another group of peptides that act as competitive antagonists are the α-conotoxins, produced by *Conus* shells. These are only 13 to 15 amino acid residues long, with just two loops of the peptide chain held in position by two disulphide links. Some of the α-conotoxins are remarkable in showing much greater action on the neuromuscular junction nAChRs of fish than on those of mammals (Myers *et al.*, 1993). Non-peptide antagonists include the alkaloid d-tubocurarine (and others of the curare complex) from the plant *Chondodendron tomentosum*, and the cyclic disulphide compound nereistoxin from the marine worm *Lumbriconereis*. The latter has been synthesized and is sold as an insecticide (Sattelle *et al.*, 1985).

Non-competitive antagonists include the alkaloid histrionicotoxin (which increases desensitization) from the arrow poison frog *Dendrobates histrionicus*, the diterpene lactone lophotoxin from gorgonian corals and the glycoside neosurugatoxin from the Japanese ivory mollusc *Babylonia japonica*, both of which block the channel pore (Adams & Swanson, 1994). In addition to its effects on dopamine receptors, the tranquillizer chlorpromazine blocks the nAChR channel pore, and has been used to help to determine the regions lining the pore (Revah *et al.*, 1990).

GABA$_A$ and glycine receptor channels

The GABA$_A$ and glycine receptor channels are concerned with fast synaptic inhibition and show selective permeability to chloride ions. We have seen in chapter 4 that they are pentameric structures showing some homology with the nAChR. Their pharmacology, however, is quite different.

The pharmacology of the GABA$_A$ receptor channel is quite complicated (Macdonald & Olsen, 1994). There are at least five different binding sites, for GABA, benzodiazepines, barbiturates, certain steroids and picrotoxin. Muscimol, a constituent of the mushroom *Amanita muscaria*, is an agonist of GABA (i.e. it binds to the GABA binding site and opens the channel). The alkaloid bicuculline and a number of synthetic compounds are competitive antagonists; they probably attach to the GABA binding site, but without opening the channel.

The benzodiazepines are a group of synthetic drugs much used in medical practice since their introduction in the 1960s. They are central nervous system depressants, producing sedation and sleep; diazepam is marketed as the tranquillizer Valium. On combination with the binding site on the GABA$_A$ receptor, they increase the effectiveness of the transmitter substance GABA. Single channel records show that this is done by increasing the frequency of channel opening, probably by enhancing the rate at which GABA binds; the single channel conductance remains the same (Rogers *et al.*, 1994). Compounds that have the opposite effect, the β-carbolines, are known as inverse agonists; they reduce the frequency of channel opening in the presence of GABA (fig. 7.10). There are also compounds that bind to the benzodiazepine site and have no action other than to prevent benzodiazepine agonists from being effective; they are known as benzodiazepine antagonists. We can describe the action of benzodiazepines on GABA binding as allosteric, since an event in one part of the molecule produces a response in another, and their effect on channel opening as modulation, since the benzodiazepine itself is not the primary agent for opening the channel.

Barbiturates are also allosteric modulators of GABA$_A$ receptor channels. They bind at a site separate from that which binds GABA. This increases the proportion of time that the channel is open when it has been activated by GABA (Macdonald *et al.*, 1989). Similar effects are produced by the steroids androsterone and pregnanolone (Twyman & Macdonald, 1992). The inhalational general anaesthetics, such as halothane and isoflurane, also potentiate the response of GABA$_A$ receptors to GABA, and this is probably the means whereby they produce anaesthesia (Franks & Lieb, 1994).

Picrotoxin is a convulsant substance from the seed of *Anamirta cocculus*, a climbing shrub from Southeast Asia. It is a non-competitive inhibitor of GABA action at GABA$_A$ receptors. It may block the channel directly by binding to a site in the channel pore, or it may be that it interferes with the allosteric link between GABA binding and channel opening so as to reduce the probability of long openings. Penicillin G reduces the average open time of GABA$_A$ receptor channels, probably by blocking the open channel (Macdonald & Olsen, 1994).

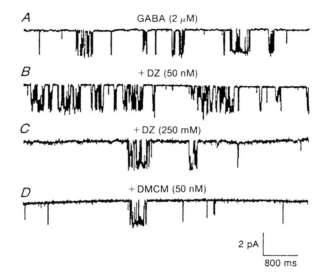

Fig. 7.10. Modulation of GABA-induced activity in GABA$_A$ receptors by benzodiazepines. All the records show single channel currents in cultured mouse spinal neurons in response to micropipette application of 2 μM GABA. Addition of diazepam (DZ) increases the rate of channel opening whereas addition of a β-carboline (DMCM) reduces it. (From Rogers et al., 1994).

The location of these various binding sites, and their relations to the different channel subunits and their subtypes, is not yet fully clear. However, some clues can be obtained from the expression of various subunit subtypes in combination in *Xenopus* oocytes and other expression systems. Thus benzodiazepines bind to α subunits, but they seem to need the γ2 subunit in order to be effective. Different benzodiazepines bind selectively to different α subunit subtypes. Since different combinations of subtypes are expressed in different regions of the nervous system, there are corresponding regional differences in drug sensitivity (Macdonald & Olsen, 1994; Vandenberg & Schofield, 1994).

Glycine receptors are blocked by the competitive antagonist strychnine. This is an alkaloid obtained from the seeds of the Indian tree *Strychnos nux-vomica*; it produces convulsions in mammals and has been used as a rat poison for centuries. It binds to the α subunits in the glycine receptor (Vandenberg & Schofield, 1994).

Glutamate receptor channels

Ionotropic glutamate receptors are classified by their pharmacological properties into three main types, as we have seen in chapter 4. The synthetic agent AMPA is a potent and highly specific synthetic agonist for one group of receptors, which are thus known as AMPA receptors. Kainic acid, obtained from the red alga *Digenea*, will bind with low affinity to AMPA receptors, but with high affinity to another group, which are thus known as kainate receptors. Domoic acid, found in the seaweed *Chondria*, is a specific agonist at kainate receptors. AMPA and kainate receptors are similar in many ways and show 35% to 40% amino acid sequence identity, so they are sometimes put together as AMPA-kainate receptors. They are both responsive to quisqualic acid, obtained from the seed of *Quisqualis indica*. A rather separate group, both in pharmacology and structure, is responsive to *N*-methyl-D-aspartate, and so they are known as NMDA receptors. Ibotenic acid, one of many neuroactive substances from the fly agaric mushroom *Amanita muscaria*, is an agonist at all three receptor types.

The synthetic compounds known as quinoxalinediones are competitive antagonists at AMPA-kainate receptors and have no effect on NMDA receptors. They include 6-cyano-7-nitroquinoxaline-2,3-dione (CNQX). A biphenyl derivative of naphthalene disulphonic acid known as Evans Blue is interesting in that it will block receptors with some subunit combinations but not others; if even more specific agents can be found from related substances, it might be possible to use them in medical practice for specific populations of glutamate receptors (Keller *et al.*, 1993).

A number of spider and wasp toxins are effective blocking agents for AMPA-kainate receptor channels. The functional reason for this is that glutamate is the neuromuscular transmitter in insects and the glutamate receptors of insect muscle are similar to the AMPA-kainate receptors of vertebrates; the spiders paralyse their prey by injecting neuromuscular blocking agents into them (see Schuster *et al.*, 1991; Usherwood & Blagbrough, 1991; Hollmann & Heinemann, 1994). Joro spider toxin, from the spider *Nephila clavata*, acts as a use-dependent channel blocker for AMPA receptors showing rectification, but not for those which do not. The Q/R site on the M2 pore loop (see fig. 5.23) is crucial here. When channels expressed in oocytes have a preponderance of their subunits with arginine (R) instead of glutamine (Q) at the Q/R site (either by using GluR2 or GluR6 subunits or mutant subunits), both rectification and block by toxin disappear (Blaschke *et al.*, 1993). Hence it seems likely that the block is caused by the spider toxin binding to a site in the channel pore near to the Q/R site.

A number of synthetic drugs are effective as NMDA channel blocking agents. They may be of some use in preventing the potentially toxic effects of receptor overactivity in conditions such as epilepsy and neurodegenerative diseases (Rogawski, 1993). The general anaesthetic ketamine probably acts by inhibiting NMDA receptors (see Franks & Lieb, 1994).

NMDA channels can be blocked by conantokin-T and conantokin-G, *Conus* peptides that have γ-carboxyglutamate in their sequences (Olivera *et al.*, 1990). The synthetic compounds D-2-amino-5-phosphonovalerate (APV) and 3-(2-carboxypiperazine-4-yl)propylphosphonate (CPP) are competitive antagonists of glutamate action at NMDA receptors. The non-competitive agents phencyclidine, MK-801 and zinc ions all act by blocking the channel pore (Hollmann & Heinemann, 1994). Magnesium ions produce a voltage-dependent block of the channel pore so that NMDA channels are normally only open when the membrane potential is depolarized. Spermine and other polyamines produce complex effects, for which the NR2 subunits are necessary, with both inhibition and stimulation of the response to glutamine being evident (Williams *et al.*, 1994). Finally, we have seen in the previous chapter that glycine potentiates NMDA receptor responses to glutamate by binding to the receptor molecule.

8 Dysfunctional channels in human disease

Ion channels are implicated in the origins or treatment of many diseases. In some autoimmune diseases the body produces antibodies to its own channels. Thus, in myasthenia gravis, antibodies to the acetylcholine receptors at the neuromuscular junction are produced, with consequent partial paralysis of many muscles (Drachman, 1994). In various neurodegenerative diseases there is a cytotoxic reaction involving excess calcium ion inflow through NMDA receptor channels (Lees, 1993). We have seen in chapter 7 how diabetes can sometimes be treated by closing the ATP-dependent potassium channels in pancreatic β cells. But in this chapter we shall restrict ourselves to just one aspect of the channel disease arena: the occurrence of channels whose structure is different from normal as a result of genetic mutation and which therefore do not function in the way they should.

The idea that inherited disorders might result from the absence or imperfection of some particular protein was first emphasized by Archibald Garrod soon after the rediscovery of Mendel's laws at the beginning of the twentieth century (Garrod, 1909, 1931). Garrod's first book listed just four such diseases, but its intellectual successor today estimates that over 4000 single-gene diseases are known (Scriver *et al.*, 1995). A number of these involve mutations of ion channels, and we can be confident that many more will be discovered in the future. Let us have a look at some examples.

Voltage-gated cation channels

Sodium channels

Three somewhat similar inherited diseases are produced by mutations in the skeletal muscle sodium channel gene (Brown, R.H., 1993; Barchi, 1995; Hoffman *et al.*, 1995). They are hyperkalaemic periodic paralysis (HYPP),

Table 8.1. *Some naturally occurring mutants of the human skeletal muscle sodium channel* α *subunit. The hereditary diseases they produce are hyperkalaemic periodic paralysis (HYPP), paramyotonia congenita (PC) and potassium-aggravated myotonia (PAM)*

Mutation	Position	Domain	Disease
Thr→Met	704	II S5	HYPP
Ala→Thr	1146	III S4–5	HYPP-PAM
Gly→Val ⎫ Gly→Glu ⎬ Gly→Ala ⎭	1306	III–IV	PAM
Thr→Met	1313	III–IV	PC
Met→Val	1360	IV S1	HYPP
Phe→Leu	1419	IV S3	HYPP (horse)
Arg→His ⎫ Arg→Cys ⎬ Arg→Pro ⎭	1448	IV S4	PC
Val→Met	1589	IV S6	PAM
Met→Val	1592	IV S6	HYPP

Data mainly from Brown, 1993; Hoffmann *et al.*, 1995.

paramyotonia congenita (PC) and potassium-aggravated myotonia (PAM). The paralysis in HYPP is brought on by heavy physical work and is associated with raised plasma potassium concentrations. The myotonias (PC and PAM) show intermittent muscle stiffness or involuntary contraction, often triggered by cold (PC) or by potassium-rich food such as bananas (PAM). Electrophysiological analysis of muscle biopsies from affected patients show that their voltage-gated sodium currents do not inactivate fully. The consequent slight depolarization might well explain both the paralysis (a rise in threshold) and the muscle stiffness (a slight activation of the contractile apparatus).

Understanding of these diseases in molecular terms began with the demonstration in 1990 that the human skeletal muscle sodium channel gene was localized on the long arm of chromosome 17 near to the GH1 growth hormone locus. In a large family with HYPP in four generations, linkage between GH1, the sodium channel gene and the occurrence of HYPP was complete (Fontaine *et al.*, 1990). Soon afterwards PC and PAM were also shown to be sodium channel defects. Sequence determination showed that the mutant genes express sodium channel proteins with single amino acid changes, as is summarized in table 8.1.

Let us look at one of these sites, the set of mutations at position 1306, which is in the cytoplasmic sequence linking domains III and IV. These mutations were discovered in a group of patients who suffered from a generalized muscle stiffness that was not influenced by cold but was much enhanced by dietary potassium intake, and hence was classified as PAM (McClatchey *et al.*, 1992; Lerche *et al.*, 1993). McClatchey and her colleagues (1992) sequenced genomic DNA from the white blood cells of a Belgian family with the disease, using the polymerase chain reaction with appropriate sodium channel primer sequences. This revealed that the Gly-1306 codon, normally the guanosine triplet GGG, had mutated to GTG, which codes for valine.

Lerche and his colleagues (1993) used single-strand conformational polymorphism (SSCP) to look for mutations in their group of patients. This is a technique for separating DNA sequences differing in just a single base; the precise way in which a piece of single-stranded DNA folds is affected by such a small change, and this in turn affects the rate at which it moves electrophoretically on a polyacrylamide gel. They found three mutations at the Gly-1306 site, as is shown in fig. 8.1: the glycine residue was changed to valine as found by McClatchey's group, but also, in different families, to alanine (coded by GCG) and to glutamic acid (coded by GAG). These three mutations all produced PAM but with differences in the severity of the symptoms. The patient with glutamic acid substituted for glycine at this point suffered from severe myotonia and needed to take tocainide (a sodium channel blocking agent) to combat the effects on the respiratory muscles. The patients (mother and son) with valine as the substitute residue had milder myotonia, which did not require medication. A patient with alanine as the substitute suffered from fluctuating levels of muscle stiffness.

The pair of glycine residues at 1306 and 1307 would fit the requirement for flexibility in the 'hinged-lid' model of sodium channel inactivation (see chapter 6) very well, so that it seems likely that this is their function. They are separated from the adjacent Ile-Phe-Met trio by just two other residues (Ile-1488 in the rat brain type IIA corresponds to Ile-1310 in the human skeletal muscle sequence). We would expect that the large charged glutamate residue would produce much larger changes in function than the smaller valine or alanine residues would, which may explain the greater severity of the disease in the patient with this mutation.

We have seen in chapter 6 that the R1448H mutant in the $S4_{IV}$ segment also affects inactivation. Figure 8.2 shows the effects of another mutation, from valine to methionine at position 1589, on patch clamp currents of sodium channels expressed in a cell line. The intermittent failure of inactivation is very clear (Mitrovic *et al.*, 1994). Similar results have been seen in cells cultured from muscle biopsies from HYPP patients (Cannon *et al.*, 1991). Effects very similar to the clinical condition have been produced in rat muscles by treatment with a sea anemone toxin; a proportion of the sodium channels entered the non-inactivating state, and the muscles were then more excitable and showed delayed relaxation as in myotonia (Cannon & Corey, 1993).

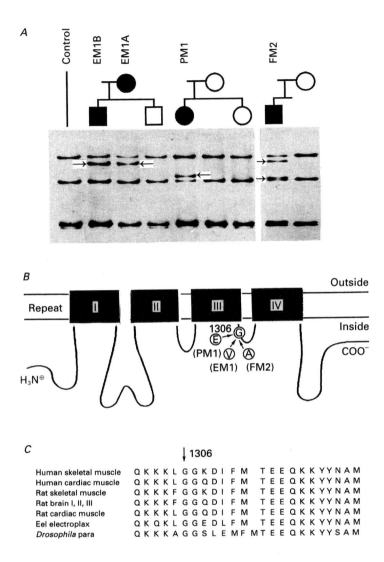

Fig. 8.1. Molecular genetics of potassium-aggravated myotonia (PAM). *A* shows gels with single-strand conformational polymorphism products for members of three different families with PAM; the part pedigrees show the relations of the family members with PAM (black symbols) and their immediate relatives (white symbols). Gels relate to the pedigree above them and (left) to one of the unrelated normal controls. Those from normal subjects with the GGG codon at the site coding for Gly-1306 show three bands, mutations are revealed by the extra bands (arrows). *B* represents the sodium channel α subunit with the position of Gly-1306 and its mutations indicated. *C* shows the amino acid sequences of residues 3 to 25 of the 53-residue III–IV cytoplasmic linker section in a number of different channels. Notice that the two glycine residues at 1306 and 1307 are completely conserved; they are thought to be important as part of the hinge in the 'hinged-lid' model of sodium channel inactivation. (From Lerche *et al.*, 1993.)

Voltage-gated calcium channels

Hypokalaemic periodic paralysis (HOPP) is an autosomal dominant skeletal muscle disorder in which episodes of muscle weakness occur. People inheriting this disease usually have their first attacks of limb weakness as teenagers. Patients have low serum potassium levels, down to 1.8 mM, in comparison with normal levels of 3.5 to 5.0 mM (HYPP patients, however, may have levels up to 7 mM). Their muscles have relatively low resting potentials and action potentials with little or no overshoot.

Genetic analysis links HOPP to the DHP receptor calcium channel at

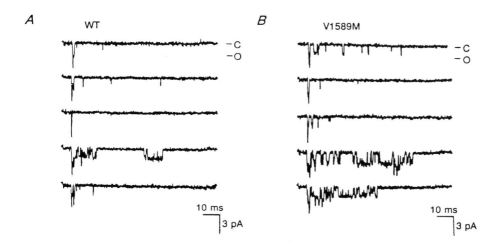

Fig. 8.2. Lack of inactivation in a hyperkalaemic periodic paralysis (HYPP) mutant sodium channel. Traces show currents elicited in cell-attached patches by depolarizations from −100 to −30 mV; C shows zero current with all channels closed; O shows current level with one channel open. Human skeletal muscle voltage-gated sodium channels were expressed in a human embryonic kidney cell line. *A* shows the responses of normal (wild type, WT) channels, *B* shows those of mutant V1589M channels. Notice the intermittent lack of inactivation in *B*. (From Mitrovic *et al.*, 1994.)

region q31-32 on chromosome 1 (Fontaine *et al.*, 1994; Ptáček *et al.*, 1994). SSCP analysis showed that patients had mutations affecting the arginine residues of the S4 segments of domains II or IV (fig. 8.3). In some families, Arg-528 in $S4_{II}$ was changed to histidine, a codon mutation from CGT to CAT, in others the Arg-1239 in $S4_{IV}$ was changed to histidine, and in others the Arg-1239 was changed to glycine (codon CCT). Myotubes cultured from HOPP patients showed calcium currents with enhanced inactivation (Sipos *et al.*, 1995). Voltage-gated calcium channels in skeletal muscle are intimately involved as voltage sensors in excitation–contraction coupling, and it may be that the rapid inactivation seen in the mutant channels interferes with this process. It is as yet not clear how the effects on membrane potentials and on serum potassium levels are produced.

Voltage-gated potassium channels

A rare hereditary neurological disease known as the episodic ataxia with myokymia syndrome is characterized by brief episodes of incoordination (ataxia) with rippling of the muscles (myokymia) at other times. Linkage studies show that the gene responsible is localized on chromosome 12p13, in the same area as a number of potassium channels. Four different point mutations of the Kv1.1 gene have been found in members of different affected families (Browne *et al.*, 1994; Kraus & McNamara, 1995). The Kv1.1 channel occurs in the cerebellum and peripheral nerves in rats, regions that might well be involved in the human disease.

A group of potassium channels related to the *ether-à-go-go* gene of *Drosophila* have been found in mammals (Warmke & Ganetzky, 1994). The human member of this subfamily, *HERG* (human *eag*-related gene) is found on chromosome 7 and codes for a delayed rectifier potassium channel of heart

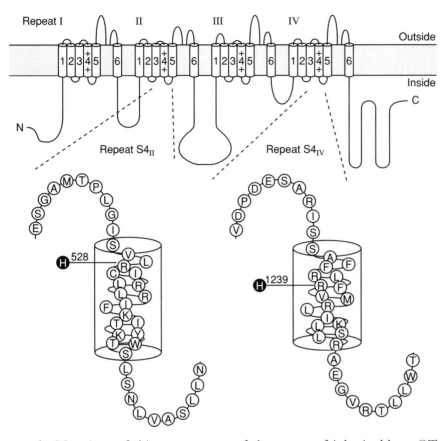

Fig. 8.3. Location in the voltage-gated calcium channel of mutations leading to hypokalaemic periodic paralysis. R528H and R1239H, and also R1239G (not shown here) occur in different families. (From Sipos *et al.*, 1995.)

muscle. Mutations of this gene are one of the causes of inherited long QT syndrome, a disorder in which the repolarization of the ventricle during the heart beat is delayed so that the QT interval in the electrocardiogram is prolonged (Curran *et al.*, 1995; Sanguinetti *et al.*, 1995).

Muscle chloride channels

Inherited myotonias (muscle stiffness) without sensitivity to cold or change in blood potassium levels are of two types. The recessive form, in which the stiffness starts in childhood in the leg muscles, and spreads to the arms, neck and facial muscles, is known as Becker's disease or generalized myotonia (GM). A less severe form, Thomsen's disease or myotonia congenita (MC), is genetically dominant. The muscles are more excitable than usual and have a reduced chloride conductance. Normally the high chloride conductance in skeletal muscle tends to hold the membrane potential near to its resting level; blockage of this with suitable drugs makes the muscle more excitable and mimics the symptoms of myotonia (Bryant & Morales-Aguilera, 1971).

Koch and her colleagues (1992) were able to clone a cDNA for the human CLC-1 channel and showed that it was physically localized on chromosome 7 at the q32-qter region. In affected families (including the descendants of J. Thomsen, who first described MC in 1876 and suffered from it himself) there was tight linkage between the disease and the channel gene. SSCP analysis of DNA from affected members of these families has shown the presence of various point mutations, two short deletions, and a base change affecting a splice site (Heine *et al.*, 1994; Hoffman *et al.*, 1995; Lorenz *et al.*, 1994). Three of the point mutations produced Thomsen-type (dominant) myotonia, the rest produced the Becker-type (recessive) disease. We have seen in chapter 4 how these mutant forms have been used to estimate the stoichiometry of the CLC-1 channel.

Cystic fibrosis and the CFTR

Cystic fibrosis is the most common fatal genetic recessive disease affecting people of north European descent. Its essential feature is a failure of chloride transport as a result of mutations to the gene coding for the CFTR (cystic fibrosis transmembrane conductance regulator) channel. The CFTR chloride channel is localized to the apical membranes of epithelial cells lining secretory tubules. Chloride movement via these CFTR channels into the lumen of the tubule is followed by water movement. Hence failure of chloride transport results in a reduction in the amount of fluid produced by the epithelial cells lining the airways of the lung and the secretory ducts of the pancreas, so that they become blocked with thick and immobile secretions. In consequence there may be difficulty in breathing, and the lungs are subject to repeated and persistent infections. Pancreatic function usually declines so that there is poor digestion of fats and proteins. There are also deficiencies of chloride transport in the gut and the sweat glands, and various other secondary effects. Sweat chloride concentrations may exceed 60 mM (the normal value is variable but typically not more than 40 mM), a feature that can be used in diagnosis. Eventually the pathological changes in the lung and pancreas are likely to be fatal; the median survival age in the USA is 29 years (Welsh *et al.*, 1995).

Unlike most of the other channels discussed in this chapter, the CFTR was discovered as a direct result of investigations into the genetic basis of the disease caused by its dysfunction. The putative structure of the CFTR is shown in fig. 4.24. It has two membrane-spanning domains, each with six putative transmembrane segments, and on the cytoplasmic surface there are two ATP-binding domains and a regulatory (R) domain.

The commonest form of cystic fibrosis, producing a severe affliction and

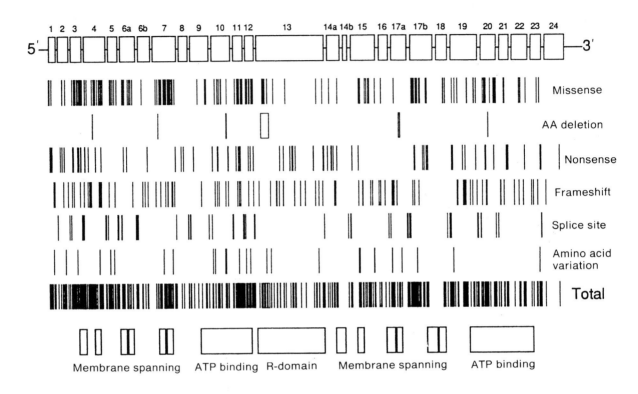

occurring in about 70% of cases overall, results from the deletion of a pheny-lalanine residue at position 508, in the first ATP binding region. About 1 in 25 north Europeans are heterozygous for this deleterious gene, and about 1 in 2500 children are homozygous and so develop cystic fibrosis (Collins, 1992). This high incidence of the ΔF508 gene suggests that there may be some advantage to be derived from possession of one copy of it. It has been suggested that heterozy-gous individuals might have a limited chloride secretion capacity in the gut, and that this would endow some resistance to cholera and other enterotoxic diar-rhoeal diseases (Hansson, 1988; Field & Semrad, 1993). However, studies with transgenic mice that possess only one copy of the normal gene have not shown any deficiency of intestinal chloride transport (Cuthbert *et al.*, 1995). An alter-native suggestion is that possession of one copy of the ΔF508 gene is protective against asthma. Statistical evidence for this was gained from a study of the rela-tives of cystic fibrosis sufferers: those who were heterozygous for ΔF508 were less likely to have asthma than those were not (Schroeder *et al.*, 1995).

More than 400 other mutations in the CFTR gene can occur, with effects that may not be as severe as ΔF508. There are also many benign polymor-phisms, non-pathological nucleotide substitutions in various parts of the CFTR gene (Tsui, 1992; Welsh *et al.*, 1995). The distribution of these muta-tions in the protein chain is shown in fig 8.4.

Fig. 8.4. (left) Distribution of cystic fibrosis mutations and amino acid (AA) polymorphisms in the CFTR protein chain. The DNA sequence is represented at the top, with the 27 exons shown as boxes (the introns between them are not drawn to scale). The boxes at the bottom show the presumed functional domains of the protein. The 'amino acid variation' set refers to presumed benign polymorphisms. Most class III and IV mutants are of the missense type, where one amino acid is substituted for another; the remaining mutation types tend to produce class I or II mutants, with incomplete protein chains that either cannot be produced or cannot be incorporated into the plasma membrane. (From Welsh *et al.*, in *The Metabolic and Molecular Bases of Inherited Disease*, ed. C.R. Scriver, A.L. Beaudet, W.S. Sly & D. Valle, vol. 3, 1995, with permission from McGraw-Hill, Inc.)

Table 8.2. *Classes of mutations that cause cystic fibrosis*

Class	Defect	Examples	Domain	Frequency (%)
I	Protein production			
	Nonsense mutations	G542X	NBD1	3.4
	Frameshift	3905 insertT	NBD2	2.1
	Splice	621+G→T	MSD1	1.3
II	Processing	ΔI507	NBD1	0.5
		ΔF508	NBD1	67.2
		S549I	NBD1	Rare
		S549R	NBD1	0.3
		N1303K	NBD2	1.8
III	Regulation	G551D	NBD1	2.4
		S1255P	NBD2	Rare
IV	Conduction	R117H	MSD1	0.8
		R334W	MSD1	0.4
		R347P	MSD1	0.5

NBD, nucleotide-binding domain; MSD, membrane-spanning domain. Simplified after Welsh & Smith 1993.

The ways in which the different mutations produce their effects have been divided into four groups, as is shown in table 8.2 (Welsh & Smith, 1993). Class I mutations lead to mRNAs that cannot produce a full protein chain. Class II mutations produce chains that are not processed properly, so they do not get incorporated into the plasma membrane; ΔF508 is of this type, so the disease is here produced by the lack of CFTR channels in the membrane (Cheng *et al.*, 1990). The remaining groups of mutations do reach the plasma membrane but do not function properly when they get there. Class III mutations affect the regulation of the channel by ATP or by phosphorylation, so that the channels are less likely to open. Class IV mutations have reduced ion flow through the open channel. Three mutations in this group affect arginine residues in putative membrane-crossing segments, and it is tempting to speculate that these large positively charged residues might be concerned with interacting with the permeant chloride ions. It is noticeable that in many patients with class IV mutations the pancreas works adequately, presumably because the pancreatic fluid secretion system is tolerant of some shortfall in chloride movement.

Standard treatment of cystic fibrosis includes draining the air passages of mucus, keeping the lungs clear of infection with antibiotics, administration of pancreatic enzymes at mealtimes (they can be contained in microspheres that release their contents in the small intestine), and helping the patients to cope socially and psychologically with their situation.

There has been much interest recently in the prospect of treatment by somatic gene therapy, by introducing DNA that will produce functional CFTR into the cells of the lungs and possibly elsewhere (Collins, 1992; Geddes, 1994). Experiments began with the demonstration in 1990 that the normal gene could be transferred into defective cells cultured from cystic fibrosis patients (Drumm *et al.*, 1990; Rich *et al.*, 1990). Then the use of transgenic mice showed that the cystic fibrosis defect could be corrected at least temporarily by gene therapy (Hyde *et al.*, 1993). The success of these experiments opened the door to short-term clinical tests on human patients.

Two methods have been employed to introduce the CFTR gene into the human respiratory tract. One uses a recombinant adenovirus vector containing cDNA for human CFTR (Zabner *et al.*, 1993; Crystal *et al.*, 1994). The other wraps up the CFTR cDNA in liposomes, very small capsules of lipid membranes (Caplen *et al.*, 1995). The majority of subjects in these experiments showed the presence of RNA derived from the transferred gene and some temporary restoration in the chloride-transporting capabilities of the cells of the respiratory passages. Problems with the viral method include some mild inflammatory response in some cases and the possibility of developing immunity to it over the longer term. Problems with the liposome method include variability in the response and its brief period of effectiveness. There is still much to be done before these promising first steps can be converted into a viable long-term method of treatment.

The glycine receptor channel

Hereditary hyperekplexia, also known as startle disease, is a rare autosomal dominant neurological disorder in which the muscles show considerable rigidity as a result of nervous action and the response to sudden sounds or touch is extremely vigorous. Linkage analysis showed that the gene involved was located on the long arm of chromosome 5. Analysis of the glycine receptor $\alpha 1$ subunit showed that the disorder was produced by a point mutation in which a leucine or glutamine residue was substituted for the arginine residue at position 271, at the extracellular end of the M2 segment (Shiang *et al.*, 1993).

The effects of these mutations have been studied by expressing mutant channels in *Xenopus* oocytes and transfected human embryonic kidney cells (Langosch *et al.*, 1994; Rajendra *et al.*, 1994). The results show a greatly reduced affinity of the receptor for glycine, coupled with a reduced chloride conductance. These effects would reduce the effectiveness of inhibitory action in the spinal cord and so lead to the symptoms of the disease.

Gap junctions

Charcot-Marie-Tooth (CMT) disease is a heterogeneous group of inherited disorders in which there is progressive degeneration of the peripheral nerves. The sex-linked form (CMTX) has been mapped to the q13.1 band of the X chromosome, a segment which also includes the connexin 32 gene. Sequencing of this gene from CMTX patients showed the presence of different point mutations in the gene in different families (Bergoffen *et al.*, 1993). The normal connexin 32 gene will produce functional gap junctions when expressed in paired *Xenopus* oocytes, but the mutant forms will not (Bruzzone *et al.*, 1994). Discovery of the mutant gene led to the demonstration that connexin 32 is well expressed in peripheral nerves. The connexin molecules form gap junctions in the Schwann cells, which provide the myelin sheaths, and are perhaps important in the transfer of ions and other small molecules between adjacent parts of these unusually shaped cells.

A proportion of CMTX patients do not show point mutations in the connexin 32 gene, and the autosomal versions of the Charcot-Marie-Tooth disorder are also not attributable to this cause; in some cases the defect has been localized to particular proteins involved in the formation of the myelin sheath (Ionasescu, 1995).

The calcium-release channel

Malignant hyperthermia is an abnormal response to general anaesthesia. The combination of potent inhalent anaesthetics such as halothane and depolarizing muscle relaxants such as succinylcholine produces a major physiological crisis in about 1 in 20 000 people. Skeletal muscles contract, the body temperature rises rapidly, ionic imbalances ensue, and the result may be fatal. The death rate from such events has fallen from around 80% to less than 10% of cases in recent years as a result of early diagnosis, better management of the crisis, and the use of the drug dantrolene, which prevents the release of calcium ions from the sarcoplasmic reticulum (MacLennan & Phillips, 1992; Kalow & Grant, 1995).

The RYR1 gene, coding for the ryanodine receptor (the calcium-release channel) of the sarcoplasmic reticulum, maps to region q13.1 of the human chromosome 19, as also does the malignant hyperthermia susceptibility locus in some families with the disorder (MacLennan *et al.*, 1990). A similar disorder occurs in domestic pigs, where halothane-sensitivity, porcine stress syndrome and the production of meat with undesirable exudates are all related to a recessive point mutation of the Arg-615 residue to cysteine in the ryanodine receptor (Fujii *et al.*, 1991). Since the mutant gene is associated with a slight increase in muscle mass in the heterozygote, it is likely to have been selected for by pig breeders in the past. Transfected cells containing the mutant gene respond to halothane by releasing calcium ions into the cytoplasm, whereas those expressing the normal ryanodine receptor do not (Otsu *et al.*, 1994).

The corresponding human mutation, R614C, is also found in a small proportion of human cases of malignant hyperthermia, but there are also other RYR1 mutations that can cause the disorder, other polymorphisms of RYR1 that seem to be harmless, and other mutations that cause malignant hyperthermia without affecting the ryanodine receptor (Ball & Johnson, 1993; MacLennan & Chen, 1993). The details of this complexity are well worth establishing so that a comprehensive genetic test for susceptibility can be developed.

9 Not the last word

We have seen in this book that the study of channels has expanded and blossomed in a most remarkable way in recent years. One way of illustrating this is to count the number of scientific papers with the word 'channel' (meaning an ion channel) in their titles in different years. There were just 3 in 1970, well over 100 in 1980, and nearly 2000 in 1990. The figure is presently approaching 3000 titles per year, with probably over 10 000 papers referring to ion channels somewhere in their content. Why has this happened?

A number of factors come together to produce a flowering of scientific endeavour in this way. Perhaps three are of particular importance: new techniques, new ideas, and the motivation that is provided by a sense of the importance of the problem.

Techniques are very important. The voltage clamp system was crucial to the separation of the sodium and potassium currents in the nerve action potential. The patch clamp system has permitted the action of individual channels to be investigated, including those of cells too small to be investigated by other methods, and has led to a whole new set of investigations on channel kinetics. Molecular genetics and recombinant DNA technology have provided detailed information about channel structure, the discovery of many new channels, and some very powerful methods of investigating the relations between channel structure and channel function.

But new techniques are of little use on their own. They have to be put to work in the context of bright ideas. It was a bright idea of Hodgkin and Huxley to set up equations describing the sodium and potassium conductances and see whether they would predict the form of the action potential. It was a bright idea of Hille to introduce the concept of the selectivity filter and to investigate its dimensions with ions of different size. It was a bright idea of Anderson and his colleagues to demonstrate that the CFTR really is a channel by showing that mutagenesis could alter its anion selectivity.

Given new techniques and good ideas, an area of scientific knowledge will advance rapidly if people feel that it is important. The fascination of looking at the behaviour of single protein molecules with the patch clamp technique, and in many cases of knowing just what the primary structure of that molecule is, have given a powerful sense of getting to grips with fundamentals. Knowledge about channels has provided a physical context for the equations of permeability studies, a molecular foundation for investigating the action of drugs. It has enabled us to make sense of a very wide variety of cellular behaviour, providing better answers to such questions as how the nervous impulse works, what controls insulin secretion, how internal calcium ion concentrations are varied, and so on. Some of this knowledge is directly relevant to the treatment of human disease, and much of it helps to provide the background of fundamental knowledge which is essential to medical science.

So what will happen next? The big question awaiting a technical breakthrough is precisely how the amino acid chains of channel proteins are arranged. If channel proteins could be crystallized, then we could have a three-dimensional map of their structure derived from X-ray diffraction studies, as is available now for proteins such as haemoglobin, lysozyme and ribonuclease. Unfortunately it is just because of their suitability for the two-dimensional lipid environment of the cell membrane that channel proteins cannot as yet be crystallized in three dimensions. Developments in nuclear magnetic resonance methods might offer an alternative approach to the problem.

Precise knowledge of channel structure, should it become available, would allow molecular dynamics simulations of ion permeation through natural channels to be attempted. These might give us some answers to the question, do the energy barriers and wells in Eyring rate theory models really correspond to particular submolecular structures? And if we can identify such structures in channels, will we be able to predict some of their permeability characteristics? Or is there a new and powerful theory of permeability waiting to burst forth?

We may expect further genetic information, from the Human Genome Project as well as from ongoing work on other genomes, to produce new channel sequences. Determination of the properties of the channel proteins for which they code should provide employment for an army of graduate students. There is also room for more information about channel action in different types of cell. There must be many more channels in plant cells than we now know about.

We can draw the focus closer and ask some particular questions about a more restricted group of channels. We may take the inward rectifier potassium channels as an example. How many pore-forming subunits are there per channel? What are the rules that govern the formation of heteromultimers?

Are there any accessory subunits? Can we get better evidence about the transmembrane topology of the polypeptide chain? Do any parts of the C-terminal section enter the membrane? Is channel selectivity provided by the H5 loop alone, or do other parts of the pore contribute significantly to it? Are all ATP-sensitive potassium channels part of this gene family? How do potassium channel openers combine with them and how do they work? What is the relationship between these potassium channels and the sulphonylurea receptor? How do the different properties of the various members of the family relate to their roles in cell function?

What is the molecular basis of the modulation of inward rectifier potassium channels by phosphorylation or by interaction with G proteins? What is the functional distribution of these channels, and how is their differential expression in different cells regulated? What determines the density and rate of turnover of the channels? How do differences between different members of the group relate to their roles in cell function? What other members of the family are to be found? What is their evolutionary relationship to other potassium channels, and to other channels with two membrane-crossing segments?

We can expect answers to many of these questions over the next few years. It will be good to see them.

References

Multiple author papers (i.e. more than two) with the same first author are arranged in chronological order

Adams, D.J., Dwyer, T.M. & Hille, B. (1980). The permeability of endplate channels to monovalent and divalent metal cations. *Journal of General Physiology* **75**, 493–510.

Adams, D.M. (1974). *Inorganic Solids*. London: John Wiley & Sons.

Adams, M.E. & Olivera, B.M. (1994). Neurotoxins: overview of an emerging research technology. *Trends in Neurosciences* **17**, 151–5.

Adams, M.E. & Swanson, G. (1994). Neurotoxins supplement. *Trends in Neurosciences* **17**(4), supplement.

Adams, P.R. & Brown, D.A. (1982). Synaptic inhibition of the M-current: slow excitatory post-synaptic potential mechanism in bullfrog sympathetic neurones. *Journal of Physiology* **332**, 263–72.

Adelman, J.P., Shen, K.-Z., Kavanaugh, M.P., Warren, R.A., Wu, Y.-N., Lagrutta, A., Bond, C.T. & North, R.A. (1992). Calcium-activated potassium channels expressed from cloned complementary DNAs. *Neuron* **9**, 209–16.

Adelman, W.J. & French, R.J. (1978). Blocking of the squid axon potassium channel by external caesium ions. *Journal of Physiology* **276**, 13–25.

Agre, P., Preston, G.M., Smith, B.M., Jung, J.S., Raina, S., Moon, C., Guggino, W.B. & Nielsen, S. (1993). Aquaporin CHIP: the archetypal molecular water channel. *American Journal of Physiology* **265**, F463–F476.

Aguilar-Bryan, L., Nichols, C.G., Wechsler, S.W., Clement, J.P., Boyd, A.E., González, G., Herrera-Sosa, H., Nguy, K., Bryan, J. & Nelson, D.A. (1995). Cloning of the β cell high-affinity sulfonylurea receptor: a regulator of insulin secretion. *Science* **268**, 423–6.

Aidley, D.J. (1990). *The Physiology of Excitable Cells*, 3rd edition. Cambridge: Cambridge University Press.

Alberts, B., Bray, D., Lewis, J., Raff, M., Roberts, K. & Watson, J.D. (1994). *Molecular Biology of the Cell*, 3rd edition. New York: Garland Publishing.

Alcayaga, C., Cecchi, X., Alvarez, O. & Latorre, R. (1989). Streaming potential measurements in Ca^{2+}-activated K^+ channels from skeletal and smooth muscle. *Biophysical Journal* **55**, 367–71.

Almers, W. (1978). Gating currents and charge movements in excitable membranes. *Reviews of Physiology, Biochemistry and Pharmacology* **82**, 96–190.

Almers, W. & McCleskey, E.W. (1984). Non-selective conductance in calcium channels of frog muscle: calcium selectivity in a single-file pore. *Journal of Physiology* **353**, 585–608.

Almers, W., McCleskey, E.W. & Palade, P.T. (1984). A non-selective cation conductance in frog muscle membrane blocked by micromolar external calcium ions. *Journal of Physiology* **353**, 565–83.

Anderson, C.R. & Stevens, C.F. (1973). Voltage clamp analysis of acetylcholine produced end-plate current fluctuations at frog neuromuscular junction. *Journal of Physiology* **235**, 655–91.

Anderson, J.A., Huprikar, S.S., Kochian, L.V., Lucas, W.J.

& Gaber, R.F. (1992). Functional expression of a possible *Arabidopsis thaliana* potassium channel in *Saccharomyces cerevisiae*. *Proceedings of the National Academy of Sciences, USA* **89**, 3736–40.

Anderson, M.P., Berger, H.A., Rich, D.P., Gregory, R.J., Smith, A.E. & Welsh, M.J. (1991a). Nucleotide triphosphates are required to open the CFTR chloride channel. *Cell* **67**, 775–84.

Anderson, M.P., Gregory, R.J., Thompson, S., Souza, D.W., Paul, S., Mulligan, R.C., Smith, A.E. & Welsh, M.J. (1991b). Demonstration that CFTR is a chloride channel by alteration of its anion selectivity. *Science* **253**, 202–5.

Armstrong, C.M. (1969). Inactivation of the potassium conductance and related phenomena caused by quaternary ammonium ion injected in squid axons. *Journal of General Physiology* **54**, 553–75.

Armstrong, C.M. (1971). Interaction of tetraethylammonium ion derivatives with the potassium channels of giant axons. *Journal of General Physiology* **58**, 413–37.

Armstrong, C.M. (1975). Ionic pores, gates, and gating currents. *Quarterly Reviews of Biophysics* **7**, 179–210.

Armstrong, C.M. (1981). Sodium channels and gating currents. *Physiological Reviews* **61**, 644–83.

Armstrong, C.M. (1992). Voltage-dependent ion channels and their gating. *Physiological Reviews* **72**, S5–S13.

Armstrong, C.M. & Bezanilla, F. (1973). Currents related to movement of the gating particles of the sodium channels. *Nature* **242**, 459–61.

Armstrong, C.M. & Bezanilla, F. (1977). Inactivation of the sodium channel. II. Gating current experiments. *Journal of General Physiology* **70**, 567–90.

Armstrong, C.M. & Gilly, W.F. (1979). Fast and slow steps in the activation of sodium channels. *Journal of General Physiology* **74**, 691–711.

Armstrong, C.M., Bezanilla, F. & Rojas, E. (1973). Destruction of sodium conductance inactivation in squid axons perfused with pronase. *Journal of General Physiology* **62**, 375–91.

Arseniev, A.S., Barsukov, I.L., Bystrov, V.F., Lomize, A.L. & Ovchinnikov, Y.A. (1985). ^{1}H-NMR study of gramicidin A transmembrane ion channel. *FEBS Letters* **186**, 168–74.

Ashcroft, F.M. & Rorsman, P. (1989). Electrophysiology of the pancreatic β cell. *Progress in Biophysics and Molecular Biology* **54**, 87–143.

Ashcroft, F.M. & Stanfield, P.R. (1982). Calcium inactivation in skeletal muscle fibres of the stick insect, *Carausius morosus. Journal of Physiology* **330**, 349–72.

Ashcroft, S.J.H. & Ashcroft, F.M. (1990). Properties and functions of ATP-sensitive K-channels. *Cellular Signalling* **2**, 197–214.

Atkinson, N.S., Robertson, G.A. & Ganetzky, B. (1991). A component of calcium-activated potassium channels encoded by the *Drosophila slo* locus. *Science* **253**, 551–5.

Attali, B., Guillemare, E., Lesage, F., Honoré, E., Romey, G., Lazdunski, M. & Barhanin, J. (1993). The protein IsK is a dual activator of K$^+$ and Cl$^-$ channels. *Nature* **365**, 850–2.

Attwell, D., Miller, B. & Sarantis, M. (1993). Arachidonic acid as a messenger in the central nervous system. *Seminars in the Neurosciences* **5**, 159–69.

Baker, M.D. & Ritchie, J.M. (1994). The action of capsaicin on type I delayed rectifier K$^+$ currents in rabbit Schwann cells. *Proceedings of the Royal Society of London* B **255**, 259–65.

Ball, F.G. & Rice, J.A. (1992). Stochastic models for ion channels: introduction and bibliography. *Mathematical Biosciences* **112**, 189–206.

Ball, S.P. & Johnson, K.J. (1993). The genetics of malignant hyperthermia. *Journal of Medical Genetics* **30**, 89–93.

Barchi, R.L. (1995). Molecular pathology of the skeletal muscle sodium channel. *Annual Review of Physiology* **57**, 355–85.

Barnard, E.A. (1992). Subunits of GABA$_A$, glycine, and glutamate receptors. In *Receptor Subunits and Complexes*, ed. A. Burgen & E.A. Barnard, pp. 163–87. Cambridge: Cambridge University Press.

Barnard, E.A., Miledi, R. & Sumikawa, K. (1982). Translation of exogenous messenger RNA coding for nicotinic acetylcholine receptors produced functional receptors in *Xenopus* oocytes. *Proceedings of the Royal Society of London* B **215**, 241–6.

Barrett, J.N., Magleby, K.L. & Pallotta, B.S. (1982). Properties of single calcium-activated potassium channels in cultured rat muscle. *Journal of Physiology* **331**, 211–30.

Barritt, G.J. (1992). *Communication within Animal Cells.* Oxford: Oxford University Press.

Bassingthwaighte, J.B., Liebovitch, L.S. & West, B.J.

(1994). *Fractal Physiology*. New York: Oxford University Press for the American Physiological Society.

Baukrowitz, T., Hwang, T.-C., Nairn, A.C. & Gadsby, D.C. (1994). Coupling of CFTR Cl⁻ channel gating to an ATP hydrolysis cycle. *Neuron* **12**, 473–82.

Bean, B.P. (1984). Nitrendipine block of cardiac calcium channels: high-affinity binding to the inactive state. *Proceedings of the National Academy of Sciences, USA* **81**, 6388–92.

Bean, B.P. (1989). Neurotransmitter inhibition of neuronal calcium currents by changes in channel voltage dependence. *Nature* **340**, 153–5.

Bean, R.C., Shepherd, W.C., Chan, H. & Eichner, J. (1969). Discrete conductance fluctuations in lipid bilayer protein membranes. *Journal of General Physiology* **53**, 741–57.

Bear, C.E., Li, C.H., Kartner, N., Bridges, R.J., Jensen, T.J., Ramjeesingh, M. & Riordan, J.R. (1992). Purification and functional reconstitution of the cystic fibrosis transmembrane conductance regulator. *Cell* **68**, 809–18.

Becker, S., Prusak-Sochaczewski, E., Zamponi, G., Beck-Sickinger, A.G., Gordon, R.D. & French, R.J. (1992). Action of derivatives of μ-conotoxin GIIIA on sodium channels: single amino acid substitutions in the toxin separately affect association and dissociation rates. *Biochemistry* **31**, 8229–38.

Begenisich, T. (1994). Permeation properties of cloned K⁺ channels. In *Handbook of Membrane Channels*, ed. C.Peracchia, pp. 17–28. San Diego, CA: Academic Press.

Beirão, P.S.L., Davies, N.W. & Stanfield, P.R. (1994). Inactivating 'ball' peptide from *Shaker* B blocks Ca²⁺-activated but not ATP-dependent K⁺ channels of rat skeletal muscle. *Journal of Physiology* **474**, 269–74.

Bekkers, J.M., Greeff, N.G. & Keynes, R.D. (1986). The conductance and density of sodium channels in the cut-open squid giant axon. *Journal of Physiology* **377**, 463–86.

Bennett, J.A. & Dingledine, R. (1995). Topology profile for a glutamate receptor: three transmembrane domains and a channel-lining reentrant membrane loop. *Neuron* **14**, 373–84.

Bennett, M.V.L, Barrio, L.C., Bargellio, T.A., Spray, D.C., Herzberg, E. & Sáez, J.C. (1991). Gap junctions: new tools, new answers, new questions. *Neuron* **6**, 305–20.

Benos, D.J., Awayda, M.S., Ismailov, I.I. & Johnson, J.P. (1995). Structure and function of amiloride-sensitive Na⁺ channels. *Journal of Membrane Biology* **143**, 1–18.

Bergoffen, J., Scherer, S.S., Wang, S., Scott, M.O., Bone, L.J., Paul, D.L., Chen, K., Lensch, M.W., Chance, P.F. & Fischbeck, K.H. (1993). Connexin mutations in X-linked Charcot-Marie-Tooth disease. *Science* **262**, 2039–42.

Berridge, M.J. (1993). Inositol trisphosphate and calcium signalling. *Nature* **361**, 315–25.

Bertrand, D., Galzi, J.L., Devillers-Thiéry, A., Bertrand, S. & Changeux, J.P. (1993). Mutations at two distinct sites within the channel domain M2 alter calcium permeability of the neuronal α7 nicotinic receptor. *Proceedings of the National Academy of Sciences, USA* **90**, 6971–5.

Beyer, E.C., Paul, D.L. & Goodenough, D.A. (1987). Connexin43: a protein from rat heart homologous to a gap junction protein from liver. *Journal of Cell Biology* **105**, 2621–9.

Bezanilla, F. & Armstrong, C.M. (1972). Negative conductance caused by entry of sodium and cesium ions into the potassium channels of squid axons. *Journal of General Physiology* **60**, 588–608.

Bezanilla, F. & Stefani, E. (1994). Voltage-dependent gating of ionic channels. *Annual Review of Biophysics* **23**, 819–46.

Bezanilla, F., Perozo, E., Papazian, D.M. & Stefani, E. (1991). Molecular basis of gating charge immobilization in Shaker potassium channels. *Science* **254**, 679–83.

Biel, M., Zong, X., Distler, M., Bosse, E., Klugbauer, N., Murakami, M., Flockerzi, V. & Hofmann, F. (1994). Another member of the cyclic nucleotide-gated channel family, expressed in testis, kidney and heart. *Proceedings of the National Academy of Sciences, USA* **91**, 3505–9.

Bishop, N.D. & Lea, E.J.A. (1994). Characterisation of the porin of *Rhodobacter capsulatus 37b4* in planar lipid bilayers. *FEBS Letters* **349**, 69–74.

Blanton, M.P. & Cohen, J.B. (1992). Mapping the lipid-exposed regions in the *Torpedo californica* nicotinic acetylcholine receptor. *Biochemistry* **31**, 3738–50.

Blanton, M.P. & Cohen, J.B. (1994). Identifying the lipid–protein interface of the *Torpedo* nicotinic acetylcholine receptor: secondary structure implications. *Biochemistry* **33**, 2859–72.

Blaschke, M., Keller, B.U., Rivosecchi, R., Hollmann, M., Heinemann, S. & Konnerth, A. (1993). A single amino acid determines the subunit-specific spider toxin block of α-amino-3-hydroxy-5-methylisoxazole-4-propionate/kainate receptor channels. *Proceedings of the National Academy of Sciences, USA* **90**, 6528–32.

Blumenthal, E.M. & Kaczmarek, L.K. (1994). The minK potassium channel exists in functional and nonfunctional forms when expressed in the plasma membrane of *Xenopus* oocytes. *Journal of Neuroscience* **14**, 3097–105.

Bogusz, S., Boxer, A. & Busath, D. (1992). An SS1–SS2 β-barrel structure for the voltage-activated potassium channel. *Protein Engineering* **5**, 285–93.

Bolotina, V.M., Najibi, S., Palacino, J.J., Pagano, P. & Cohen, R.A. (1994). Nitric oxide directly activates calcium-dependent potassium channels in vascular smooth muscle. *Nature* **368**, 850–3.

Bond, C.T., Pessia, M., Xia, X.M., Lagrutta, A., Kavanaugh, M.P. & Adelman, J.P. (1994). Cloning and expression of a family of inward rectifier potassium channels. *Receptors and Channels* **2**, 183–91.

Bormann, J., Hamill, O.P. & Sakmann, B. (1987). Mechanism of anion permeation through channels gated by glycine and gamma-aminobutyric acid in mouse cultured spinal neurones. *Journal of Physiology* **385**, 243–86.

Brake, A.J., Wagenbach, M.J. & Julius, D. (1994). New structural motif for ligand-gated channels defined by an ionotropic ATP receptor. *Nature* **371**, 519–23.

Branden, C. & Tooze, J. (1991). *Introduction to Protein Structure*. New York: Garland Publishing.

Bredt, D.S. & Snyder, S.H. (1994). Nitric oxide: a physiologic messenger molecule. *Annual Review of Biochemistry* **63**, 175–95.

Brehm, P. & Eckert, R. (1978). Calcium entry leads to inactivation of calcium channels in *Paramecium*. *Science* **202**, 1203–6.

Brisson, A. & Unwin, P.N.T. (1985). Quaternary structure of the acetylcholine receptor. *Nature* **315**, 474–7.

Brown, A.M. (1993). Functional bases for interpreting amino acid sequences of voltage-dependent K^+ channels. *Annual Review of Biophysics* **22**, 173–98.

Brown, A.M., Drewe, J.A., Hartmann, H.A., Taglialatella, M., De Biasi, M., Soman, K. & Kirsch, G.E. (1993). The potassium pore and its regulation. *Annals of the New York Academy of Sciences* **707**, 74–80.

Brown, D.A. (1990). G-proteins and potassium currents in neurons. *Annual Review of Physiology* **52**, 215–42.

Brown, D.A. & Adams, P.R. (1980). Muscarinic supression of a novel voltage-sensitive K^+-current in a vertebrate neurone. *Nature* **283**, 673–6.

Brown, R.H. (1993). Ion channel mutations in periodic paralysis and related myotonic diseases. *Annals of the New York Academy of Sciences* **707**, 305–16.

Browne, D.L., Gancher, S.T., Nutt, J.G., Brunt, E.R.P., Smith, E.A., Kramer, P. & Litt, M. (1994). Episodic ataxia/myokymia syndrome is associated with point mutations in the human potassium channel gene, *KCNA1*. *Nature Genetics* **8**, 136–40.

Brüggemann, A., Pardo, L.A., Stühmer, W. & Pongs, O. (1993). *Ether-à-go-go* encodes a voltage-gated channel permeable to K^+ and Ca^{2+} and modulated by cAMP. *Nature* **365**, 445–8.

Bruzzone, R., White, T.W., Scherer, S.S., Fischbeck, K.H. & Paul, D.L. (1994). Null mutations of connexin32 in patients with X-linked Charcot-Marie-Tooth disease. *Neuron* **13**, 1253–60.

Bryant, S.H. & Morales-Aguilera, A. (1971). Chloride conductance in normal and myotonic muscle fibres and the action of monocarboxylic aromatic acids. *Journal of Physiology* **219**, 367–83.

Buckley, J.T. (1992). Crossing three membranes: channel formation by aerolysin. *FEBS Letters* **307**, 30–3.

Burgess, J. (1978). *Metal Ions in Solution*. Chichester: Ellis Horwood.

Burnashev, N., Monyer, H., Seeburg, P.H. & Sakmann, B. (1992a). Divalent ion permeability of AMPA receptor channels is dominated by the edited form of a single subunit. *Neuron* **8**, 189–98.

Burnashev, N., Schoepfer, R., Monyer, H., Ruppersberg, J.P., Günther, W., Seeburg, P.H. & Sakmann, B. (1992b). Control by asparagine residues of calcium permeability and magnesium blockade in the NMDA receptor. *Science* **257**, 1415–19.

Busath, D.D. (1993). The use of physical methods in determining gramicidin channel structure and function. *Annual Review of Physiology* **55**, 473–501.

Butler, A., Wei, A., Baker, K. & Salkoff, L. (1989). A family of putative potassium channel genes in *Drosophila*. *Science* **243**, 943–7.

Butler, A., Tsunoda, S., McCobb, D.P., Wei, A. & Salkoff, L. (1993). *mSlo*, a complex mouse gene encoding "maxi" calcium-activated potassium channels. *Science* **261**, 221–4.

Cafiso, D.S. (1994). Alamethicin: a peptide model for voltage gating and protein-membrane interactions. *Annual Review of Biophysics* **23**, 141–65.

Cahalan, M. & Neher, E. (1992). Patch clamp techniques: an overview. *Methods in Enzymology* **207**, 3–14.

Cannessa, C.M., Horisberger, J.-D. & Rossier, B.C. (1993). Epithelial sodium channel related to proteins involved in neurodegeneration. *Nature* **361**, 467–70.

Cannessa, C.M., Schild, L., Buell, G., Thorens, B., Gautschi, I., Horisberger, J.-D. & Rossier, B.C. (1994). Amiloride-sensitive epithelial Na^+ channel is made of three homologous subunits. *Nature* **367**, 463–7.

Cannon, S.C. & Corey, D.P. (1993). Loss of Na^+ channel inactivation by anemone toxin (ATX II) mimics the myotonic state in hyperkalaemic periodic paralysis. *Journal of Physiology* **466**, 501–20.

Cannon, S.C., Brown, R.H. & Corey, D.P (1991). A sodium channel defect in hyperkalemic periodic paralysis: potassium-induced failure of inactivation. *Neuron* **6**, 619–26.

Caplen, N.J., Alton, E.W.F.W., Middleton, P.G., Dorin, J.R., Stevenson, B.J., Gao, X., Durham, S.R., Jeffery, P.K., Hodson, M.E., Coutelle, C., Huang, L., Porteous, D.J., Williamson, R. & Geddes, D.M. (1995). Liposome-mediated *CFTR* gene transfer to the nasal epithelium of patients with cystic fibrosis. *Nature Medicine* **1**, 39–46.

Castle, N.A. & Haylett, D.G. (1987). Effect of channel blockers on potassium efflux from metabolically exhausted frog skeletal muscle. *Journal of Physiology* **383**, 31–43.

Castle, N.A., Haylett, D.G. & Jenkinson, D.H. (1989). Toxins in the characterization of potassium channels. *Trends in Neurosciences* **12**, 59–65.

Catterall, W.A. (1986). Voltage-dependent gating of sodium channels: correlating structure and function. *Trends in Neurosciences* **9**, 7–10.

Catterall, W.A. (1988). Structure and function of voltage-sensitive ion channels. *Science* **242**, 50–61.

Catterall, W.A. (1992). Cellular and molecular biology of voltage-gated sodium channels. *Physiological Reviews* **72**, S15–S48.

Catterall, W.A. (1993). Structure and function of voltage-gated ion channels. *Trends in Neurosciences* **16**, 500–6.

Catterall, W.A. & Striessnig, J. (1992). Receptor sites for Ca^{2+} channel antagonists. *Trends in Pharmacological Sciences* **13**, 256–62.

Caulfield, M.P., Jones, S., Vallis, Y., Buckley, N.J., Kim, G-D., Milligan, G. & Brown, D.A. (1994). Muscarinic M-current inhibition via $G_{\alpha q/11}$ and α-adrenoceptor inhibition of Ca^{2+} current via $G_{\alpha o}$ in rat sympathetic neurones. *Journal of Physiology* **477**, 415–22.

Chabala, L.D. (1984). The kinetics of recovery and development of potassium channel inactivation in perfused squid (*Loligo pealii*) giant axons. *Journal of Physiology* **356**, 193–220.

Chad, J.E. & Eckert, R. (1986). An enzymic mechanism for calcium current inactivation in dialysed *Helix* neurones. *Journal of Physiology* **378**, 31–51.

Chahine, M., George, A.L., Zhou, M., Ji, S., Sun, W., Barchi, R.L. & Horn, R. (1994). Sodium channel mutations in paramyotonia congenita uncouple inactivation from activation. *Neuron* **12**, 281–94.

Chandy, K.G. Douglas, J., Gutman, G.A., Jan, L., Joho, R., Kalzmavek, L., McKinnon, D., North, R.A., Numa, S., Philipson, L., Ribera, A.B., Rudy, B., Salkoff, L., Swanson, R., Steiner, D., Tanouye, M. & Tempel, B.L. (1991). Simplified gene nomenclature. *Nature* **352**, 26.

Changeux, J.-P. (1985). *Neuronal Man*. Oxford: Oxford University press.

Changeux, J.-P., Kasai, M. & Lee, C.Y. (1970). The use of a snake venom toxin to characterize the cholinergic receptor protein. *Proceedings of the National Academy of Sciences, USA* **67**, 1241–7.

Changeux, J.-P., Galzi, J.-L., Devillers-Thiéry, A. & Bertrand, D. (1992). The functional architecture of the acetylcholine nicotinic receptor explored by affinity labelling and site-directed mutagenesis. *Quarterly Review of Biophysics* **25**, 395–432.

Chaudhari, N. & Hahn, W.E. (1983). Genetic expression in the developing brain. *Science*, **220**, 924–8.

Chavez, R.A. & Hall, Z.W. (1991). The transmembrane topology of the amino terminus of the α subunit of the nicotinic acetylcholine receptor. *Journal of Biological Chemistry* **266**, 15532–8.

Chavez, R.A. & Hall, Z.W. (1992). Expression of fusion proteins of the nicotinic acetylcholine receptor from mammalian muscle identifies the membrane-spanning regions in the α and δ subunits. *Journal of Cell Biology* **116**, 385–93.

Chen, T.-Y., Peng, Y.-W., Dhallan, R.S., Ahamed, B., Reed, R.R. & Yau, K.-W. (1993). A new subunit of the cyclic nucleotide-gated cation channel in retinal rods. *Nature* **362**, 764–7.

Chen, Y.-H. & DeHaan, R.L. (1993). Multiple channel conductance states in gap junctions. *Progress in Cell Research* **3**, 97–103.

Cheng, S.H., Gregory, R.J., Marshall, J., Paul, S., Souza, D.W., White, G.A., O'Riordan, C.R. & Smith, A.E. (1990). Defective intracellular transport and processing of CFTR is the molecular basis of most cystic fibrosis. *Cell* **63**, 827–34.

Chepelinsky, A.B. (1994). The MIP transmembrane channel gene family. In *Handbook of Membrane Channels*, ed. C.Peracchia, pp. 413–32. San Diego, CA: Academic Press.

Chiu, S.-W., Subramaniam, S., Jakobsson, E. & McCammon, J.A. (1989). Water and polypeptide conformations in the gramicidin channel. A molecular dynamics study. *Biophysical Journal* **56**, 253–61.

Chiu, S.-W., Novotny, J.A. & Jakobsson, E. (1993). The nature of ion and water barrier crossings in a simulated ion channel. *Biophysical Journal* **64**, 98–108.

Choi, K.L., Mossman, C., Aubé, J. & Yellen, G. (1993). The internal quaternary ammonium receptor site of *Shaker* potassium channels. *Neuron* **10**, 533–41.

Christensen, O. (1987). Mediation of cell volume regulation by Ca^{2+} influx through stretch-activated channels. *Nature* **330**, 66–8.

Clapham, D.E. (1995). Calcium signaling. *Cell* **80**, 259–68.

Clark, A.J. (1926). The reaction between acetyl choline and muscle cells. *Journal of Physiology* **61**, 530–46.

Clark, A.J. (1933). *The Mode of Action of Drugs on Cells.* London: Edward Arnold.

Claudio, T. (1989). Molecular genetics of acetylcholine receptor-channels. In *Molecular Neurobiology*, ed. D.M. Glover & B.D. Hames, pp. 63–142. Oxford: IRL Press.

Claudio, T., Ballivet, M., Patrick, J. & Heinemann, S. (1983). Nucleotide and deduced amino acid sequences of *Torpedo californica* acetylcholine receptor γ subunit. *Proceedings of the National Academy of Sciences, USA* **80**, 1111–15.

Claudio, T., Green, W.N., Hartman, D.S., Hayden, D., Paulson, H.L., Sigworth, F.J., Sine, S.M. & Swedlund, A. (1987). Genetic reconstitution of functional acetylcholine receptor channels in mouse fibroblasts. *Science* **238**, 1688–94.

Cohen, P. (1988). Protein phosphorylation and hormone action. *Proceedings of the Royal Society of London* B **234**, 115–44.

Cole, K.S. (1968). *Membranes, Ions and Impulses.* Berkeley, CA: University of California Press.

Collingridge, G.L. & Lester, R.A.J. (1989). Excitatory amino acid receptors in the vertebrate central nervous system. *Pharmacological Reviews* **41**, 143–210.

Collins, F.S. (1992). Cystic fibrosis: molecular biology and therapeutic implications. *Science* **256**, 774–9.

Colquhoun, D. (1994). Practical analysis of single channel records. In *Microelectrode Techniques: The Plymouth Workshop Handbook*, 2nd edition, ed. D. Ogden, pp. 101–39. Cambridge: Company of Biologists.

Colquhoun, D. & Hawkes, A.G. (1977). Relaxation and fluctuations of membrane currents that flow through drug-operated ion channels. *Proceedings of the Royal Society of London* B **199**, 231–62.

Colquhoun, D. & Hawkes, A.G. (1981). On the stochastic properties of single ion channels. *Proceedings of the Royal Society of London* B **211**, 205–35.

Colquhoun, D. & Hawkes, A.G. (1982). On the stochastic properties of bursts of single ion channel openings and of clusters of bursts. *Philosophical Transactions of the Royal Society of London* B **300**, 1–59.

Colquhoun, D. & Hawkes, A.G. (1994). The interpretation of single channel recordings. In *Microelectrode Techniques: The Plymouth Workshop Handbook*, 2nd edition, ed. D. Ogden, pp. 141–88. Cambridge: Company of Biologists.

Colquhoun, D. & Hawkes, A.G. (1995). The principles

of the stochastic interpretation of ion-channel mechanisms. In *Single-Channel Recording*, 2nd edition, ed. B. Sakmann & E. Neher, pp. 397–482. New York: Plenum Press.

Colquhoun, D. & Sakmann, B. (1985). Fast events in single-channel currents activated by acetylcholine and its analogues at the frog muscle end-plate. *Journal of Physiology* **369**, 501–57.

Colquhoun, D. & Sigworth, F.J. (1983, 1995). Fitting and statistical analysis of single-channel records. In *Single-Channel Recording*, ed. B. Sakmann & E. Neher, pp. 191–263 and (2nd edition) 483–587. New York: Plenum Press.

Cooper, E., Couturier, S. & Ballivet, M. (1991). Pentameric structure and subunit stoichiometry of a neuronal acetylcholine receptor. *Nature* **350**, 235–8.

Cooper, K.E., Gates, P.Y. & Eisenberg, R.S. (1988). Diffusion theory and discrete rate constants in ion permeation. *Journal of Membrane Biology* **106**, 95–105.

Coronado, R. & Latorre, R. (1983). Phospholipid bilayers made from monolayers on patch-clamp electrodes. *Biophysical Journal* **43**, 231–6.

Coronado, R., Rosenberg, R. & Miller, C. (1980). Ionic selectivity, saturation and block in a K^+-selective channel from sarcoplasmic reticulum. *Journal of General Physiology* **76**, 425–46.

Couturier, S., Bertrand, D., Matter, J.M., Hernandez, M.-C., Bertrand, S., Millar, N., Valera, S., Barkas, T. & Ballivet, M. (1990). Neuronal nicotinic acetylcholine receptor subunit ($\alpha 7$) is developmentally regulated and forms a homo-oligomeric channel blocked by α-BTX. *Neuron* **5**, 847–56.

Covarrubias, M., Wei, A. & Salkoff, L. (1991). *Shaker, Shal, Shab* and *Shaw* express independent K^+ current systems. *Neuron* **7**, 763–73.

Cowan, S.W., Schirmer, T., Rummel, G., Steiert, M., Ghosh, R., Pauptit, R.A., Jansonius, J.N. & Rosenbusch, J.P. (1992). Crystal structures explain functional properties of two *E. coli* porins. *Nature* **358**, 727–33.

Creighton, T.E. (1993). *Proteins: Structures and Molecular Properties*. New York: W.H. Freeman & Co.

Criado, M., Hochschwender, S., Sarin, V., Fox, J.L. & Lindstrom, J. (1985). Evidence for unpredicted transmembrane domains in acetylcholine receptor subunits. *Proceedings of the National Academy of Sciences, USA* **82**, 2004–8.

Crystal, R.G., McElvaney, N.G., Rosenfeld, M.A., Chu, C.S., Mastrangeli, A., Hay, J.G., Brody, S.L., Jaffe, H.A., Eissa, N.T. & Danel, C. (1994). Administration of an adenovirus containing the human CFTR cDNA to the respiratory tract of individuals with cystic fibrosis. *Nature Genetics* **8**, 42–51.

Curran, M.E., Splawski, I., Timothy, K.W., Vincent, G.M., Green, E.D. & Keating, M.T. (1995). A molecular basis for cardiac arrythmia: *HERG* mutations cause long QT syndrome. *Cell* **80**, 795–803.

Curtis, H.J & Cole, K.S. (1942). Membrane resting and action potentials in giant fibers of squid nerve. *Journal of Cellular and Comparative Physiology* **19**, 135–44.

Cuthbert, A.W., Halstead, J., Ratcliff, R., Colledge, W.H. & Evans, M.J. (1995). The genetic advantage hypothesis in cystic fibrosis heterozygotes: a murine study. *Journal of Physiology* **482**, 449–54.

Daly, J.W., Myers, C.W., Warnick, J.E. & Albuquerque, E.X. (1980). Levels of batrachotoxin and lack of sensitivity to its action in poison-dart frogs (*Phyllobates*). *Science* **208**, 1383–5.

Dani, J.A. (1989). Open channel structure and ion binding sites of the nicotinic acetylcholine receptor channel. *Journal of Neuroscience* **9**, 884–92.

Dani, J.A. & Levitt, D.G. (1990). Diffusion and kinetic approaches to describe permeation in ionic channels. *Journal of Theoretical Biology* **146**, 289–301.

Davies, N.W., Spruce, A.E., Standen, N.B. & Stanfield, P.R. (1989). Multiple blocking mechanisms of ATP-sensitive potassium channels of frog skeletal muscle by tetraethylammonium ions. *Journal of Physiology* **413**, 31–48.

Davies, N.W., Standen, N.B. & Stanfield, P.R. (1991). ATP-dependent potassium channels of muscle cells: their properties, regulation and possible functions. *Journal of Bioenergetics and Biomembranes* **23**, 509–35.

Davies, N.W., Standen, N.B. & Stanfield, P.R. (1992). The effect of intracellular pH on ATP-dependent potassium channels of frog skeletal muscle. *Journal of Physiology* **445**, 549–68.

Debye, P.J.W. & Hückel, E. (1923). Zur Theorie der Elektrolyte. *Physikalische Zeitschrift* **24**, 185–206 and 305–25.

DeFelice, L.J. (1981). *Introduction to Membrane Noise*. New York: Plenum Press.

Deisenhofer, J., Epp, O., Miki, K., Huber, R. & Michel, H. (1985). Structure of the protein subunits in the photosynthetic reaction centre of *Rhodopseudomonas viridis* at 3Å resolution. *Nature* **318**, 618–21.

Del Castillo, J. & Katz, B. (1957). Interaction at end-plate receptors between different choline derivatives. *Proceedings of the Royal Society of London* B **146**, 369–81.

Demo, S.D. & Yellen, G. (1991). The inactivation gate of the *Shaker* K^+ channel behaves like an open-channel blocker. *Neuron* **7**, 743–53.

Dempsey, C.E. (1990). The actions of mellitin on membranes. *Biochimica et Biophysica Acta* **1031**, 143–61.

Dennis, M., Giraudat, J., Kotzyba-Hibert, F., Goeldner, M., Hirth, C., Chang, J.Y., Lazure, C., Chrétien, M. & Changeux, J.-P. (1988). Amino acids of the *Torpedo marmorata* acetylcholine receptor α subunit labelled by a photoaffinity ligand for the acetylcholine binding site. *Biochemistry* **27**, 2346–57.

Derkach, V., Surprenant, A. & North, R.A. (1989). 5–HT_3 receptors are membrane ion channels. *Nature* **339**, 706–9.

Devillers-Thiéry, A., Giraudat, J., Bentaboulet, M. & Changeux, J.-P. (1983). Complete mRNA coding sequence of the acetylcholine-binding α-subunit of *Torpedo marmorata* acetylcholine receptor: a model for the transmembrane organization of the polypeptide chain. *Proceedings of the National Academy of Sciences, USA* **80**, 2067–71.

Dhallan, R.S., Yau, K.-Y., Schrader, K. & Reed, R.R. (1990). Primary structure and functional expression of a cyclic nucleotide-activated channel from olfactory neurons. *Nature* **347**, 184–7.

Dionne, V.E., Steinbach, J.H. & Stevens, C.F. (1978). An analysis of the dose-response relationship at voltage-clamped frog neuromuscular junctions. *Journal of Physiology* **281**, 421–44.

DiPaola, M., Czajkowski, C. & Karlin, A. (1989). The sidedness of the COOH terminus of the acetylcholine receptor δ-subunit. *Journal of Biological Chemistry* **264**, 15457–63.

Dolphin, A.C. (1995). Voltage-dependent calcium channels and their modulation by neurotransmitters and G proteins. *Experimental Physiology* **80**, 1–36.

Donoghue, M.J. (1992). Homology. In *Keywords in Evolutionary Biology*, ed. E.F. Keller & E.A. Lloyd, pp. 170–9. Cambridge, MA: Harvard University Press.

Doupnik, C.A., Davidson, N. & Lester, H.A. (1995). The inward rectifier potassium channel family. *Current Opinion in Neurobiology* **5**, 268–77.

Drachman, D.B. (1994). Myasthenia gravis. *New England Journal of Medicine* **330**, 1797–810.

Drumm, M.L., Pope, H.A., Cliff, W.H., Rommens, J.M., Marvin, S.A., Tsui, L.-C., Collins, F.S., Frizell, R.A. & Wilson, J.M. (1990). Correction of the cystic fibrosis defect in vitro by retrovirus-mediated gene transfer. *Cell* **62**, 1227–33.

Dryer, S.E., Fujii, J.T. & Martin, A.R. (1989). A Na^+-activated K^+ current in cultured brain stem neurones from chicks. *Journal of Physiology* **410**, 283–96.

Duclohier, H., Molle, G. & Spach, G. (1989). Antimicrobial peptide magainin I from *Xenopus* skin forms anion-permeable channels in planar lipid bilayers. *Biophysical Journal* **56**, 1017–21.

Dumbacher, J.P., Beehler, B.M., Spande, T.F., Garaffo, H.M. & Daly, J.W. (1992). Homobatrachotoxin in the genus *Pitohui*: chemical defense in birds? *Science* **258**, 799–801.

Dunlap, K. & Fischbach, G.D. (1981). Neurotransmitters decrease the calcium conductance activated by depolarization of embryonic chick sensory neurones. *Journal of Physiology* **317**, 519–35.

Dunn, R.J., Hackett, N.R., McCoy, J.M., Chao, B.H., Kimura, K. & Khorana, H.G. (1987). Structure–function studies on bacteriorhodopsin. 1. Expression of the bacterio-opsin gene in *Esherichia coli*. *Journal of Biological Chemistry* **262**, 9246–54.

Durell, S.R. & Guy, H.R. (1992). Atomic scale structure and functional models of voltage-gated potassium channels. *Biophysical Journal* **62**, 238–50.

Dwyer, T.M., Adams, D.J. & Hille, B. (1980). The permeability of the endplate channel to organic cations in frog muscle. *Journal of General Physiology* **75**, 469–92.

Edsall, J.T. & McKenzie, H.A. (1978). Water and proteins. I. The significance and structure of water; its interactions with electrolytes and nonelectrolytes. *Advances in Biophysics* **10**, 137–207.

Edwards, F.A., Gibb, A.J. & Colquhoun, D. (1992). ATP

receptor-mediated synaptic currents in the central nervous system. *Nature* **359**, 144–7.

Edwards, G. & Weston, A.H. (1990). Structure–activity relationships of K$^+$ channel openers. *Trends in Pharmacological Sciences* **11**, 417–22.

Edwards, G. & Weston, A.H. (1993). The pharmacology of ATP-sensitive potassium channels. *Annual Review of Pharmacology and Toxicology* **33**, 597–637.

Ehrenstein, G. & Lecar, H. (1977). Electrically gated ionic channels in lipid bilayers. *Quarterly Review of Biophysics* **10**, 1–34.

Ehring, G.R., Zampighi, G., Horwitz, J., Bok, D. & Hall, J.E. (1990). Properties of channels reconstituted from the major intrinsic protein of lens fiber membranes. *Journal of General Physiology* **96**, 631–64.

Einstein, A. (1926). *Investigations on the Theory of the Brownian Movement*, ed. R. Fürth, transl. A.D. Cowper. London: Methuen. Republished in 1956 by Dover Publications, Inc.

Eisenman, G., Latorre, R. & Miller, C. (1986). Multi-ion conductance and selectivity in the high-conductance Ca^{2+}-activated K$^+$ channel from skeletal muscle. *Biophysical Journal* **50**, 1025–34.

Elgoyhen, A.B., Johnson, D.S., Boulter, J., Vetter, D.E. & Heinemann, S. (1994). α9: an acetylcholine receptor with novel pharmacological properties expressed in rat cochlear hair cells. *Cell*, **79**, 705–15.

Ellinor, P.T., Zhang, J.-F., Horne, W.A. & Tsien, R.W. (1994). Structural determinants of the blockade of N-type calcium channels by a peptide neurotoxin. *Nature* **372**, 272–5.

Elmslie, K.S. & Jones, S.W. (1994). Concentration dependence of neurotransmitter effects on calcium current kinetics in frog sympathetic neurones. *Journal of Physiology* **481**, 35–46.

Elmslie, K.S., Zhou, W. & Jones, S.W. (1990). LHRH and GTP-γ-S modify calcium current activation in bullfrog sympathetic neurons. *Neuron* **5**, 75–80.

Eyring, H., Lumry, R. & Woodbury, J.W. (1949). Some applications of modern rate theory to physiological systems. *Record of Chemical Progress* **100**, 100–14.

Fakler, B., Brändle, U., Bond, C., Glowatzki, S., König, C., Adelman, J.P., Zenner, H.-P. & Ruppersberg, J.P. (1994). A structural determinant of differential sensitivity of cloned inward rectifier

K$^+$ channels to intracellular spermine. *FEBS Letters* **356**, 199–203.

Fakler, B., Brändle, U., Glowatzki, S., Weidemann, S., Zenner, H.-P. & Ruppersberg, J.P. (1995). Strong voltage-dependent inward rectification of inward rectifier K$^+$ channels is caused by intracellular spermine. *Cell* **80**, 149–54.

Faraday, M. (1834). Experimental researches on electricity. Seventh Series. *Philosophical Transactions of the Royal Society of London* **124**, 77–122.

Fatt, P. & Katz, B. (1951). An analysis of the end-plate potential recorded with an intracellular electrode. *Journal of Physiology* **115**, 320–69.

Fertuck, H.C. & Salpeter, M.M. (1974). Localization of acetylcholine receptor by ^{125}I-labelled α-bungarotoxin binding at mouse motor end-plates. *Proceedings of the National Academy of Sciences, USA* **71**, 1376–8.

Ficker, E., Taglialatela, M., Wible, B.A., Henley, C.M. & Brown, A.M. (1994). Spermine and spermidine as gating molecules for inward rectifier K$^+$ channels. *Science* **266**, 1068–72.

Field, M. & Semrad, C.E. (1993). Toxigenic diarrheas, congenital diarrheas, and cystic fibrosis: disorders of intestinal ion transport. *Annual Review of Physiology* **55**, 631–55.

Finer-Moore, J. & Stroud, R.M. (1984). Amphipathic analysis and possible formation of the ion channel in an acetylcholine receptor. *Proceedings of the National Academy of Sciences, USA* **81**, 155–9.

Finkelstein, A. (1984). Water movement through membrane channels. *Current Topics in Membranes and Transport* **21**, 295–308.

Fleckenstein, A. (1983). History of calcium antagonists. *Circulation Research* **52,** supplement I, 3–16.

Fontaine, B., Khurana, T.S., Hoffman, E.P., Bruns, G.A.P., Haines, J.L., Trofatter, J.A., Hanson, M.P., Rich, J., McFarlane, H., Yasek, D.M., Romano, D., Gusella, J.F. & Brown, R.H. (1990). Hyperkalemic periodic paralysis and the adult muscle sodium channel α-subunit gene. *Science* **250**, 1000–2.

Fontaine, B., Vale-Santos, J., Jurkat-Rott, K., Reboul, J., Plassart, E., Rime, C.-S., Elbaz, A., Heine, R., Guimaraes, J., Weissenbach, J., Baumann, N., Fardeau, M. & Lehmann-Horn, F. (1994). Mapping of the hypokalaemic periodic paralysis (HypoPP) locus to chromosome 1q31–32 in three European families. *Nature Genetics* **6**, 267–72.

Forsythe, I.D., Linsdell, P. & Stanfield, P.R. (1992). Unitary A-currents of rat locus coeruleus neurones grown in cell culture: rectification caused by internal Mg^{2+} and Na^{+}. *Journal of Physiology* **451**, 553–83.

Fox, R.O. & Richards, F.M. (1982). A voltage-gated ion channel model inferred from the crystal structure of alamethicin at 1.5 Å resolution. *Nature* **300**, 325–30.

Franciolini, F. & Nonner, W. (1987). Anion and cation permeability of a chloride channel in rat hippocampal neurons. *Journal of General Physiology* **90**, 453–78.

Frank, H.S. (1958). Covalency in the hydrogen bond and properties of water and ice. *Proceedings of the Royal Society of London* A **247**, 481–92.

Frankenhaeuser, B. & Hodgkin, A.L. (1957). The action of calcium on the electrical properties of squid axons. *Journal of Physiology* **137**, 218–44.

Franks, N.P. & Lieb, W.R. (1994). Molecular and cellular mechanisms of general anaesthesia. *Nature* **367**, 607–14.

Franzini-Armstrong, C. (1980). Structure of the sarcoplasmic reticulum. *Federation Proceedings* **39**, 2403–9.

Freifelder, D. (1985). *Principles of Physical Chemistry with Applications to the Biological Sciences.* Boston, MA: Jones and Bartlett.

French, R.J. & Shoukimas, J.J. (1981). Blockage of squid axon potassium conductance by internal tetra-*N*-alkylammonium ions of various sizes. *Biophysical Journal* **34**, 271–91.

French, R.J. & Shoukimas, J.J. (1985). An ion's view of the potassium channel. *Journal of General Physiology* **85**, 669–98.

Fujii, J., Otsu, K., Zorzato, F., de Leon, S., Khanna, V.K., Weiler, J.E., O'Brien, P.J. & MacLennan, D.H. (1991). Identification of a mutation in porcine ryanodine receptor associated with malignant hyperthermia. *Science* **253**, 448–51.

Furuichi, T., Yoshikawa, S., Miyawaki, A., Wada, K., Maeda, N. & Mikoshiba, K. (1989). Primary structure and functional expression of the inositol 1,4,5-trisphosphate-binding protein P_{400}. *Nature* **342**, 32–8.

Furuichi, T., Kohda, K., Miyawaki, A. & Mikoshiba, K. (1994). Intracellular channels. *Current Opinion in Neurobiology* **4**, 294–303.

Fushimi, K., Uchida, S., Hara, Y., Hirata, Y., Marumo, F. & Sasaki, S. (1993). Cloning and expression of apical membrane water channel of rat kidney collecting tubule. *Nature* **361**, 549–52.

Gadsby, D.C. & Nairn, A.C. (1994). Regulation of CFTR channel gating. *Trends in Biochemical Sciences* **19**, 513–18.

Galzi, J.-L., Revah, F., Black, D., Goeldner, M., Hirth, C. & Changeux, J.-P. (1990). Identification of a novel amino acid α-Tyr93 within the active site of the acetylcholine receptor by photoaffinity labelling: additional evidence for a three-loop model of the acetylcholine binding site. *Journal of Biological Chemistry* **265**, 10430–7.

Galzi, J.-L., Bertrand, D., Devillers-Thiéry, A., Revah, F., Bertrand, S. & Changeux, J.-P. (1991). Functional significance of aromatic amino acids from three peptide loops of the α7 neuronal nicotinic receptor site investigated by site-directed mutagenesis. *FEBS Letters* **294**, 198–202.

Galzi, J.-L., Devillers-Thiéry, A., Hussy, N., Bertrand, S., Changeux, J.-P. & Bertrand, D. (1992). Mutations in the channel domain of a neuronal nicotinic receptor convert ion selectivity from cationic to anionic. *Nature* **359**, 500–5.

Garcia, M.L., Galvez, A., Garcia-Calvo, M., King, V.F., Vazquez, J. & Kaczorowski, G.J. (1991). Use of toxins to study potassium channels. *Journal of Bioenergetics and Biomembranes* **23**, 615–45.

Garrod, A.E. (1909). *Inborn Errors of Metabolism.* London: Frowde, Hodder & Stoughton. (Reprinted in 1963 as *Garrod's Inborn Errors of Metabolism* with a supplement by H. Harris. London: Oxford University Press.)

Garrod, A.E. (1931). *Inborn Factors in Disease.* Oxford: Clarendon Press. (Reprinted in 1989 as *Garrod's Inborn Factors in Disease* with annotations by C.R. Scriver and B. Childs. Oxford: Oxford University Press.)

Garthwaite, J. & Boulton, C.L. (1995). Nitric oxide signaling in the central nervous system. *Annual Review of Physiology* **57**, 683–706.

Gautron, S., Dos Santos, G., Pinto-Henrique, D., Koulakoff, A. & Berwald-Netter, Y. (1992). The glial voltage-gated sodium channel: cell and tissue-specific mRNA expression. *Proceedings of the National Academy of Sciences, USA* **89**, 7272–6.

Gay, L.A. & Stanfield, P.R. (1977). Cs$^+$ causes a voltage-dependent block of inward K currents in resting skeletal muscle fibres. *Nature* **267**, 169–70.

Geddes, D. (1994). Cystic fibrosis: gene therapy trials come to the UK. *MRC News* Spring 1994, 13–16.

George, A.L., Knittle, T.J. & Tamkun, M.M. (1992). Molecular cloning of an atypical voltage-gated sodium channel expressed in human heart and uterus: evidence for a distinct gene family. *Proceedings of the National Academy of Sciences, USA* **89**, 4893–7.

Gibb, A.J. (1993). Receptor pharmacology. In *Medicinal Chemistry: The Role of Organic Chemistry in Drug Research*, second edition, ed. C.R. Ganellin & S.M. Roberts, pp. 37–59. London: Academic Press.

Gilman, A.G. (1987). G proteins: transducers of receptor-generated signals. *Annual Review of Biochemistry* **56**, 615–49.

Glasstone, S.K., Laidler, K.J. & Eyring, H. (1941). *The Theory of Rate Processes*. New York: McGraw-Hill.

Goldman, D.E. (1943). Potential, impedance, and rectification in membranes. *Journal of General Physiology* **27**, 37–60.

Gordon, L.G.M. & Haydon, D.A. (1975). Potential-dependent conductances in lipid membranes containing alamethicin. *Philosophical Transactions of the Royal Society of London* B **270**, 433–47.

Gorin, M.B., Yancey, S.B., Cline, J., Revel, J.-P. & Horwitz, J. (1984). The major intrinsic protein (MIP) of the bovine lens fiber membrane: characterization and structure based on cDNA cloning. *Cell* **39**, 49–59.

Gorman, A.L.F. & Thomas, M.V. (1980). Potassium conductance and internal calcium accumulation in a molluscan neurone. *Journal of Physiology* **308**, 287–313.

Görne-Tschelnokow, U., Strecker, A., Kaduk, C., Naumann, D. & Hucho, F. (1994). The transmembrane domains of the nicotinic acetylcholine receptor contain α-helical and β structures. *EMBO Journal* **13**, 338–41.

Goulding, E.H., Tibbs, G.R., Liu, D. & Siegelbaum, S.A. (1993). Role of H5 domain in determining pore diameter and ionic permeation through cyclic nucleotide-gated channels. *Nature* **364**, 61–4.

Goulding, E.H., Tibbs, G.R. & Siegelbaum, S.A. (1994). Molecular mechanism of cyclic-nucleotide-gated channel activation. *Nature* **372**, 369–74.

Grant, T.L., Ohnmacht, C.J. & Howe, B.B. (1994). Anilide tertiary carbinols: a novel series of K$^+$ channel openers. *Trends in Pharmacological Sciences* **15**, 402–4.

Green, W.N. & Andersen, O.S. (1991). Surface charges and ion channel function. *Annual Review of Physiology* **53**, 341–59.

Green, W.N. & Claudio, T. (1993). Acetylcholine receptor assembly: subunit folding and oligomerization occur sequentially. *Cell* **74**, 57–69.

Grissmer, S., Nguyen, A.N., Aiyar, J., Hanson, D.C., Mather, R.J., Gutman, G.A., Karmilowicz, M.J., Auperin, D.D. & Chandy, K.G. (1994). Pharmacological characterization of five cloned voltage-gated K$^+$ channels, types Kv1.1, 1.2, 1.3, 1.5, and 3.1, stably expressed in mammalian cell lines. *Molecular Pharmacology* **45**, 1227–34.

Gronemeier, M., Condie, A., Prosser, J., Steinmeyer, K., Jentsch, T.J. & Jockusch, H. (1994). Nonsense and missense mutations in the muscular chloride channel gene Clc-1 of myotonic mice. *Journal of Biological Chemistry* **269**, 5963–7.

Gros, P. & Buschman, E. (1993). The multidrug resistance transport protein: identification of functional domains. In *Molecular Biology and Function of Transport Proteins* (Society of General Physiologists Series, vol. 48), ed. L. Reuss, J.M. Russell & M.L. Jennings, pp. 95–117. New York: Rockefeller University Press.

Gründer, S., Thiemann, A., Pusch, M. & Jentsch, T.J. (1992). Regions involved in the opening of ClC-2 chloride channel by voltage and cell volume. *Nature* **360**, 759–63.

Guharay, F. & Sachs, F. (1984). Stretch-activated single ion channel currents in tissue-cultured embryonic skeletal muscle. *Journal of Physiology* **352**, 685–701.

Guy, H.R. & Seetharamulu, P. (1986). Molecular model of the action potential sodium channel. *Proceedings of the National Academy of Sciences, USA* **83**, 508–12.

Haase, H., Karczewski, P., Beckert, R. & Krause, E.G. (1993). Phosphorylation of the L-type calcium channel β subunit is involved in β-adrenergic signal transduction in canine myocardium. *FEBS Letters* **335**, 217–22.

Hadley, R.W. & Lederer, W.J. (1991). Ca^{2+} and voltage inactivate Ca^{2+} channels in guinea-pig ventricular myocytes through independent mechanisms. *Journal of Physiology* **444**, 257–68.

Hagiwara, S., Miyazaki, S. & Rosenthal, N.P. (1976). Potassium current and the effect of cesium on this current during anomalous rectification of the egg cell membrane of a starfish. *Journal of General Physiology* **67**, 621–38.

Hagiwara, S., Miyazaki, S., Krasne, S. & Ciani, S. (1977). Anomalous permeabilities of the egg cell membrane of a starfish in K^+–Tl^+ mixtures. *Journal of General Physiology* **70**, 269–81.

Hakamata, Y., Nakai, J., Takeshima, H. & Imoto, K. (1992). Primary structure and distribution of a novel ryanodine receptor/calcium release channel from rabbit brain. *FEBS Letters* **312**, 229–35.

Halliwell, J.V., Plant, T.D., Robbins, J. & Standen, N.B. (1994). Voltage clamp techniques. In *Microelectrode Techniques: The Plymouth Workshop Handbook*, ed. D. Ogden, pp. 17–35. Cambridge: Company of Biologists.

Hamill, O.P., Marty, A., Neher, E., Sakmann, B. & Sigworth, F.J. (1981). Improved patch-clamp techniques for high-resolution current recording from cells and cell-free membrane patches. *Pflügers Archiv* **391**, 85–100.

Hansson, C.C. (1988). Cystic fibrosis and chloride secreting diarrhoea. *Nature* **333**, 711.

Harrison, L.J. (1991). Poisonous marine morsels. *Journal of the Florida Medical Association* **78**, 219–21.

Hartmann, H.A., Kirsch, G.E., Drewe, J.A., Taglialatela, M., Joho, R.H. & Brown, A.M. (1991). Exchange of conduction pathways between two related K^+ channels. *Science* **251**, 942–4.

Hartzell, H.C., Méry, P.-F., Fischmeister, R. & Szabo, G. (1991). Sympathetic regulation of cardiac calcium current is due exclusively to cAMP-dependent phosphorylation. *Nature* **351**, 573–6.

Heginbotham, L. & MacKinnon, R. (1992). The aromatic binding site for tetraethylammonium ion on potassium channels. *Neuron* **8**, 483–91.

Heginbotham, L. & MacKinnon, R. (1993). Conduction properties of the cloned *Shaker* K^+ channel. *Biophysical Journal* **65**, 2089–96.

Heginbotham, L., Abramson, T. & MacKinnon, R. (1992). A functional connection between the pores of

distantly related ion channels as revealed by mutant K^+ channels. *Science* **258**, 1152–5.

Heidmann, O., Buonanno, A., Geoffroy, B., Robert, B., Guenet, J.-L., Merlie, J.P. & Changeux, J.-P. (1986). Chromosomal localization of muscle nicotinic acetylcholine receptor genes in the mouse. *Science* **234**, 866–8.

Heine, R., George, A.L., Pika, U., Deymeer, F., Rüdel, R. & Lehmann-Horn, F. (1994). Proof of a non-functional muscle chloride channel in recessive myotonia congenita (Becker) by detection of a 4 base pair deletion. *Human Molecular Genetics* **3**, 1123–8.

Heinemann, S.H. (1995). Guide to data acquisition and analysis. In *Single-Channel Recording*, 2nd edition, ed. B. Sakmann & E. Neher, pp. 53–91. New York: Plenum Press.

Heinemann, S.H., Terlau, H., Stühmer, W., Imoto, K. & Numa, S. (1992). Calcium channel characteristics conferred on the sodium channel by single mutations. *Nature* **356**, 441–3.

Heitler, W.J. (1992) *NeuroSim: A Neurophysiology Simulation Package for IBM PC*. Cambridge: Biosoft.

Henderson, R. & Unwin, P.N.T. (1975). Three-dimensional model of purple membrane obtained by electron microscopy. *Nature* **257**, 28–32.

Henderson, R., Baldwin, J.M., Ceska, T.A., Zemlin, F., Beckmann, E. & Downing, K.H. (1990). Model for the structure of bacteriorhodopsin based on high-resolution electron cryomicroscopy. *Journal of Molecular Biology* **213**, 899–929.

Hess, P., Lansman, J.B. & Tsien, R.W. (1984). Different modes of Ca channel gating behaviour favoured by dihydropyridine Ca agonists and antagonists. *Nature* **311**, 538–44.

Hidalgo, P. & MacKinnon, R. (1995). Revealing the architecture of a K^+ channel pore through mutant cycles with a peptide inhibitor. *Science* **268**, 307–10.

Higashida, H. & Brown, D.A. (1986). Two polyphosphatidylinositide metabolites control two K^+ channels in a neuronal cell. *Nature* **323**, 333–5.

Higgins, C.F. (1992). ABC transporters: from microorganisms to man. *Annual Review of Cell Biology* **8**, 67–113.

Hill, A.V. (1909). The mode of action of nicotine and curare determined by the form of the contraction curve and the method of temperature coefficients. *Journal of Physiology* **39**, 361–73.

Hille, B. (1968). Pharmacological modifications of the sodium channels of frog nerve. *Journal of General Physiology* **51**, 199–219.

Hille, B. (1971). The permeability of the sodium channel to organic cations in myelinated nerve. *Journal of General Physiology* **58**, 599–619.

Hille, B. (1972). The permeability of the sodium channel to metal cations in myelinated nerve. *Journal of General Physiology* **59**, 637–58.

Hille, B. (1973). Potassium channels in myelinated nerve: selective permeability to small cations. *Journal of General Physiology* **61**, 669–86.

Hille, B. (1975). Ionic selectivity, saturation, and block in sodium channels: a four-barrier model. *Journal of General Physiology* **66**, 535–60.

Hille, B. (1977a). The pH-dependent rate of action of local anesthetics on the node of Ranvier. *Journal of General Physiology* **69**, 475–96.

Hille, B. (1977b). Local anesthetics: hydrophilic and hydrophobic pathways for the drug–receptor reaction. *Journal of General Physiology* **69**, 497–515.

Hille, B. (1992a). *Ionic Channels of Excitable Membranes*, 2nd edition. Sunderland, MA: Sinauer Associates.

Hille, B. (1992b). G protein-coupled mechanisms and nervous signaling. *Neuron* **9**, 187–95.

Hille, B. (1994). Modulation of ion-channel function by G-protein-coupled receptors. *Trends in Neurosciences* **17**, 531–6.

Hille, B. & Schwarz, W. (1978). Potassium channels as multi-ion single-file pores. *Journal of General Physiology* **72**, 409–42.

Hille, B., Woodhull, A.M. & Shapiro, B.I. (1975). Negative surface charge near sodium channels of nerve: divalent ions, monovalent ions, and pH. *Philosophical Transactions of the Royal Society of London* B **270**, 301–18.

Hladky, S.B. & Haydon, D.A. (1970). Discreteness of conductance change in bimolecular lipid membranes in the presence of certain antibiotics. *Nature* **225**, 451–3.

Hladky, S.B. & Haydon, D.A. (1984). Ion movements in gramicidin channels. *Current Topics in Membranes and Transport* **21**, 327–72.

Ho, K., Nichols, C.G., Lederer, W.J., Lytton, J., Vassilev, P.M., Kanazirska, M.V. & Hebert, S.C. (1993). Cloning and expression of an inwardly rectifying ATP-regulated potassium channel. *Nature* **362**, 31–8.

Hodgkin, A.L. (1958). Ionic movements and electrical activity in giant nerve fibres. *Proceedings of the Royal Society of London* B **126**, 87–121.

Hodgkin, A.L. (1992). *Chance and Design*. Cambridge: Cambridge University Press.

Hodgkin, A.L. & Huxley, A.F. (1939). Action potentials recorded from inside a nerve fibre. *Nature* **140**, 710–11.

Hodgkin, A.L. & Huxley, A.F. (1952a). Currents carried by sodium and potassium ions through the membrane of the giant axon of *Loligo*. *Journal of Physiology* **116**, 449–72.

Hodgkin, A.L. & Huxley, A.F. (1952b). A quantitative description of membrane current and its application to conduction and excitation in nerve. *Journal of Physiology* **117**, 500–44.

Hodgkin, A.L. & Katz, B. (1949). The effect of sodium ions on the electrical activity of the giant axon of the squid. *Journal of Physiology* **108**, 37–77.

Hodgkin, A.L. & Keynes, R.D. (1955). The potassium permeability of a giant nerve fibre. *Journal of Physiology* **128**, 61–88.

Hoffman, E.P., Lehmann-Horn, F. & Rüdel, R. (1995). Overexcited or inactive: ion channels in muscle disease. *Cell* **80**, 681–6.

Hoffman, P.W., Ravindran, A. & Huganir, R.L. (1994). Role of phosphorylation in desensitization of acetylcholine receptors. *Journal of Neuroscience* **14**, 4185–95.

Hofmann, F., Biel, M. & Flockerzi, V. (1994). Molecular basis for Ca^{2+} channel diversity. *Annual Review of Neuroscience* **17**, 399–418.

Hollmann, M. & Heinemann, S. (1994). Cloned glutamate receptors. *Annual Review of Neuroscience* **17**, 31–108.

Hollmann, M., Maron, C. & Heinemann, S. (1994). N-glycosylation site tagging suggests a three transmembrane domain topology for the glutamate receptor GluR1. *Neuron* **13**, 1331–43.

Horn, R. (1987). Statistical methods for model discrimination: applications to gating kinetics and permeation of the acetylcholine receptor channel. *Biophysical Journal* **51**, 255–63.

Horn, R. & Lange, K. (1983). Estimating kinetic constants from single channel data. *Biophysical Journal* **43**, 207–23.

Hoshi, T., Zagotta, W.N. & Aldrich, R.W. (1990). Biophysical and molecular mechanisms of *Shaker* potassium channel inactivation. *Science* **250**, 533–8.

Hoshi, T., Zagotta, W.N. & Aldrich, R.W. (1991). Two types of inactivation in *Shaker* K^+ channels: effects of alterations in the carboxy-terminal region. *Neuron* **7**, 547–56.

Hoshi, T., Zagotta, W.N. & Aldrich, R.W. (1994). *Shaker* potassium channel gating. I: Transitions near the open state. *Journal of General Physiology* **103**, 249–78.

Howard, J., Roberts, W.M. & Hudspeth, A.J. (1988). Mechanoelectrical transduction by hair cells. *Annual Review of Biophysics* **17**, 99–124.

Hucho, F., Görne-Tschelnokow, U. & Strecker, A. (1994). β-structure in the membrane-spanning part of the nicotinic acetylcholine receptor (or how helical are transmembrane helices?). *Trends in Biochemical Sciences* **19**, 383–7.

Hudspeth, A.J. (1989). How the ear's works work. *Nature* **341**, 397–404.

Huganir, R. (1988). Regulation of the nicotinic acetylcholine receptor channel by protein phosphorylation. *Current Topics in Membranes and Transport* **33**, 147–63.

Huganir, R.L., Delcour, A.H., Greengard, P. & Hess, G.P. (1986). Phosphorylation of the nicotinic acetylcholine receptor regulates its rate of desensitization. *Nature* **321**, 774–6.

Hume, R.I., Dingledine, R. & Heinemann, S.F. (1991). Identification of a site in glutamate receptor subunits that controls calcium permeability. *Science* **253**, 1028–31.

Hunter, T. (1995). Protein kinases and phosphatases: the yin and yang of protein phosphorylation and signaling. *Cell* **80**, 225–36.

Hwang, T.-C., Nagel, G., Nairn, A.C. & Gadsby, D. (1994). Regulation of the gating of cystic fibrosis transmembrane conductance regulator Cl channels by phosphorylation and ATP hydrolysis. *Proceedings of the National Academy of Sciences, USA* **91**, 4698–702.

Hyde, S.C., Gill, D.R., Higgins, C.F., Trezise, A.E.O., MacVinish, L.J., Cuthbert, A.W., Ratcliff, R., Evans, M.J. & Colledge, W.H. (1993). Correction of the ion transport defect in cystic fibrosis transgenic mice by gene therapy. *Nature* **362**, 250–5.

Imoto, K. (1993). Molecular aspects of ion permeation through channels. *Annals of the New York Academy of Sciences* **707**, 38–50.

Imoto, K., Busch, C., Sakmann, B., Mishina, M., Konno, T., Nakai, J., Bujo, H., Mori, Y., Fukuda, K. & Numa, S. (1988). Rings of negatively charged amino acids determine the acetylcholine receptor channel conductance. *Nature* **335**, 645–8.

Imoto, K., Konno, T., Nakai, J., Wang, F., Mishina, M. & Numa, S. (1991). A ring of uncharged polar amino acids as a component of channel constriction in the nicotinic acetylcholine receptor. *FEBS Letters* **289**, 193–200.

Inagaki, N., Tsuura, Y., Namba, N., Masuda, K., Gonoi, T., Horie, M., Seino, Y., Mizuta, M. & Seino, S. (1995). Cloning and functional characterization of a novel ATP-sensitive potassium channel ubiquitously expressed in rat tissues, including pancreatic islets, pituitary, skeletal muscle, and heart. *Journal of Biological Chemistry* **270**, 5691–4.

Inoue, M., Nakajima, S. & Nakajima, Y. (1988). Somatostatin induces an inward rectification in rat locus coeruleus neurones through a pertussis toxin-sensitive mechanism. *Journal of Physiology* **407**, 177–98.

Inui, M., Saito, A. & Fleischer, S. (1987). Purification of the ryanodine receptor and identity with feet structures of the junctional terminal cisternae of sarcoplasmic reticulum from fast skeletal muscle. *Journal of Biological Chemistry* **262**, 1740–7.

Ionasescu, V.V. (1995). Charcot-Marie-Tooth neuropathies: from clinical description to molecular genetics. *Muscle and Nerve* **18**, 267–75.

Isacoff, E.Y., Jan, Y.N. & Jan, L.Y. (1990). Evidence for the formation of heteromultimeric potassium channels in *Xenopus* oocytes. *Nature* **345**, 530–4.

Isacoff, E.Y., Jan, Y.N. & Jan, L.Y. (1991). Putative receptor for the cytoplasmic inactivation gate in the *Shaker* K^+ channel. *Nature* **353**, 86–90.

Ishihara, K., Mituiye, T., Noma, A. & Takano, M. (1989). The Mg^{2+} block and intrinsic gating underlying inward rectification of the K^+ current in guinea-pig cardiac myocytes. *Journal of Physiology* **419**, 297–320.

Ishii, T., Moriyoshi, K., Sigijara, H., Sakurada, K.,

Kadotani, H., Yokoi, M., Akazawa, C., Shigemoto, R., Mizuno, N., Masu, M., & Nakanishi, S. (1993). Molecular characterization of the family of the N-methyl-D-aspartate receptor subunits. *Journal of Biological Chemistry* **268**, 2836–43.

Isom, L.L., De Jongh, K.S., Patton, D.E., Reber, B.F.X., Offord, J., Charbonneau, H., Walsh, K., Goldin, A.L. & Catterall, W.A. (1992). Primary structure and functional expression of the β_1 subunit of the rat brain sodium channel. *Science* **256**, 839–42.

Isom, L.L., De Jongh, K.S. & Catterall, W.A. (1994). Auxiliary subunits of voltage-gated ion channels. *Neuron* **12**, 1183–94.

Jackson, M.B. (1986). Kinetics of unliganded acetylcholine receptor channel gating. *Biophysical Journal* **49**, 663–72.

Jackson, M.B. (1988). Dependence of acetylcholine receptor channel kinetics on agonist concentration in cultured mouse muscle fibres. *Journal of Physiology* **397**, 555–83.

Jackson, M.B. (1989). Perfection of a synaptic receptor: kinetics and energetics of the acetylcholine receptor. *Proceedings of the National Academy of Sciences, USA* **86**, 2199–203.

Jackson, M.B. (1994). Single channel currents in the nicotinic acetylcholine receptor: a direct demonstration of allosteric transitions. *Trends in Biochemical Sciences* **19**, 396–9.

Jackson, M.B., Imoto, K., Mishina, M., Konno, T., Numa, S. & Sakmann, B. (1990). Spontaneous and agonist-induced openings of the acetylcholine receptor channel composed of bovine muscle α-, β- and δ-subunits. *Pflügers Archiv* **417**, 129–35.

Jan, L.Y. & Jan, Y.N. (1990). A superfamily of ion channels. *Nature* **345**, 672.

Jan, L.Y. & Jan, Y.N. (1992). Tracing the roots of ion channels. *Cell* **69**, 715–18.

Jentsch, T.J., Steinmeyer, K. & Schwarz, G. (1990). Primary structure of *Torpedo marmorata* chloride channel isolated by expression cloning in *Xenopus* oocytes. *Nature* **348**, 510–14.

Johnson, J.W. & Ascher, P. (1987). Glycine potentiates the NMDA response in cultured mouse brain neurons. *Nature* **325**, 529–31.

Johnson, L.N. & Barford, D. (1990). Glycogen phos-

phorylase: the structural basis of the allosteric response and comparison with other allosteric proteins. *Journal of Biological Chemistry* **265**, 2409–12.

Kalow, W. & Grant, D.M. (1995). Pharmacogenetics. In *The Metabolic and Molecular Bases of Inherited Disease*, ed. C.R. Scriver, A.L. Beaudet, W.S. Sly & D. Valle, vol. 1, pp 293–326. New York: McGraw-Hill.

Karlin, A. (1993). Structure of nicotinic acetylcholine receptors. *Current Opinion in Neurobiology* **3**, 299–309.

Karplus, M. & Petsko, G.A. (1990). Molecular dynamics simulations in biology. *Nature* **347**, 631–9.

Katz, B. & Miledi, R. (1972). The statistical nature of the acetylcholine potential and its molecular components. *Journal of Physiology* **224**, 665–700.

Katzung, B.G. (1995). *Basic and Clinical Pharmacology*, 6th edition. East Norwalk, CN: Appleton & Lange.

Kaupp, U.B., Niidome, T., Tanabe, T., Terada, S., Bönigk, W., Stühmer, W., Cook, N.J., Kangawa, K., Matsuo, H., Hirose, T., Miyata, T. & Numa, S. (1989). Primary structure and functional expression from complementary DNA of the rod photoreceptor cyclic GMP-gated channel. *Nature* **342**, 762–6.

Kayano, T., Noda, M., Flockerzi, V., Takahashi, H. & Numa, S. (1988). Primary structure of rat brain sodium channel III deduced from the cDNA sequence. *FEBS Letters* **228**, 187–94.

Keller, B.U., Blaschke, M., Rivosecchi, R., Hollmann, M., Heinemann, S. & Konnerth, A. (1993). Identification of a subunit-specific antagonist of α-amino-3-hydroxy-5-methylisoxazole-4-propionate/kainate receptor channels. *Proceedings of the National Academy of Sciences, USA* **90**, 605–9.

Kenakin, T.P. (1984). The classification of drugs and drug receptors in isolated tissues. *Pharmacological Reviews* **36**, 165–222.

Kerem, B., Rommens, J.M., Buchanan, J.A., Markiewicz, D., Cox, T.K., Chakravarti, A., Buchwald, M. & Tsui, L.-C. (1989). Identification of the cystic fibrosis gene: genetic analysis. *Science* **245**, 1073–80.

Keynes, R.D. (1990). A series-parallel model of the voltage-gated sodium channel. *Proceedings of the Royal Society of London* B **240**, 425–32.

Keynes, R.D. (1992). A new look at the mechanism of activation and inactivation of voltage-gated ion channels. *Proceedings of the Royal Society of London* B **249**, 107–12.

Keynes, R.D. (1994). The kinetics of voltage-gated ion channels. *Quarterly Review of Biophysics* **27**, 339–434.

Keynes, R.D. & Aidley, D.J. (1991). *Nerve and Muscle*, 2nd edition. Cambridge: Cambridge University Press.

Keynes, R.D. & Ritchie, J.M. (1984). On the binding of labelled saxitoxin to the squid giant axon. *Proceedings of the Royal Society of London* B **222**, 147–53.

Keynes, R.D. & Rojas, E. (1974). Kinetics and steady state properties of the charge system controlling sodium conductance in the squid giant axon. *Journal of Physiology* **239**, 393–434.

Keynes, R.D., Greef, N.G. & Forster, I.C. (1992). Activation, inactivation and recovery in the sodium channels of the squid giant axon dialysed with different solutions. *Philosophical Transactions of the Royal Society of London* B **337**, 471–84.

Khodorov, B.I. (1985). Batrachotoxin as a tool to study voltage-sensitive sodium channels of excitable membranes. *Progress in Biophysics and Molecular Biology* **45**, 57–148.

Kieferle, S., Fong, P., Bens, M., Vandewalle, A. & Jentsch, T.J. (1994). Two highly homologous members of the ClC chloride channel family in both rat and human kidney. *Proceedings of the National Academy of Sciences, USA* **91**, 6943–7.

Kienker, P. (1989). Equivalence of aggregated Markov models of ion-channel gating. *Proceedings of the Royal Society of London* B **236**, 269–309.

Kirsch, G.E., Drewe, J.A., Talialatela, M., Jojo, R.H., DeBiasi, M., Hartmann, H.A. & Brown, A.M. (1992). A single nonpolar residue in the deep pore of related K^+ channels acts as a K^+:Rb^+ conductance switch. *Biophysical Journal* **62**, 136–44.

Knepper, M.A. (1994). The aquaporin family of molecular water channels. *Proceedings of the National Academy of Sciences, USA* **91**, 6255–8.

Knowles, M.R., Stutts, M.J., Spock, A., Fischer, N., Gatzy, J.T. & Boucher, R.C. (1983). Abnormal ion permeation through cystic fibrosis respiratory epithelium. *Science* **221**, 1067–70.

Koch, M.C., Steinmeyer, K., Lorenz, C., Ricker, K., Wolf, F., Otto, M., Zoll, B., Lehmann-Horn, F., Grzeschik, K.-H. & Jentsch, T.J. (1992). The skeletal muscle chloride channel in dominant and recessive human myotonia. *Science* **257**, 797–800.

Köhler, M., Burnashev, N., Sakmann, B. & Seeburg, P.H. (1993). Determinants of Ca^{2+} permeability in both TM1 and TM2 of high affinity kainate receptor channels: diversity by RNA editing. *Neuron* **10**, 491–500.

Konno, T., Busch, C., von Kitzing, E., Imoto, K., Wang, F., Nakai, J., Mishinia, M., Numa, S. & Sakmann, B. (1991). Rings of anionic acids as structural determinants of ion selectivity in the acetylcholine receptor channel. *Proceedings of the Royal Society of London* B **244**, 69–79.

Koryta, J. (1982). *Ions, Electrodes, and Membranes.* Chichester: John Wiley & Sons.

Krapivinsky, G.P., Ackermann, M.J., Gordon, E.A., Krapivinsky, L.D. & Clapham, D.E. (1994). Molecular characterization of a swelling-induced chloride conductance regulatory protein, pI_{Cln}. *Cell* **76**, 439–48.

Krapivinsky, G., Gordon, E.A., Wickman, K., Velimivoric, L., Krapivinsky, L. & Clapham, D.E. (1995). The G-protein-gated atrial K^+ channel I_{KACh} is a heteromultimer of two inwardly rectifying K^+-channel proteins. *Nature* **374**, 135–41.

Kraus, J.E. & McNamara, J.O. (1995). Clinical relevance of defects in signalling pathways. *Current Opinion in Neurobiology* **5**, 358–66.

Krebs, E.G. (1983). Historical perspectives on protein phosphorylation and a classification system for protein kinases. *Philosophical Transactions of the Royal Society of London* B **302**, 3–11.

Kreienkamp, H.-J., Maeda, R.K., Sine, S.M. & Taylor, P. (1995). Intersubunit contacts governing assembly of the nicotinic acetylcholine receptor. *Neuron* **14**, 635–44.

Kubo, Y., Baldwin, T.J., Jan, Y.N. & Jan, L.Y. (1993a). Primary structure and functional expression of a mouse inward rectifier potassium channel. *Nature* **362**, 127–33.

Kubo, Y., Reuveny, E., Slesinger, P.A., Jan, Y.N. & Jan, L.Y. (1993b). Primary structure and functional expression of a rat G-protein-coupled potassium channel. *Nature* **364**, 802–6.

Kuhse, J., Becker, C.-M., Schmieden, V., Hoch, W., Pribilla, I., Langosch, D., Malosio, M.-L., Muntz, M. & Betz, H. (1991). Heterogeneity of the inhibitory glycine receptor. *Annals of the New York Academy of Sciences* **625**, 129–35.

Kullberg, R., Owens, J.L., Camacho, P., Mandel, G. & Brehm, P. (1990). Multiple conductance classes of mouse nicotinic acetylcholine receptors expressed in *Xenopus* oocytes. *Proceedings of the National Academy of Sciences, USA* **87**, 2067–71.

Kumpf, R.A. & Dougherty, D.A. (1993). A mechanism for ion selectivity in potassium channels: computational studies of cation-π interactions. *Science* **261**, 1708–10.

Kürz, L.L., Zuhlke, R.D., Zhang, H.J. & Joho, R.H. (1995). Side-chain accessibilities in the pore of a K channel probed by sulfhydryl-specific reagents after cysteine-scanning mutagenesis. *Biophysical Journal* **68**, 900–5.

Kyte, J. & Doolittle, R.F. (1982). A simple method for displaying the hydropathic character of a protein. *Journal of Molecular Biology* **157**, 105–32.

Labarca, P., Coronado, R. & Miller, C. (1980). Thermodynamic and kinetic studies of the gating behaviour of a K^+-selective channel from the sarcoplasmic reticulum membrane. *Journal of General Physiology* **76**, 397–424.

Lamb, R.A., Zebedee, S.L. & Richardson, C.D. (1985). Influenza virus M_2 protein is an integral membrane protein expressed on the infected-cell surface. *Cell* **40**, 627–33.

Langmuir, I. (1918). The adsorption of gases on plane surfaces of glass, mica and platinum. *Journal of the American Chemical Society* **40**, 1361–403.

Langosch, D., Thomas, L. & Betz, H. (1988). Conserved quaternary structure of ligand-gated ion channels: the postsynaptic glycine receptor is a pentamer. *Proceedings of the National Academy of Sciences, USA* **85**, 7394–8.

Langosch, D., Laube, B., Rundström, N., Schmieden, V., Bormann, J. & Betz, H. (1994). Decreased agonist affinity and chloride conductance of mutant glycine receptors associated with human hereditary hyperekplexia. *EMBO Journal* **13**, 4223–8.

Latorre, R. (1994). Molecular workings of large conductance (maxi) Ca^{2+}-activated K^+ channels. In *Handbook of Membrane Channels*, ed. C. Peracchia, pp. 79–102. San Diego, CA: Academic Press.

Latorre, R., Oberhauser, A., Labarca, P. & Alvarez, O. (1989). Varieties of calcium-activated potassium channels. *Annual Review of Physiology* **51**, 385–99.

Läuger, P. (1982). Microscopic calculation of ion-transport rates in membrane channels. *Biophysical Chemistry* **15**, 89–100.

Le Novère, N. & Changeux, J.-P. (1995). Molecular evolution of the nicotinic acetylcholine receptor: an example of multigene family in excitable cells. *Journal of Molecular Evolution* **40**, 155–72.

Lear, J.D., Wasserman, Z.R. & DeGrado, W.F. (1988). Synthetic amphiphilic peptide models for protein ion channels. *Science* **240**, 1177–81.

Lees, G.J. (1993). Contributary mechanisms in the causation of neurodegenerative disorders. *Neuroscience* **54**, 287–322.

Lei, S.Z., Pan, Z.-H., Aggarwal, S.K., Chen, H.-S.V., Hartman, J., Sucher, N.J. & Lipton, S.A. (1992). Effect of nitric oxide production on the redox modulatory site of the NMDA receptor-channel complex. *Neuron* **8**, 1087–99.

Lerche, H., Heine, R., Pika, U., George, A.L., Mitrovic, N., Browatzki, M., Weiss, T., Rivet-Bastide, M., Franke, C., Lomonaco, M., Ricker, K. & Lehmann-Horn, F. (1993). Human sodium channel myotonia: slowed channel inactivation due to substitutions for a glycine within the III–IV linker. *Journal of Physiology* **470**, 13–22.

Lester, H.A. (1992). The permeation pathway of neurotransmitter-gated ion channels. *Annual Review of Biophysics* **21**, 267–92.

Levis, R.A. & Rae, J.L. (1992). Constructing a patch clamp setup. *Methods in Enzymology* **207**, 14–66.

Levitan, I.B. (1985). Phosphorylation of ion channels. *Journal of Membrane Biology* **87**, 177–90.

Levitan, I.B. (1994). Modulation of ion channels by protein phosphorylation and dephosphorylation. *Annual Review of Physiology* **56**, 193–212.

Levitt, D.G. (1986). Interpretation of biological ion channel flux data – reaction-rate versus continuum theory. *Annual Review of Biophysics* **15**, 29–57.

Lewis, C.A. (1979). Ion-concentration dependence of the reversal potential and the single-channel conductance of ion channels at the frog neuromuscular junction. *Journal of Physiology* **286**, 417–45.

Lewis, C.A. & Stevens, C.F. (1979). Mechanism of ion permeation through channels in a postsynaptic membrane. In *Membrane Transport Processes*, vol. 3 *Ion Permeation through Membrane Channels*, ed. C.F. Stevens & R.W. Tsien, pp. 133–51. New York: Raven Press.

Li, M., Jan, Y.N. & Jan, L.Y. (1992a). Specification of subunit assembly by the hydrophilic amino-terminal domain of the Shaker potassium channel. *Science* **257**, 1225–30.

Li, M., West, J.W., Lai, Y., Scheuer, T. & Catterall, W.A. (1992b). Functional modulation of brain sodium channels by cAMP-dependent phosphorylation. *Neuron* **8**, 1151–9.

Liebovitch, L.S. & Toth, T.I. (1991). A model of ion channel kinetics using deterministic chaotic rather than stochastic processes. *Journal of Theoretical Biology* **148**, 243–67.

Liman, E.R., Tytgat, J. & Hess, P. (1992). Subunit stoichiometry of a mammalian K^+ channel determined by construction of multimeric cDNAs. *Neuron* **9**, 861–71.

Linder, M.E. & Gilman, A.G. (1992). G proteins. *Scientific American* **267**(1), 36–43.

Ling, G. & Gerard, R.W. (1949). The normal membrane potential of frog sartorius fibers. *Journal of Cellular and Comparative Physiology* **34**, 383–96.

Lingle, C.J., Maconochie, D. & Steinbach, J.H. (1992). Activation of skeletal muscle nicotinic acetylcholine receptors. *Journal of Membrane Biology* **126**, 195–217.

Liu, Y. & Dilger, J.P. (1991). Opening rate of acetylcholine receptor channels. *Biophysical Journal* **60**, 424–32.

Lodish, H., Baltimore, D., Berk, A., Zipursky, S.L., Matsudaira, P. & Darnell, J. (1995). *Molecular Cell Biology*, 3rd edition. New York: Scientific American Books.

Loewi, O. (1936). The chemical transmission of nerve action. Nobel lecture. Reprinted (1965) in *Nobel Lectures: Physiology or Medicine 1922–1941*, pp. 416–29. Amsterdam: Elsevier.

Logothetis, D.E., Kurachi, Y., Galper, J., Neer, E.J. & Clapham, D.E. (1987). The $\beta\gamma$ subunits of GTP-binding proteins activate the muscarinic K^+ channel in heart. *Nature* **325**, 321–6.

Lopatin, A.N., Makhina, E.N. & Nichols, C.G. (1994). Potassium channel block by cytoplasmic polyamines as the mechanism of intrinsic rectification. *Nature* **372**, 366–9.

Lopez, G.A., Jan, Y.N. & Jan, L.Y. (1991). Hydrophobic substitution mutations in the S4 sequence alter voltage-dependent gating in *Shaker* K^+ channels. *Neuron* **7**, 327–36.

Lopez-Barneo, J., Hoshi, T., Heinemann, S.H. & Aldrich, R.W. (1993). Effects of external cations and mutations in the pore region on C-type inactivation of *Shaker* potassium channels. *Receptors and Channels* **1**, 61–71.

Lorenz, C., Meyer-Kleine, C., Steinmeyer, K., Koch, M.C. & Jentsch, T.J. (1994). Genomic organization of the human muscle chloride channel ClC-1 and analysis of novel mutations leading to Becker-type myotonia. *Human Molecular Genetics* **3**, 941–6.

Loutrari, H., Tzartos, S.J. & Claudio, T. (1992). Use of *Torpedo*-mouse hybrid acetylcholine receptors reveals immunodominance of the alpha subunit in myasthenia gravis antisera. *European Journal of Immunology* **22**, 2949–56.

Lü, Q. & Miller, C. (1995). Silver as a probe of pore-forming residues in a potassium channel. *Science* **268**, 304–7.

Lu, Z. & MacKinnon, R. (1994). Electrostatic tuning of Mg^{2+} affinity in an inward-rectifier K^+ channel. *Nature* **371**, 243–6.

MacDermott, A.B., Mayer, M.L., Westbrook, M., Smith, S.J. & Barker, J.L. (1986). NMDA-receptor activation increases cytoplasmic calcium concentration in cultured spinal cord neurones. *Nature* **321**, 519–22.

Macdonald, R.L. & Olsen, R.W. (1994). $GABA_A$ receptor channels. *Annual Review of Neuroscience* **17**, 569–602.

Macdonald, R.L., Rogers, C.J. & Twyman, R.E. (1989). Barbiturate regulation of kinetic properties of the $GABA_A$ receptor channel of mouse spinal neurones in culture. *Journal of Physiology* **417**, 483–500.

MacKinnon, R. (1991). Determination of the subunit stoichiometry of a voltage-activated potassium channel. *Nature* **350**, 232–5.

MacKinnon, R. & Miller, C. (1988). Mechanism of charybdotoxin block of the high-conductance, Ca^{2+}-activated K^+ channel. *Journal of General Physiology* **91**, 335–49.

MacKinnon, R. & Yellen, G. (1990). Mutations affecting TEA blockade and ion permeation in voltage-activated K$^+$ channels. *Science* **250**, 276–9.

MacKinnon, R., Heginbotham, L. & Abramson, T. (1990). Mapping the receptor site for charybdotoxin, a pore-blocking potassium channel inhibitor. *Neuron* **5**, 767–71.

MacLennan, D.H. & Chen, S.R.W. (1993). The role of calcium release channel of skeletal muscle sarcoplasmic reticulum in malignant hyperthermia. *Annals of the New York Academy of Sciences* **707**, 294–304.

MacLennan, D.H. & Phillips, M.S. (1992). Malignant hyperthermia. *Science* **256**, 789–94.

MacLennan, D.H., Duff, C., Fujii, J., Phillips, M., Korneluk, R.G., Frodis, W., Britt, B.A. & Worton, R.G. (1990). Ryanodine receptor gene is a candidate for predisposition to malignant hyperthermia. *Nature* **343**, 559–61.

Maconochie, D.J., Fletcher, G.H. & Steinbach, J.H. (1995). The conductance of the muscle nicotinic receptor channel changes rapidly upon gating. *Biophysical Journal* **68**, 483–90.

Magleby, K.L. & Stevens, C.F. (1972). The effect of voltage on the time course of end-plate currents. *Journal of Physiology* **223**, 151–71.

Makowski, L., Caspar, D.L., Phillips, W.C. & Goodenough, D.A. (1977). Gap junction structures. II. Analysis of the X-ray diffraction data. *Journal of Cell Biology* **74**, 629–45.

Manella, C.A. (1992). The 'ins' and 'outs' of mitochondrial membrane channels. *Trends in Biochemical Sciences* **17**, 315–20.

Maricq, A.V., Peterson, A.S., Brake, A.J., Myers, R.M. & Julius, D. (1991). Primary structure and functional expression of the 5HT$_3$ receptor, a serotonin-gated ion channel. *Science* **254**, 432–6.

Massefski, W., Jr, Redfield, A.G., Hare, D.R. & Miller, C. (1990). Molecular structure of charybdotoxin, a pore-directed inhibitor of potassium ion channels. *Science* **249**, 521–4.

Matsuda, H. (1988). Open-state substructure of inwardly rectifying potassium channels revealed by magnesium block in guinea-pig heart cells. *Journal of Physiology* **397**, 237–58.

Matsuda, H., Saigusa, A. & Irisawa, H. (1987). Ohmic conductance through the inwardly rectifying K channel and blocking by internal Mg^{2+}. *Nature* **325**, 156–9.

Mattera, R., Graziano, M.P., Yatani, A., Zhou, Z., Graf, R., Codina, J., Birnbaumer, L., Gilman, A.G. & Brown, A.M. (1989). Splice variants of the α subunit of the G protein G$_s$ activate both adenylyl cyclase and calcium channels. *Science* **243**, 804–7.

McBain, C.J. & Mayer, M.L. (1994). N-methyl-D-aspartic acid receptor structure and function. *Physiological Reviews* **74**, 723–60.

McClatchey, A.I., Van den Bergh, P., Pericak-Vance, M.A., Raskind, W., Verellen, C., McKenna-Yasek, D., Rao, K., Haines, J.L., Bird, T., Brown, R.H. & Gusella, J.F. (1992). Temperature-sensitive mutations in the III–IV cytoplasmic loop region of the skeletal muscle sodium channel gene in paramyotonia congenita. *Cell* **68**, 769–74.

McCleskey, E.W. (1994). Calcium channels: cellular roles and molecular mechanisms. *Current Opinion in Neurobiology* **4**, 304–12.

McCormack, K., Tanouye, M.A., Iverson, L.E., Lin, J.-W., Ramaswami, M., McCormack, T., Campanelli, J.T., Mathew, M.K. & Rudy, B. (1991). A role for hydrophobic residues in the voltage-dependent gating of Shaker K$^+$ channels. *Proceedings of the National Academy of Sciences, USA* **88**, 2931–5.

McCrea, P.D., Engelman, D.M. & Popot, J.-L. (1988). Topography of integral membrane proteins: hydrophobicity analysis vs. immunolocalization. *Trends in Biochemical Sciences* **13**, 289–90.

McManus, O.B. & Magleby, K.L. (1988). Kinetic states and modes of single large-conductance calcium-activated potassium channels in cultured rat skeletal muscle. *Journal of Physiology* **402**, 79–120.

McPhee, J.C., Ragsdale, D.S., Scheuer, T. & Catterall, W.A. (1994). A mutation in segment IVS6 disrupts fast inactivation of sodium channels. *Proceedings of the National Academy of Sciences, USA* **91**, 12346–50.

Meech, R.W. & Standen, N.B. (1975). Potassium activation in *Helix aspersa* neurones under voltage clamp; a component mediated by calcium influx. *Journal of Physiology* **249**, 211–39.

Meech, R.W. & Thomas, R.C. (1980). Effect of measured calcium chloride injections on the membrane potential and internal pH of snail neurones. *Journal of Physiology* **298**, 111–29.

Meves, H. (1994). Modulation of ion channels by arachidonic acid. *Progress in Neurobiology* **43**, 175–86.

Mignery, G.A. & Sudhof, T.C. (1990). The ligand binding site and transduction mechanism in the inositol 1,4,5–trisphosphate receptor. *EMBO Journal* **9**, 3893–8.

Mignery, G.A., Newton, C.L., Archer, B.T. & Sudhof, T.C. (1990). Structure and expression of the rat inositol 1,4,5–trisphosphate receptor. *Journal of Biological Chemistry* **265**, 12679–85.

Mikoshiba, K. (1993). Inositol 1,4,5–trisphosphate receptor. *Trends in Pharmacological Sciences* **14**, 86–9.

Milks, L., Kumar, N.M., Houghten, R., Unwin, N. & Gilula, N.B. (1988). Topology of the 32–kd liver gap junction protein determined by site-directed antibody localizations. *EMBO Journal* **7**, 2967–75.

Miller, C. (1982a). Coupling of water and ion fluxes in a K^+-selective channel of sarcoplasmic reticulum. *Biophysical Journal* **38**, 227–30.

Miller, C. (1982b). *Bis*-quaternary ammonium blockers as structural probes of the sarcoplasmic reticulum K^+ channel. *Journal of General Physiology* **79**, 869–91.

Miller, C. (1983). Integral membrane channels: studies in model membranes. *Physiological Reviews* **63**, 1209–42.

Miller, C. (ed.) (1986). *Ion Channel Reconstitution*. New York: Plenum Press.

Miller, C. (1995). The charybdotoxin family of K^+ channel-blocking peptides. *Neuron* **15**, 5–10.

Miller, C., Bell, J.E. & Garcia, A.M. (1984). The potassium channel of the sarcoplasmic reticulum. *Current Topics in Membranes and Transport* **21**, 99–132.

Millhauser, G.L., Salpeter, E.E. & Oswald, R.E. (1988). Diffusion models of ion-channel gating and the origin of power-law distributions from single-channel recording. *Proceedings of the National Academy of Sciences, USA* **85**, 1502–7.

Mintz, I.M., Venema, V.J., Swiderek, K.M., Lee, T.D., Bean, B.P. & Adams, M.E. (1992). P-type calcium channels blocked by the spider toxin ω-Aga-IVA. *Nature* **355**, 827–29.

Mishina, M., Kurosaki, T., Tobimatsu, T., Morimoto, Y., Noda, M., Yamamoto, T., Tereo, M., Lindstrom, J., Takahashi, T., Kuno, M. & Numa, S. (1984). Expression of functional acetylcholine receptor from cloned cDNAs. *Nature* **307**, 604–8.

Mishina, M., Tobimatsu, T., Imoto, K., Tanaka, K., Fujita, Y., Fukuda, K., Kurasaki, M., Takahashi, H., Morimoto, Y., Hirose, T., Inayama, S., Takahashi, T., Kuno, M. & Numa, S. (1985). Location of functional regions of acetylcholine receptor α-subunit by site-directed mutagenesis. *Nature* **313**, 364–9.

Mishina, M., Takai, T., Imoto, K., Noda, M., Takahashi, T., Numa, S., Methfessel, C. & Sakmann, B. (1986). Molecular distinction between fetal and adult forms of muscle acetylcholine receptor. *Nature* **321**, 406–11.

Mitrovic, N., George, A.L., Heine, R., Wagner, S., Pika, U., Hartlaub, U., Zhou, M., Lerche, H., Fahlke, C. & Lehmann-Horn, F. (1994). K^+-aggravated myotonia: destabilization of the inactivated state of the human muscle Na^+ channel by the V1589M mutation. *Journal of Physiology* **478**, 395–402.

Moczydlowski, E. (1992). Analysis of drug action at single-channel level. *Methods in Enzymology* **207**, 791–806.

Monaghan, D.T., Bridges, R.J. & Cotman, C.W. (1989). The excitatory amino acid receptors. *Annual Review of Pharmacology and Toxicology* **29**, 365–402.

Moncada, S., Palmer, R.M.J. & Higgs, E.A. (1991). Nitric oxide: physiology, pathophysiology, and pharmacology. *Pharmacological Reviews* **43**, 109–42.

Monod, J., Wyman, J. & Changeux, J.-P. (1965). On the nature of allosteric transitions: a plausible model. *Journal of Molecular Biology* **12**, 88–118.

Montal, M. (1990). Molecular anatomy and molecular design of channel proteins. *FASEB Journal* **4**, 2623–35.

Montal, M. & Mueller, P. (1972). Formation of bimolecular membranes from lipid monolayers and a study of their electrical properties. *Proceedings of the National Academy of Sciences, USA* **69**, 3561–6.

Monyer, H., Sprengel, R., Schoepfer, R., Herb, A., Higuchi, M., Lomeli, H., Burnashev, N., Sakmann, B. & Seeburg, P.H. (1992). Heteromeric NMDA receptors: molecular and functional distinction of subtypes. *Science* **256**, 1217–21.

Moorman, J.R., Palmer, C.J., John, J.E., Durieux, M.E. & Jones, L.R. (1992). Phospholemman expression induces a hyperpolarization-activated chloride current in *Xenopus* oocytes. *Journal of Biological Chemistry* **267**, 14551–4.

Mori, Y. (1994). Molecular biology of voltage-dependent calcium channels. In *Handbook of Membrane Channels*, ed. C. Peracchia, pp. 163–75. San Diego, CA: Academic Press.

Morishige, K., Takahashi, N., Jahangir, A., Yamada, M., Koyama, H., Zanelli, J.S. & Kurachi, Y. (1994). Molecular cloning and functional expression of a novel brain-specific inward rectifier potassium channel. *FEBS Letters* **346**, 251–6.

Moriyoshi, K., Masu, M., Ishii, T., Shigemoto, R., Mizuno, N. & Nakanishi, S. (1991). Molecular cloning and characterization of the rat NMDA receptor. *Nature* **354**, 31–7.

Morley, J. (1994). K^+ channel openers and suppression of airway hyperreactivity. *Trends in Pharmacological Sciences* **15**, 463–8.

Morris, C.E. (1990). Mechanosensitive ion channels. *Journal of Membrane Biology* **113**, 93–107.

Myers, R.A., Cruz, L.J., Rivier, J.E. & Olivera, B.M. (1993). *Conus* peptides as chemical probes for receptors and ion channels. *Chemical Reviews* **93**, 1923–36.

Nakai, J., Imagawa, T., Hakamata, Y., Shigekawa, M., Takeshima, H. & Numa, S. (1990). Primary structure and functional expression from cDNA of the cardiac ryanodine receptor/calcium release channel. *FEBS Letters* **271**, 169–77.

Nakamura, T. & Gold, G.H. (1987). A cyclic nucleotide-gated conductance in olfactory receptor cilia. *Nature* **325**, 442–4.

Nakanishi, S. (1992). Molecular diversity of glutamate receptors and implications for brain function. *Science* **258**, 597–603.

Nakanishi, S. & Masu, M. (1994). Molecular diversity and functions of glutamate receptors. *Annual Review of Biophysics* **23**, 319–48.

Nakatani, K. & Yau, K.-W. (1988). Calcium and magnesium fluxes across the plasma membrane of the toad rod outer segment. *Journal of Physiology* **395**, 695–729.

Narahashi, T., Moore, J.W. & Scott, W.R. (1964). Tetrodotoxin blockage of sodium conductance increase in lobster giant axons. *Journal of General Physiology* **47**, 965–74.

Narahashi, T., Hass, H.G. & Therrien, E.F. (1967).

Saxitoxin and tetrodotoxin: comparison of nerve blocking mechanism. *Science* **157**, 1441–2.

Nathan, C. (1992). Nitric oxide as a secretory product of mammalian cells. *FASEB Journal* **6**, 3051–64.

Neer, E.J. (1995). Heterotrimeric G proteins: organizers of transmembrane signals. *Cell* **80**, 249–57.

Neher, E. (1992). Ion channels for communication between and within cells. *Science* **256**, 498–502. Also published in *Les Prix Nobel 1991*, pp. 120–35. Stockholm: Almquist & Wiksell.

Neher, E. & Sakmann, B. (1976). Single-channel currents recorded from membrane of denervated frog muscle fibres. *Nature* **260**, 799–802.

Neher, E. & Stevens, C.F. (1977). Conductance fluctuations and ionic pores in membranes. *Annual Review of Biophysics* **6**, 345–81.

Nernst, W. (1888). Zur Kinetik der in Lösung befindlichen Körper. Theorie der Diffusion. *Zeitschrift für Physikalische Chemie* **2**, 613–37.

Neyton, J. & Trautmann, A. (1985). Single-channel currents of an intercellular junction. *Nature* **317**, 331–5.

Nicholl, D.S.T. (1994). *An Introduction to Genetic Engineering*. Cambridge: Cambridge Unversity Press.

Nichols, C.G., Lederer, W.J. & Cannell, M.B. (1991). ATP dependence of K_{ATP} channel kinetics in isolated membrane patches from rat ventricle. *Biophysical Journal* **60**, 1164–77.

Nikaido, H. (1994). Porins and specific diffusion channels in bacterial outer membranes. *Journal of Biological Chemistry* **269**, 3905–8.

Noda, M., Takahashi, H., Tanabe, T., Toyosato, M., Furutani, Y., Hirose, T., Asai, M., Inayama, S., Miyata, T. & Numa, S. (1982). Primary structure of α-subunit precursor of *Torpedo californica* acetylcholine receptor deduced from cDNA sequence. *Nature* **299**, 793–7.

Noda, M., Takahashi, H., Tanabe, T., Toyosato, M., Kikyotani, S., Hirose, T., Asai, M., Takashima, H., Inayama, S., Miyata, T. & Numa, S. (1983a). Primary structures of β- and δ-subunit precursors of *Torpedo californica* acetylcholine receptor deduced from cDNA sequences. *Nature* **301**, 251–5.

Noda, M., Takahashi, H., Tanabe, T., Toyosato, M., Kikyotani, S., Furutani, Y., Hirose, T., Takashima, H., Inayama, S., Miyata, T. & Numa, S. (1983b). Structural homology of *Torpedo californica* acetylcholine receptor subunits. *Nature* **302**, 528–32.

Noda, M., Furutani, Y., Takahashi, H., Toyosato, M., Tanabe, T., Shimizu, S., Kikyotani, S., Kayano, T., Hirose, T., Inayama, S. & Numa, S. (1983c). Cloning and sequence analysis of calf cDNA and human genomic DNA encoding α-subunit precursor of muscle acetylcholine receptor. *Nature* **305**, 818–23.

Noda, M., Shimizu, S., Tanabe, T., Takai, T., Kayano, T., Ikeda, T., Takahashi, H., Nakayama, H., Kanaoka, Y., Minamino, N., Kangawa, K., Matsuo, H., Raftery, M.A., Hirose, T., Inayama, S., Hayashida, H., Miyati, T. & Numa, S. (1984). Primary structure of *Electrophorus electricus* sodium channel deduced from cDNA sequence. *Nature* **312**, 121–7.

Noda, M., Ikeda, T., Kayano, T., Suzuki, H., Takeshima, H., Kurasaki, M., Takahashi, H. & Numa, S. (1986). Existence of distinct sodium channel messenger RNAs in rat brain. *Nature* **320**, 188–92.

Noda, M., Suzuki, H., Numa, S. & Stühmer, W. (1989). A single point mutation confers tetrodotoxin and saxitoxin insensitivity on the sodium channel II. *FEBS Letters* **259**, 213–16.

Noma, A. (1983). ATP-regulated K^+ channels in cardiac muscle. *Nature* **305**, 147–8.

Norton, R.S. (1991). Structure and structure-function relationships of sea anemone proteins that interact with the sodium channel. *Toxicon* **29**, 1051–84.

Nowak, L., Bregestovski, P., Ascher, P., Herbet, A. & Prochiantz, A. (1984). Magnesium gates glutamate-activated channels in mouse central neurones. *Nature* **307**, 462–5.

Numa, S. (1986). Evolution of ionic channels. *Chemica Scripta* **26B**, 173–8.

Numann, R., Catterall, W.A. & Scheuer, T. (1991). Functional modulation of brain sodium channels by protein kinase C phosphorylation. *Science* **254**, 115–18.

Oblatt-Montal, M., Buhler, L.K., Iwamoto, T., Tomich, J.M. & Montal, M. (1993a). Synthetic peptides and four-helix bundle proteins as model systems for the pore-forming structure of channel proteins. I. Transmembrane segment M2 of the nicotinic acetylcholine receptor channel is a key pore-lining structure. *Journal of Biological Chemistry* **268**, 14601–7.

Oblatt-Montal, M., Iwamoto, T., Tomich, J.M. & Montal, M. (1993b). Design, synthesis and functional characterization of a pentameric channel protein that mimics the presumed pore structure of the nicotinic cholinergic receptor. *FEBS Letters* **320**, 261–6.

Oblatt-Montal, M., Reddy, G.L., Iwamoto, T., Tomich, J.M. & Montal, M. (1994). Identification of an ion channel-forming motif in the primary structure of CFTR, the cystic fibrosis chloride channel. *Proceedings of the National Academy of Sciences, USA* **91**, 1495–9.

O'Connell, A.M., Koeppe, R.E. & Andersen, O.S. (1990). Kinetics of gramicidin channel formation in lipid bilayers: transmembrane monomer association. *Science* **250**, 1256–9.

Ogden, D. (ed.) (1994a). *Microelectrode Techniques: The Plymouth Workshop Handbook*, 2nd edition. Cambridge: Company of Biologists.

Ogden, D. (1994b). Microelectrode electronics. In *Microelectrode Techniques: The Plymouth Workshop Handbook*, 2nd edition, ed. D. Ogden, pp. 407–37. Cambridge: Company of Biologists.

Ogden, D. & Stanfield, P.R. (1994). Patch clamp techniques for single channel and whole-cell recording. In *Microelectrode Techniques: The Plymouth Workshop Handbook*, 2nd edition, ed. D. Ogden, pp. 53–78. Cambridge: Company of Biologists.

Ogura, Y. (1971). Fugu (puffer-fish) poisoning and the pharmacology of crystalline tetrodotoxin in poisoning. In *Neuropoisons: Their Pathophysiological Actions* vol. 1 *Poisons of Animal Origin*, ed. L.L. Simpson, pp. 139–58. New York: Plenum Press.

Oiki, S., Danho, W. & Montal, M. (1988a). Channel protein engineering: synthetic 22–mer peptide from the primary structure of the voltage-sensitive sodium channel forms ionic channels in lipid bilayers. *Proceedings of the National Academy of Sciences, USA* **85**, 2393–7.

Oiki, S., Danho, W., Madison, V. & Montal, M. (1988b). M2δ, a candidate for the structure lining the ionic channel of the nicotinic cholinergic receptor. *Proceedings of the National Academy of Sciences, USA* **85**, 8703–7.

Old, R.W. & Primrose, S.B. (1994). *Principles of Gene Manipulation*, 5th edition. Oxford: Blackwell Scientific Publications.

Olivera, B.M., Rivier, J., Clark, C., Ramilo, C., Corpuz,

G.P., Abogadie, F.C., Mena, E.E., Woodward, S.R., Hillyard, D.R. & Cruz, L.J. (1990). Diversity of *Conus* neuropeptides. *Science* **249**, 257–63.

Olivera, B.M., Rivier, J., Scott, J.K., Hillyard, D.R. & Cruz, L.J. (1991). Conotoxins. *Journal of Biological Chemistry* **266**, 22067–70.

Olivera, B.M., Miljanich, G.P., Ramachandran, J. & Adams, M.E. (1994). Calcium channel diversity and neurotransmitter release: the ω-conotoxins and ω-agatoxins. *Annual Review of Biochemistry* **63**, 823–67.

Ordway, R.W., Walsh, J.V. & Singer, J.J. (1989). Arachidonic acid and other fatty acids directly activate potassium channels in smooth muscle cells. *Science* **244**, 1176–9.

Ordway, R.W., Singer, J.J. & Walsh, J.V. (1991). Direct regulation of ion channels by fatty acids. *Trends in Neurosciences* **14**, 96–100.

O'Shea, E.K., Klemm, J.D., Kim, P.S. & Alber, T. (1991). X-ray structure of the GCN4 leucine zipper, a two-stranded, parallel coiled-coil. *Science* **254**, 539–44.

Oswald, R.E., Millhauser, G.L. & Carter, A.A. (1991). Diffusion model in ion channel gating: extension to agonist-activated ion channels. *Biophysical Journal* **59**, 1136–42.

Otsu, K., Nishida, K., Kimura, Y., Kuzuya, T., Hori, M., Kamada, T. & Tada, M. (1994). The point mutation Arg615→Cys in the Ca^{2+} release channel of the skeletal sarcoplasmic reticulum is responsible for hypersensitivity to caffeine and halothane in malignant hyperthermia. *Journal of Biological Chemistry* **269**, 9413–15.

Pallotta, B.S. (1991). Single ion channel's view of classical receptor theory. *FASEB Journal* **5**, 2035–43.

Papazian, D.M., Schwarz, T.L., Tempel, B.L., Jan, Y.N & Jan, L.Y. (1987). Cloning of genomic and complementary DNA from *Shaker*, a putative potassium channel gene from *Drosophila*. *Science* **237**, 749–53.

Papazian, D.M., Timpe, L.C., Jan, Y.N. & Jan, L.Y. (1991). Alteration of voltage-dependence of *Shaker* potassium channel by mutations in the S4 sequence. *Nature* **349**, 305–10.

Papazian, D.M., Shao, X.M., Seoh, S.-A., Mock, A.F., Huang, Y. & Wainstock, D.H. (1995). Electrostatic interactions of S4 voltage sensor in Shaker K^+ channel. *Neuron* **14**, 1293–301.

Park, C.S. & Miller, C. (1992a). Interaction of charybdotoxin with permeant ions inside the pore of a K^+ channel. *Neuron* **9**, 307–13.

Park, C.S. & Miller, C. (1992b). Mapping function to structure in a channel-blocking peptide: electrostatic mutants of charybdotoxin. *Biochemistry* **31**, 7749–55.

Parker, M.W., Pattus, F., Tucker, A.D. & Tseroglou, D. (1989). Structure of the membrane-pore-forming fragment of colicin A. *Nature* **337**, 93–6.

Parker, M.W., Buckley, J.T., Postma, J.P.M., Pattus, F. & Tsernoglou, D. (1994). Structure of the *Aeromonas* toxin proaerolysin in its water-soluble and membrane-channel states. *Nature* **367**, 292–5.

Partenskii, M.B. & Jordan, P.C. (1992). Theoretical perspectives on ion-channel electrostatics. *Quarterly Review of Biophysics* **25**, 477–510.

Pascual, J.M., Shieh, C.-C., Kirsch, G.E. & Brown, A.M. (1995). K^+ pore structure revealed by reporter cysteines at inner and outer surfaces. *Neuron* **14**, 1055–63.

Patlak, J. (1991). Molecular kinetics of voltage-dependent Na^+ channels. *Physiological Reviews* **71**, 1047–80.

Patlak, J., Rovner, A., Lieberman, M. & Hirschberg, B. (1995). A movement of 10–11 electronic charges is suggested by the voltage dependence of the steady-state open probability in non-inactivating sodium channels. *Biophysical Journal* **68**, A156.

Patterson, C. (1988). Homology in classical and molecular biology. *Molecular Biology and Evolution* **5**, 603–25.

Patton, D.E., West, J.W., Catterall, W.A. & Goldin, A.L. (1993). A peptide segment critical for sodium channel inactivation functions as an inactivation gate in a potassium channel. *Neuron* **11**, 967–74.

Pattus, F., Lakey, J. & Dargent, B. (1988). Bacterial proteins which form membrane pores: a biophysical approach. *Microbiological Science* **5**, 207–10.

Paul, D.L. (1986). Molecular cloning of cDNA for rat liver gap junction protein. *Journal of Cell Biology* **103**, 123–34.

Pauling, L. (1927). The sizes of ions and the structure of ionic crystals. *Journal of the American Chemical Society* **49**, 769–90.

Pauling, L. (1960). *The Nature of the Chemical Bond*, 3rd edition. Ithaca, NY: Cornell University Press.

Paulmichl, M., Li, Y., Wickman, K., Ackermann, M.,

Peralta, E. & Clapham, D.E. (1992). New mammalian chloride channel identified by expression cloning. *Nature* **365**, 238–41.

Pellegrino, R.G. & Ritchie, J.M. (1984). Sodium channels in the axolemma of normal and degenerating rabbit optic nerve. *Proceedings of the Royal Society of London* B **222**, 155–60.

Peng, S., Blachly-Dyson, E., Forte, M. & Colombini, M. (1992). Large scale rearrangement of protein domains is associated with voltage gating of the VDAC channel. *Biophysical Journal* **62**, 123–35.

Peracchia, C., Lazrak, A. & Peracchia, L.L. (1994). Molecular models of channel interaction and gating in gap junctions. In *Handbook of Membrane Channels*, ed. C. Peracchia, pp. 361–77. San Diego, CA: Academic Press.

Perutz, M.F. (1989). Mechanisms of cooperativity and allosteric regulation in proteins. *Quarterly Review of Biophysics* **22**, 139–236.

Petrou, S., Ordway, R.W., Hamilton, J.A., Walsh, J.V. & Singer, J.J. (1994). Structural requirements for charged lipid molecules to directly increase or suppress K^+ channel activity in smooth muscle cells. *Journal of General Physiology* **103**, 471–86.

Pinto, L.H., Holsinger, L.J. & Lamb, R.A. (1992). Influenza virus M_2 protein has ion channel activity. *Cell* **69**, 517–28.

Polymeropoulos, E.E. & Brickmann, J. (1985). Molecular dynamics of ion transport through transmembrane model channels. *Annual Review of Biophysics* **14**, 315–30.

Pongs, O. (1992). Molecular biology of voltage-dependent potassium channels. *Physiological Reviews* **72**, S69–S88.

Popot, J.-L. (1993). Integral membrane protein structure: transmembrane α-helices as autonomous folding domains. *Current Biology* **3**, 532–40.

Pottosin, I.I. (1992). Probing of pore in the *Chara gymnophylla* K^+ channel by blocking cations and by streaming potential measurements. *FEBS Letters* **298**, 253–6.

Poxleitner, M., Seitz-Beywl, J. & Heinzinger, K. (1993). Ion transport through gramicidin A. Water structure and functionality. *Zeitschrift für Naturforschung* **48c**, 654–65.

Preston, G.M., Carroll, T.P., Guggino, W.B. & Agre, P. (1992). Appearance of water channels in *Xenopus* oocytes expressing red cell CHIP28 protein. *Science* **256**, 385–7.

Ptáček, L.J., Tawil, R., Criggs, R.C., Engel, A.G., Layzer, R.B., Kwiecinski, H., McManis, P.G., Santiago, L., Moore, M., Fouad, G., Bradley, P. & Leppert, M.F. (1994). Dihydropyridine receptor mutations cause hypokalemic periodic paralysis. *Cell* **77**, 863–8.

Pusch, M. & Jentsch, T.J. (1994). Molecular physiology of voltage-gated chloride channels. *Physiological Reviews* **74**, 813–27.

Pusch, M., Ludewig, U., Rehfeldt, A. & Jentsch, T.J. (1995). Gating of the voltage-dependent chloride channel ClC-0 by the permeant anion. *Nature* **373**, 527–30.

Qin, D., Takano, M. & Noma, A. (1989). Kinetics of ATP-sensitive K^+ channel revealed with oil-gate concentration jump method. *American Journal of Physiology* **257**, H1624–H1633.

Quinton, P.M. (1983). Chloride impermeability in cystic fibrosis. *Nature* **301**, 421–2.

Quinton, P.M. (1990). Cystic fibrosis: a disease of electrolyte transport. *FASEB Journal* **4**, 2709–17.

Raftery, M.A., Hunkapiller, M.W., Strader, C.D. & Hood, L.E. (1980). Acetylcholine receptor: complex of homologous subunits. *Science* **208**, 1454–7.

Rajendra, S., Lynch, J.W., Pierce, K.D., French, C.R., Barry, P.H. & Schofield, P.R. (1994). Startle disease mutations reduce the agonist sensitivity of the human inhibitory glycine receptor. *Journal of Biological Chemistry* **269**, 18739–42.

Ramaswami, M. & Tanouye, M.A. (1989). Two sodium-channel genes in *Drosophila*: implications for channel diversity. *Proceedings of the National Academy of Sciences, USA* **86**, 2029–82.

Rauch, G., Gambale, F. & Montal, M. (1990). Tetanus toxin channel in phosphatidylserine planar bilayers: conductance states and pH dependence. *European Biophysical Journal* **18**, 79–83.

Rayner, M.D., Starkus, J.G., Ruben, P.C. & Alicata, D.A. (1992). Voltage-sensitive and solvent-sensitive processes in ion channel gating. *Biophysical Journal* **61**, 96–108.

Rehm, H. (1991). Molecular aspects of neuronal

voltage-dependent K$^+$ channels. *European Journal of Biochemistry* **202**, 701–13.

Reizer, J., Reizer, A. & Saier, M.H. (1993). The MIP family of integral membrane channel proteins: sequence comparisons, evolutionary relationships, reconstructed pathway of evolution, and proposed functional differentiation of the two repeated halves of the proteins. *Critical Reviews in Biochemistry and Molecular Biology* **28**, 235–57.

Rettig, J., Heinemann, S.H., Wunder, F., Lorra, C., Parcej, D.N., Dolly, J.O. & Pongs, O. (1994). Inactivation properties of voltage-gated K$^+$ channels altered by presence of β-subunit. *Nature* **369**, 289–94.

Reuveny, E., Slesinger, P.A., Inglese, J., Morales, J.M., Iñiguez-Lluhi, J.A., Lefkowitz, R.J., Bourne, H.R., Jan, Y.N. & Jan, L.Y. (1994). Activation of the cloned muscarinic potassium channel by G protein βγ subunits. *Nature* **370**, 143–6.

Revah, F., Galzi, J.-L., Giraudat, J., Haumont, P.-Y., Lederer, F. & Changeux, J.-P. (1990). The noncompetitive blocker [^{3}H]chlorpromazine labels three amino acids of the acetylcholine receptor gamma subunit: implications for the α-helical organization of regions MII and for the structure of the ion channel. *Proceedings of the National Academy of Sciences, USA* **87**, 4675–9.

Revel, J.-P. & Karnovsky, M.J. (1967). Hexagonal array of subunits in intercellular junctions of the mouse heart and liver. *Journal of Cell Biology* **33**, C7–C12.

Rich, D.P., Anderson, M.P., Gregory, R.J., Cheng, S.H., Paul, S., Jefferson, D.M., McCann, J.D., Klinger, K.W., Smith, A.E. & Welsh, M.J. (1990). Expression of cystic fibrosis transmembrane conductance regulator corrects defective chloride channel regulation in cystic fibrosis airway epithelial cells. *Nature* **347**, 358–63.

Riordan, J.R., Rommens, J.M., Kerem, B.-S., Alon, N., Rozmahel, R., Grzelczak, Z., Zielenski, J., Lok, S., Plavsik, N., Chou, J.-L., Drumm, M.L., Ianuzzi, M.C., Collins, F.S. & Tsui, L.-C. (1989). Identification of the cystic fibrosis gene: cloning and characterization of complementary DNA. *Science* **245**, 1066–73.

Ritchie, J.M., Rogart, R.B. & Strichartz, G.R. (1976). A new method for labelling saxitoxin and its binding to non-myelinated fibres of rabbit vagus, lobster walking leg, and garfish olfactory nerve. *Journal of Physiology* **261**, 477–94.

Robertson, D.W. & Steinberg, M.I. (1990). Potassium channel modulators: scientific applications and therapeutic promise. *Journal of Medicinal Chemistry* **33**, 1529–41.

Rogart, R.B., Cribbs, L.L., Muglia, L.K., Kephart, D.D. & Kaiser, M.W. (1989). Molecular cloning of a putative tetrodotoxin-resistant rat heart sodium channel isoform. *Proceedings of the National Academy of Sciences, USA* **86**, 8170–4.

Rogawski, M.A. (1993). Therapeutic potential of excitatory amino acid antagonists: channel blockers and 2,3-benzodiazepines. *Trends in Pharmacological Sciences* **14**, 325–31.

Rogers, C.J., Twyman, R.E. & Macdonald, R.L. (1994). Benzodiazepine and β-carboline regulation of single GABA$_A$ receptor channels of mouse spinal neurones in culture. *Journal of Physiology* **475**, 69–82.

Röhrkasten, A., Meyer, H.E., Nastainczyk, W., Sieber, M. & Hofmann, F. (1988). cAMP-dependent protein kinase rapidly phosphorylates serine-687 of the skeletal muscle receptor for calcium channel blockers. *Journal of Biological Chemistry* **263**, 15325–9.

Role, L.W. (1992). Diversity in primary structure and function of neuronal nicotinic acetylcholine receptor channels. *Current Opinion in Neurobiology* **2**, 254–62.

Rommens, J.M., Ianuzzi, M.C., Kerem, B.-S., Drumm, M.L., Melmer, G., Dean, M., Rozmahel, R., Cole, J.L., Kennedy, D., Hidaka, N., Zsiga, M., Buchwald, M., Riordan, J.R., Tsui, L.-C. & Collins, F.S. (1989). Identification of the cystic fibrosis gene: chromosome walking and jumping. *Science* **245**, 1059–65.

Rosenberg, P.A. & Finkelstein, A. (1978). Interaction of ions and water in gramicidin A channels. *Journal of General Physiology* **72**, 327–40.

Rossie, S. & Catterall, W.A. (1989). Phosphorylation of the α subunit of rat brain sodium channels by cAMP-dependent protein kinase at a new site containing Ser^{686} and Ser^{687}. *Journal of Biological Chemistry* **264**, 14220–4.

Rossie, S., Gordon, D. & Catterall, W.A. (1987). Identification of an intracellular domain of the sodium channel having multiple cAMP-dependent phosphorylation sites. *Journal of Biological Chemistry* **262**, 17530–5.

Roux, B. & Karplus, M. (1994). Molecular dynamics simulations of the gramicidin channel. *Annual Review of Biophysics* **23**, 731–61.

Rudy, B. & Iverson, L.E. (eds.) (1992). *Ion Channels (Methods in Enzymology* **207**). San Diego: Academic Press.

Ruppersberg, J.P., Schröter, K.H., Sakmann, B., Stocker, M., Sewing, S. & Pongs, O. (1990). Heteromultimeric channels formed by rat brain potassium-channel proteins. *Nature* **345**, 535–7.

Ruppersberg, J.P., Frank, R., Pongs, O. & Stocker, M. (1991). Cloned neuronal $I_K(A)$ channels reopen during recovery from inactivation. *Nature* **353**, 657–60.

Sackin, H. (1995). Mechanosensitive channels. *Annual Review of Physiology* **57**, 333–53.

Sakmann, B. (1992). Elementary steps in synaptic transmission revealed by currents through single ion channels. *Science* **256**, 503–12. Also published in *Les Prix Nobel 1991*, pp. 141–69. Stockholm: Almquist & Wiksell.

Sakmann, B. & Neher, E. (eds.) (1983). *Single-Channel Recording*. New York: Plenum Press.

Sakmann, B. & Neher, E. (eds.) (1995). *Single-Channel Recording*, 2nd edition. New York: Plenum Press.

Sakmann, B., Methfessel, C., Mishina, M., Takahashi, T., Takai, T., Kurasaki, M., Fukuda, K. & Numa, S. (1985). Role of acetylcholine receptor subunits in gating of the channel. *Nature* **318**, 538–43.

Sakmann, B., Edwards, F.A., Konnerth, A. & Takahashi, T. (1989). Patch clamp techniques used for studying synaptic transmission in slices of mammalian brain. *Quarterly Journal of Experimental Physiology* **74**, 1107–18.

Salkoff, L. (1983). Genetic and voltage-clamp analysis of a *Drosophila* potassium channel. *Cold Spring Harbor Symposium on Quantitative Biology* **48**, 221–31.

Salkoff, L., Baker, K., Butler, A., Covarrubias, M., Pak, M.D. & Wei, A. (1992). An essential 'set' of K^+ channels conserved in flies, mice and humans. *Trends in Neurosciences* **15**, 161–6.

Sanguinetti, M.C., Jiang, C., Curran, M.E. & Keating, M.T. (1995). A mechanistic link between an inherited and an acquired cardiac arrhythmia: *HERG* encodes the I_{Kr} potassium channel. *Cell* **81**, 299–307.

Sansom, M.S.P. (1991). The biophysics of peptide models of ion channels. *Progress in Biophysics and Molecular Biology* **55**, 139–235.

Sansom, M.S.P. (1993). Alamethicin and related peptabiols – model ion channels. *European Biophysical Journal* **22**, 105–24.

Sansom, M.S.P. & Kerr, I.D. (1993). Influenza virus M2 protein: a molecular modelling study of the ion channel. *Protein Engineering* **6**, 65–74.

Sansom, M.S.P., Ball, F.G., Kerry, C.J., McGee, R., Ramsey, R.L. & Usherwood, P.N.R. (1989). Markov, fractal, diffusion, and related models of ion channel gating: a comparison with experimental data from two ion channels. *Biophysical Journal* **56**, 1229–43.

Santis, G. (1995). Basic molecular genetics. In *Cystic Fibrosis*, ed. M.E. Hodson & D.M. Geddes, pp. 15–39. London: Chapman & Hall Medical.

Sargent, P.B. (1993). The diversity of neuronal nicotinic acetylcholine receptors. *Annual Review of Neuroscience* **16**, 403–43.

Sather, W.A., Yang, J. & Tsien, R.W. (1994). Structural basis of ion channel permeation and selectivity. *Current Opinion in Neurobiology* **4**, 313–23.

Satin, J., Kyle, J.W., Chen, M., Bell, P., Cribbs, L.L., Fozzard, H.A. & Rogart, R.B. (1992). A mutant of TTX-resistant cardiac sodium channels with TTX-sensitive properties. *Science* **256**, 1202–5.

Sattelle, D.B., Harrow, I.D., David, J.A., Pelhate, M., Callec, J.J., Gepner, J.I. & Hall, L.M. (1985). Nereistoxin: actions on a CNS acetylcholine receptor/ion channel in the cockroach *Periplaneta americana*. *Journal of Experimental Biology* **118**, 37–52.

Schindler, H. (1980). Formation of planar bilayers from artificial or native membrane vesicles. *FEBS Letters* **122**, 77–9.

Schindler, H. & Quast, U. (1980). Functional acetylcholine receptor from *Torpedo marmorata* in planar membranes. *Proceedings of the National Academy of Sciences, USA* **77**, 3052–6.

Schofield, P.R., Darlison, M.G., Fujita, N., Burt, D. Stephenson, F.A., Rodriguez.H., Rhee, L.M., Ramachandran, J., Reale, V., Glencorse, T.A., Seeburg, P.H. & Barnard, E.A. (1987). Sequence and functional expression of the $GABA_A$ receptor shows a ligand-gated receptor super-family. *Nature* **328**, 221–7.

Schoppa, N.E., McCormack, K., Tanouye, M.A. & Sigworth, F.J. (1992). The size of the gating charge in wild-type and mutant *Shaker* potassium channels. *Science* **255**, 1712–15.

Schroeder, J.I., Ward, J.M. & Gassmann, W. (1994). Perspectives on the physiology and structure of inward-rectifying K$^+$ channels in higher plants. *Annual Review of Biophysics* **23**, 441–71.

Schroeder, S.A., Gaughan, D.M. & Swift, M. (1995). Protection against bronchial asthma by the *CFTR* ΔF508 mutation: a heterozygote advantage in cystic fibrosis. *Nature Medicine* **1**, 703–5.

Schubert, B., VanDongen, A.M.J., Kirsch, G.E. & Brown, A.M. (1989). β-adrenergic inhibition of cardiac sodium channels by dual G-protein pathways. *Science* **245**, 516–19.

Schulz, G.E. (1993). Bacterial porins: structure and function. *Current Opinion in Cell Biology* **5**, 701–7.

Schuster, C.M., Ultsch, A., Schloss, P., Cox, J.A., Schmitt, B. & Betz, H. (1991). Molecular cloning of an invertebrate glutamate receptor subunit expressed in *Drosophila* muscle. *Science* **254**, 112–14.

Schwarz, W., Palade, P.T. & Hille, B. (1977). Local anesthetics: effect of pH on use-dependent block of sodium channels in frog muscle. *Biophysical Journal* **20**, 343–68.

Schwarzmann, G., Wiegand, H., Rose, B., Zimmerman, D., Ben-Haim, D. & Loewenstein, W.R. (1981). Diameter of cell-to-cell junctional membrane channels as probed with neutral molecules. *Science* **213**, 551–3.

Scriver, C.R., Beaudet, A.L., Sly, W.S. & Valle, D. (eds.) (1995). *The Metabolic and Molecular Bases of Inherited Disease*. New York: McGraw-Hill.

Seeburg, P.H. (1993). The molecular biology of mammalian glutamate receptor channels. *Trends in Neurosciences* **16**, 359–65.

Séguéla, P., Wadiche, J., Dineley-Miller, K., Dani, J.A. & Patrick, J.W. (1993). Molecular cloning, functional properties, and distribution of rat brain α$_7$: a nicotinic cation channel highly permeable to calcium. *Journal of Neuroscience* **13**, 596–604.

Sentenac, H., Bonneaud, N., Minet, M., Lacroute, F., Salmon, J.-M., Gaymard, F. & Grignon, C. (1992). Cloning and expression in yeast of a plant potassium ion transport system. *Science* **256**, 663–5.

Shannon, R.D. (1976). Revised effective radii and systematic studies of interatomic distances in halides and chalcogenides. *Acta Crystallographica* **A32**, 751–67.

Shen, N.V. & Pfaffinger, P.J. (1995). Molecular recognition and assembly sequences involved in the subfamily-specific assembly of voltage-gated K$^+$ channel subunit proteins. *Neuron* **14**, 625–33.

Shen, N.V., Chen, X., Boyer, M.M. & Pfaffinger, P.J. (1993). Deletion analysis of K$^+$ channel assembly. *Neuron* **11**, 67–76.

Sheng, M., Liao, Y.J., Jan, Y.N. & Jan, L.Y. (1993). Presynaptic A-current based on heteromultimeric K$^+$ channels detected *in vivo*. *Nature* **365**, 72–5.

Shiang, R., Ryan, S.G., Zhu, Y.-Z., Hahn, A.F., O'Connell, P. & Wasmuth, J.J. (1993). Mutations in the α$_1$ subunit of the inhibitory glycine receptor cause the dominant neurologic disorder, hyperekplexia. *Nature Genetics* **5**, 351–8.

Shibahara, S., Kubo, T., Perski, H.J., Takahashi, H., Noda, M. & Numa, S. (1985). Cloning and sequence analysis of human DNA encoding gamma subunit precursor of muscle acetylcholine receptor. *European Journal of Biochemistry* **146**, 15–22.

Sigworth, F.J. (1980a). The variance of sodium current fluctuations at the node of Ranvier. *Journal of Physiology* **307**, 97–129.

Sigworth, F.J. (1980b). The conductance of sodium channels under conditions of reduced current at the node of Ranvier. *Journal of Physiology* **307**, 131–42.

Sigworth, F.J. (1994). Voltage gating of ion channels. *Quarterly Review of Biophysics* **27**, 1–40.

Sigworth, F.J. (1995). Electronic design of the patch clamp. In *Single-Channel Recording*, 2nd edition, ed. B. Sakmann & E. Neher, pp. 95–127. New York: Plenum Press.

Sigworth, F.J. & Sine, S.M. (1987). Data transformations for improved display and fitting of single-channel dwell time histograms. *Biophysical Journal* **52**, 1047–54.

Silverman, J.A., Mindell, J.A., Zhan, H., Finkelstein, A. & Collier, R.J. (1994). Structure-function relationships in diphtheria toxin channels: I. Determining a minimal channel-forming domain. *Journal of Membrane Biology* **137**, 17–28.

Simon, M.I., Strathmann, M.P. & Gautam, N. (1991).

Diversity of G proteins in signal transduction. *Science* **252**, 802–8.

Sine, S.M. & Steinbach, J.H. (1984). Agonists block currents through acetylcholine receptor channels. *Biophysical Journal* **46**, 277–84.

Sine, S.M., Claudio, T. & Sigworth, F.J. (1990). Activation of *Torpedo* acetylcholine receptors expressed in mouse fibroblasts. *Journal of General Physiology* **96**, 395–437.

Sipos, I., Jurkat-Rott, K., Harasztosi, C., Fontaine, B., Kovacs, L., Melzer, W. & Lemann-Horn, F. (1995). Skeletal muscle DHP receptor mutations alter calcium currents in human hypokalaemic periodic paralysis myotubes. *Journal of Physiology* **483**, 299–306.

Sixma, T.K., Pronk, S.E., Kalk, K.H., Wartna, E.S., van Zenten, B.A.M., Witholt, B. & Hol, W.G.J. (1991). Crystal structure of a cholera toxin-related heat-labile enterotoxin for *E. coli. Nature* **351**, 371–7.

Skinner, F.K., Ward, C.A. & Bardakjian, B.L. (1993). Permeation in channels: a statistical rate theory approach. *Biophysical Journal* **65**, 618–29.

Skou, J.C. (1989). Sodium–potassium pump. In *Membrane Transport: People and Ideas*, ed. D.C. Tosteson, pp. 155–85. Bethesda, MD: American Physiological Society.

Slatin, S.L., Qiu, X.-Q., Jakes, K.S. & Finkelstein, A. (1994). Identification of a translocated protein segment in a voltage-dependent channel. *Nature* **371**, 158–61.

Slesinger, P.A., Jan, Y.N. & Jan, L.Y. (1993). The S4–S5 loop contributes to the ion-selective pore of potassium channels. *Neuron* **11**, 739–49.

Smith, C.M. & Reynard, A.M. (1992). *Textbook of Pharmacology*. Philadelphia, PA: W.B. Saunders Co.

Smith, M. (1994). Synthetic DNA and biology. Nobel lecture. *Angewandte Chemie International Edition* **33**, 1214–21.

Snutch, T.P. & Reiner, P.B. (1992). Ca^{2+} channels: diversity of form and function. *Current Opinion in Neurobiology* **2**, 247–53.

Snutch, T.P., Tomlinson, W.J., Leonard, J.P. & Gilbert, M.M. (1991). Distinct calcium channels are generated by alternative splicing and are differentially expressed in the mammalian CNS. *Neuron* **7**, 45–57.

Sommer, B. & Seeburg, P.H. (1992). Glutamate receptor channels: novel properties and new clones. *Trends in Pharmacological Sciences* **13**, 291–6.

Sommer, B., Keinänen, K., Verdoorn, T.A., Wisden, W., Burnashev, N., Herb, A., Köhler, M., Takagi, T., Sakmann, B. & Seeburg, P.H. (1990). Flip and flop: a cell-specific functional switch in glutamate-operated channels of the CNS. *Science* **249**, 1580–5.

Sommer, B., Köhler, M., Sprengel, R. & Seeburg, P.H. (1991). RNA editing in brain controls a determinant of ion flow in glutamate-gated channels. *Cell* **67**, 11–19.

Soong, T.W., Stea, A., Hodson, C.D., Dubel., S.J., Vincent, S.R. & Snutch, T.P. (1993). Structure and functional expression of a member of the low voltage-activated calcium channel family. *Science* **260**, 1133–6.

Sorrentino, V. & Volpe, P. (1993). Ryanodine receptors: how many, where and why? *Trends in Pharmacological Sciences* **14**, 98–103.

Spalding, B.C., Senyk, O., Swift, J.G. & Horowicz, P. (1981). Unidirectional flux ratio for potassium ions in depolarized frog skeletal muscle. *American Journal of Physiology* **241**, C68–C75.

Spedding, M & Paoletti, R. (1992). Classification of calcium channels and the sites of action of drugs modifying channel function. *Pharmacological Reviews* **44**, 363–76.

Sprang, S.R., Acharya, K.R., Goldsmith, E.J., Stuart, D.I., Varvill, K., Fletterick, R.J., Madsen, N.B. & Johnson, L.N. (1988). Structural changes in glycogen phosphorylase induced by phosphorylation. *Nature* **336**, 215–21.

Spray, D.C. (1990). Electrophysiological properties of gap junction channels. In *Parallels in Cell to Cell Junctions in Plants and Animals*, ed. A.W. Robards, W.J. Lucas, J.D. Pitts, H.J. Jongsma & D.C. Spray, pp. 63–85. Berlin: Springer-Verlag.

Spruce, A.E., Standen, N.B. & Stanfield, P.R. (1987a). Studies of the unitary properties of adenosine-5'-triphosphate-regulated potassium channels of frog skeletal muscle. *Journal of Physiology* **382**, 213–36.

Spruce, A.E., Standen, N.B. & Stanfield, P.R. (1987b). The action of external tetraethylammonium ions on unitary delayed rectifier potassium channels of frog skeletal muscle. *Journal of Physiology* **393**, 467–478.

Standen, N.B. (1992). Potassium channels, metabolism and muscle. *Experimental Physiology* **77**, 1–25.

Standen, N.B. & Stanfield, P.R. (1992). Patch clamp methods for single channel and whole cell recording. In *Monitoring Neuronal Activity: A Practical Approach*, ed. J.A. Stamford, pp. 59–83. Oxford: IRL Press.

Stanfield, P.R. (1983). Tetraethylammonium ions and the potassium permeability of excitable cells. *Reviews of Physiology, Biochemistry and Pharmacology* **97**, 1–67.

Stanfield, P.R. (1986). Voltage-dependent calcium channels of excitable membranes. *British Medical Bulletin* **42**, 359–67.

Stanfield, P.R. (1988). Intracellular Mg^{2+} may act as a co-factor in ion channel function. *Trends in Neurosciences* **11**, 475–7.

Stanfield, P.R., Nakajima, Y. & Yamaguchi, K. (1985). Substance P raises neuronal membrane excitability by reducing inward rectification. *Nature* **315**, 498–501.

Stanfield, P.R., Davies, N.W., Shelton, P.A., Khan, I.A., Brammar, W.J., Standen, N.B. & Conley, E.C. (1994a). The intrinsic gating of inward rectifier K^+ channels expressed from the murine IRK1 gene depends on voltage, K^+ and Mg^{2+}. *Journal of Physiology* **475**, 1–7.

Stanfield, P.R., Davies, N.W., Shelton, P.A., Sutcliffe, M.J., Khan, I.A., Brammar, W.J. & Conley, E.C. (1994b). A single aspartate residue is involved in both intrinsic gating and blockage by Mg^{2+} of the inward rectifier, IRK1. *Journal of Physiology* **478**, 1–6.

Stein, P.E., Boodhoo, A., Tyrrell, G.J., Brunton, J.L. & Read, R.J. (1992). Crystal structure of the cell-binding B oligomer of verotoxin-1 from *E. coli*. *Nature* **355**, 748–50.

Steinberg, T.H., Civitelli, R., Geist, S.T., Robertson, A.J., Hick, E., Veenstra, R.D., Wang, H.-Z., Warlow, P.M., Westphale, E.M., Laing, J.G. & Beyer, E.C. (1994). Connexin43 and connexin45 form gap junctions with different molecular permeabilities in osteoblastic cells. *EMBO Journal* **13**, 744–50.

Steinmeyer, K., Ortland, C. & Jentsch, T.J. (1992). Primary structure and functional expression of a developmentally regulated skeletal muscle chloride channel. *Nature* **354**, 301–4.

Steinmeyer, K., Lorenz, C., Pusch, M., Koch, M.C. & Jentsch, T.J. (1994). Multimeric structure of ClC-1

chloride channel revealed by mutations in dominant myotonic congenita. *EMBO Journal* **13**, 737–43.

Stocker, M., Stühmer, W., Wittka, R., Wang, S., Müller, R., Ferrus, A. & Pongs, O. (1990). Alternative *Shaker* transcripts express either rapidly inactivating or non-inactivating K^+ channels. *Proceedings of the National Academy of Sciences, USA* **87**, 8903–7.

Strichartz, G.R. (1973). The inhibition of sodium currents in myelinated nerve by quaternary derivatives of lidocaine. *Journal of General Physiology* **62**, 37–57.

Strichartz, G.R., Rando, T. & Wang, G.K. (1987). An integrated view of the molecular toxinology of sodium channel gating in excitable cells. *Annual Review of Neuroscience* **10**, 237–67.

Strong, M., Chandy, K.G. & Gutman, G.A. (1993). Molecular evolution of voltage-sensitive ion channel genes: on the origins of electrical excitability. *Molecular Biology and Evolution* **10**, 221–42.

Stryer, L. (1991). Visual excitation and recovery. *Journal of Biological Chemistry* **266**, 10711–14.

Stühmer, W., Conti, F., Suzuki, H., Wang, X., Noda, M., Yahagi, N., Kubo, H. & Numa, S. (1989). Structural parts involved in activation and inactivation of the sodium channel. *Nature* **339**, 597–603.

Super, M. (1992). Milestones in cystic fibrosis. *British Medical Bulletin* **48**, 717–37.

Surprenant, A., Buell, G. & North, R.A. (1995). P_{2X} receptors bring new structure to ligand-gated ion channels. *Trends in Neurosciences* **18**, 224–9.

Sutcliffe, M.J. & Stanfield, P.R. (1994). Pore region of K^+ channel RACTK1. *Nature* **369**, 616.

Sutherland, E.W. (1971). Studies on the mechanism of hormone action. In *Les Pris Nobel 1971*, pp. 240–57. Stockholm: Almquist & Wiksell.

Sutherland, E.W. & Rall, T.W. (1960). Relation of adenosine 3′,5′-phosphate and phosphorylase to the action of catecholamines and other hormones. *Pharmacological Reviews* **12**, 265–99.

Suzuki, M., Takahashi, K., Ikeda, M., Hayakawa, H., Ogawa, A., Kawaguchi, Y. & Sakai, O. (1994). Cloning of pH-sensitive K^+ channel possessing two trans-membrane segments. *Nature* **367**, 642–5.

Taglialatela, M., Drewe, J.A., Kirsch, G.E., de Biasi, M., Hartmann, H.A. & Brown, A.M. (1993). Regulation of K^+/Rb^+ selectivity and internal TEA blockade by mutations at a single site in K^+ pores. *Pflügers Archiv* **423**, 104–12.

Takahashi, N., Morishige, K., Jahangir, A., Yamada, M., Findlay, I., Koyama, H. & Kurachi, Y. (1994). Molecular cloning and functional expression of cDNA encoding a second class of inward rectifier potassium channels in mouse brain. *Journal of Biological Chemistry* **269**, 23274–9.

Takai, T., Noda, M., Mishina, M., Shimizu, S., Furutani, Y., Kayano, T., Ikeda, T., Kubo, T., Takahashi, H., Takahashi, T., Kuno, M. & Numa, S. (1985). Cloning, sequencing and expression of cDNA for a novel subunit of acetylcholine receptor from calf muscle. *Nature* **315**, 761–4.

Takano, K., Stanfield, P.R., Nakajima, S. & Nakajima, Y. (1995). Protein kinase C-mediated inhibition of an inward rectifier potassium channel by substance P in nucleus basalis neurons. *Neuron* **14**, 999–1008.

Takeshima, H., Nishimura, S., Matsumoto, T., Ishida, H., Kangawa, K., Minamino, N., Matsuo, H., Ueda, M., Hanaoka, M., Hirose, T., Numa, S. (1989). Primary structure and expression from complementary DNA of skeletal muscle ryanodine receptor. *Nature* **339**, 439–45.

Takumi, T. (1993). A protein with a single transmembrane domain forms an ion channel. *News in Physiological Sciences* **8**, 175–8.

Takumi, T., Ohkubo, H. & Nakanishi, S. (1988). Cloning of a membrane protein that induces a slow voltage-gated potassium current. *Science* **242**, 1042–5.

Tanabe, T., Takeshima, H., Mikami, A., Flockerzi, V., Takahashi, H., Kangawa, K., Kojima, M., Matsuo, H., Hirose, T. & Numa, S. (1987). Primary structure of the receptor for calcium channel blockers from skeletal muscle. *Nature* **328**, 313–18.

Tanabe, T., Beam, K.G., Adams, B.A., Niidome, T. & Numa, S. (1990). Regions of the skeletal muscle dihydropyridine receptor critical for excitation–contraction coupling. *Nature* **346**, 567–9.

Tanguy, J. & Yeh, J.Z. (1991). BTX modification of Na channels in squid axons. I. State dependence of BTX action. *Journal of General Physiology* **97**, 499–519.

Tank, D.W & Miller, C. (1983). Patch-clamped liposomes. In *Single-Channel Recording*, ed. B. Sakmann & E. Neher, pp. 91–105. New York: Plenum Press.

Taylor, S.S. (1989). cAMP-dependent protein kinase. *Journal of Biological Chemistry* **264**, 8443–6.

Tempel, B.L., Papazian, D.M., Schwarz, T.L., Jan, Y.N & Jan, L.Y. (1987). Sequence of a probable potassium channel component encoded at *Shaker* locus of *Drosophila*. *Science*, **237**, 770–5.

Thiemann, A., Gründer, S., Pusch, M. & Jentsch, T.J. (1992). A chloride channel widely expressed in epithelial and non-epithelial cells. *Nature* **356**, 57–60.

Thomsen, W.J. & Catterall, W.A. (1989). Localization of the receptor site for alpha-scorpion toxins by antibody mapping: implications for sodium channel topology. *Proceedings of the National Academy of Sciences, USA* **86**, 10161–5.

Thuringer, D. & Escande, D. (1989). Apparent competition between ATP and the potassium channel opener RP-49356 on ATP-sensitive K^+ channels of cardiac myocytes. *Molecular Pharmacology* **36**, 897–902.

Tillotson, D. (1979). Inactivation of Ca conductance dependent on entry of Ca ions in molluscan neurons. *Proceedings of the National Academy of Sciences, USA* **76**, 1497–500.

Tomaselli, G.F., McLaughlin, J.T., Jurman, M.E., Hawrot, E. & Yellen, G. (1991). Mutations affecting agonist sensitivity of the nicotinic acetylcholine receptor. *Biophysical Journal* **60**, 721–7.

Tomlins, B. & Williams, A.J. (1986). Solubilisation and reconstitution of the rabbit skeletal muscle sarcoplasmic reticulum K^+ channel into liposomes suitable for patch clamp studies. *Pflügers Archiv* **407**, 341–7.

Toro, L., Stefani, E. & Latorre, R. (1992). Internal blockade of a Ca^+-activated K^+ channel by *Shaker* B inactivating "ball" peptide. *Neuron* **9**, 237–45.

Toyoshima, C. & Unwin, P.N.T. (1988). Ion channel of acetylcholine receptor reconstructed from images of postsynaptic membranes. *Nature* **336**, 247–50.

Trautwein, W. & Hescheler, J. (1990). Regulation of cardiac L-type calcium current by phosphorylation and G proteins. *Annual Review of Physiology* **52**, 257–74.

Triggle, D.J. (1979). Receptor theory. In *Receptors in Pharmacology*, pp. 1–65. New York: Marcel Dekker.

Trimmer, J.S. & Agnew, W.S. (1989). Molecular diversity of voltage-sensitive Na channels. *Annual Review of Physiology* **51**, 401–18.

Trimmer, J.S., Cooperman, S.S., Tomiko, S.A., Zhou, J., Crean, S.M.., Boyle, M.B., Kallen., R.G., Sheng, Z., Barchi, R.L., Sigworth, F.J., Goodman, R.H., Agnew, W.S. & Mandel, G. (1989). Primary structure and functional expression of a mammalian skeletal muscle sodium channel. *Neuron* **3**, 33–49.

Tsien, R.W., Hess, P., McCleskey, E.W. & Rosenberg, R.W. (1987). Calcium channels: mechanisms of selectivity, permeation and block. *Annual Review of Biophysics* **16**, 265–90.

Tsien, R.W., Lipscombe, D., Madison, D.V., Bley, K.R. & Fox, A.P. (1988). Multiple types of neuronal calcium channels and their selective modulation. *Trends in Neurosciences* **11**, 431–8.

Tsui, L.-C. (1992). The spectrum of cystic fibrosis mutations. *Trends in Genetics* **8**, 392–8.

Twyman, R.E. & Macdonald, R.L. (1992). Neurosteroid regulation of GABA$_A$ receptor single-channel kinetic properties of mouse spinal cord neurons in culture. *Journal of Physiology* **456**, 215–45.

Uchitel, O.D., Protti, D.A., Sanchez, V., Cherksey, B.D., Sugimori, M. & Llinás, R. (1992). Voltage-dependent calcium channel mediates presynaptic calcium influx and transmitter release in mammalian synapses. *Proceedings of the National Academy of Sciences, USA* **89**, 3330–3.

Unwin, N. (1989). The structure of ion channels in membranes of excitable cells. *Neuron* **3**, 665–76.

Unwin, N. (1993a). Nicotinic acetylcholine receptor at 9 Å resolution. *Journal of Molecular Biology* **229**, 1101–24.

Unwin, N. (1993b). Neurotransmitter action: opening of ligand-gated ion channels. *Cell* **72**/*Neuron* **10** supplement, 31–41.

Unwin, N. (1995). Acetylcholine receptor channel imaged in the open state. *Nature* **373**, 37–43.

Unwin, P.N.T. & Ennis, P.D. (1984). Two configurations of a channel-forming membrane protein. *Nature* **307**, 609–13.

Urry, D.W., Jing, N., Trapane, T.L., Luan, C.-H. & Waller, M. (1988). Ion interactions with the gramicidin A transmembrane channel: cesium-133 and calcium-43 NMR studies. *Current Topics in Membranes and Transport* **33**, 51–90.

Usherwood, P.N.R. & Blagbrough, I.S. (1991). Spider toxins affecting glutamate receptors: polyamines in therapeutic neurochemistry. *Pharmacology and Therapeutics* **52**, 245–68.

Ussing, H. H. (1949). The distinction by means of tracers between active transport and diffusion. *Acta Physiologica Scandinavica* **19**, 43–56.

Valera, S.., Hussy, N., Evans, R.J., Adami, N., North, R.A., Surprenant, A. & Buell, G. (1994). A new class of ligand-gated ion channel defined by P$_{2X}$ receptor for extracellular ATP. *Nature* **371**, 516–19.

Vandenberg, C.A. (1987). Inward rectification of a potassium channel in cardiac ventricular cells depends upon internal magnesium ions. *Proceedings of the National Academy of Sciences, USA* **84**, 2560–4.

Vandenberg, R.J. & Schofield, P.R. (1994). Inhibitory ligand-gated channel receptors: molecular biology and pharmacology of GABA$_A$ and glycine receptors. In *Handbook of Membrane Channels*, ed. C. Peracchia, pp. 317–33. San Diego, CA: Academic Press.

Varadi, G., Mori, Y., Mikala, G. & Schwartz, A. (1995). Molecular determinants of Ca^{2+} channel function and drug action. *Trends in Pharmacological Sciences* **16**, 43–9.

Varnum, M.D., Maylie, J., Busch, A. & Adelman, J.P. (1995). Persistent activation of min K channels by chemical crosslinking. *Neuron* **14**, 407–12.

Velimirovic, B.M., Koyano, K., Nakajima, S. & Nakajima, Y. (1995). Opposing mechanisms of regulation of a G-protein-coupled inward rectifier K$^+$ channel in rat brain neurons. *Proceedings of the National Academy of Sciences, USA* **92**, 1590–4.

Vernino, S., Amador, M., Luetje, C.W., Patrick, J. & Dani, J.A. (1992). Calcium modulation and high calcium permeability of neuronal nicotinic acetylcholine receptors. *Neuron* **8**, 127–34.

Vestergaard-Bogind, B., Stampe, P. & Christophersen, P. (1985). Single-file diffusion through the Ca^{2+}-activated K$^+$ channel of human red cells. *Journal of Membrane Biology* **88**, 67–75.

Villaroel, A. & Sakmann, B. (1992). Threonine in the selectivity filter of the acetylcholine receptor channel. *Biophysical Journal* **62**, 196–208.

Villaroel, A., Alvarez, O., Oberhauser, A. & Latorre, R. (1988). Probing a Ca^{2+}-activated K$^+$ channel with quaternary ammonium ions. *Pflügers Archiv* **413**, 118–26.

Viviani, R. (1992). Eutrophication, marine biotoxins, human health. *Science of the Total Environment* Supplement 1992, 631–61.

Voilley, N., Lingueglia, E., Champigny, G., Mattéi, M.-G., Waldmann, R., Lazdunski, M. & Barbry, P. (1994). The lung amiloride-sensitive Na$^+$ channel: biophysical properties, pharmacology, ontogenesis, and molecular cloning. *Proceedings of the National Academy of Sciences, USA* **91**, 247–51.

Wada, K., Ballivet, M., Boulter, J., Connolly, J., Wada, E., Deneris, E.S., Swanson, L.W., Heinemann, S. & Patrick, J. (1988). Functional expression of a new pharmacological subtype of brain nicotinic acetylcholine receptor. *Science* **240**, 330–4.

Wagoner, P.K. & Oxford, G.S. (1987). Cation permeation through the voltage-dependent potassium channel in the squid axon. *Journal of General Physiology* **90**, 261–90.

Wallace, B.A. (1990). Gramicidin channels and pores. *Annual Review of Biophysics* **19**, 127–57.

Wang, H., Kunkel, D.D., Martin, T.M., Schwartzkroin, P.A. & Tempel, B.L. (1993). Heteromultimeric K$^+$ channels in terminal and juxtaparanodal regions of neurons. *Nature* **365**, 75–9.

Wang, J. & Best, P.M. (1994). Characterization of the potassium channel from frog skeletal muscle sarcoplasmic reticulum membrane. *Journal of Physiology* **477**, 279–90.

Warmke, J.W. & Ganetzky, B. (1994). A family of potassium channel genes related to *eag* in *Drosophila* and mammals. *Proceedings of the National Academy of Sciences, USA* **91**, 3438–42.

Warmke, J., Drysdale, R. & Ganetzky, B. (1991). A distinct potassium channel polypeptide encoded by the *Drosophila eag* locus. *Science* **252**, 1560–2.

Watson, J.D., Gilman, M., Witkowski, J. & Zoller, M. (1992). *Recombinant DNA*, 2nd edition. New York: Scientific American Books.

Wei, A., Covarrubias, M., Butler, A., Baker, K., Pak, M. & Salkoff, L. (1990). K$^+$ current diversity is produced by an extended gene family conserved in *Drosophila* and the mouse. *Science* **248**, 599–603.

Wei, A., Solaro, C., Lingle, C. & Salkoff, L. (1994). Calcium sensitivity of BK-type K$_{Ca}$ channels determined by a separable domain. *Neuron* **13**, 671–81.

Weiss, M.S., Wacker, T., Weckesser, J., Welte, W. & Schulz, G.E. (1990). The three-dimensional structure of porin from *Rhodobacter capsulatus* at 3 Å resolution. *FEBS Letters* **267**, 268–72.

Weiss, M.S., Abele, U., Weckesser, J., Welte, W., Schiltz, E. & Schulz, G.E. (1991). Molecular architecture and electrostatic properties of a bacterial porin. *Science* **254**, 1627–30.

Welsh, M.J. & Smith, A.E. (1993). Molecular mechanisms of CFTR chloride channel dysfunction in cystic fibrosis. *Cell* **73**, 1251–4.

Welsh, M.J., Anderson, M.P., Rich, D.P., Berger, H.A., Denning, G.M., Ostedgaard, L.S., Sheppard, D.N., Cheng, S.H., Gregory, R.J. & Smith, A.E. (1992). Cystic fibrosis transmembrane conductance regulator: a chloride channel with novel regulation. *Neuron* **8**, 821–9.

Welsh, M.J., Tsui, L.-C., Boat, T.F. & Beaudet, A.L. (1995). Cystic fibrosis. In *The Metabolic and Molecular Bases of Inherited Disease*, ed. C.R. Scriver, A.L. Beaudet, W.S. Sly & D. Valle, vol. 3, pp. 3799–876. New York: McGraw-Hill.

West, J.B. (1991). *The Physiological Basis of Medical Practice*, 12th edition. Baltimore, MD: Williams and Wilkins.

West, J.W., Numann, R., Murphy, B.J., Scheuer, T. & Catterall, W.A. (1991). A phosphorylation site in the Na$^+$ channel required for modulation by protein kinase C. *Science* **254**, 866–8.

West, J.W., Patton, D.E., Scheuer, T., Wang, Y., Goldin, A.L. & Catterall, W.A. (1992). A cluster of hydrophobic amino acid residues required for fast Na$^+$ channel inactivation. *Proceedings of the National Academy of Sciences, USA* **89**, 10910–14.

Westbrook, G.L. & Jahr, C.E. (1989). Glutamate receptors in excitatory neurotransmission. *Seminars in the Neurosciences* **1**, 103–14.

White, M.M. & Miller, C. (1979). A voltage-gated anion channel from electric organ of *Torpedo californica*. *Journal of Biological Chemistry* **254**, 10161–6.

White, R.E., Schonbrunn, A. & Armstrong, D.L. (1991). Somatostatin stimulates Ca^{2+}-activated K$^+$ channels through protein dephosphorylation. *Nature* **351**, 570–3.

White, T.W., Bruzzone, R., Wolfram, S., Paul, D.L. & Goodenough, D.A. (1994). Selective interactions among the multiple connexin proteins expressed in the vertebrate lens: the second extracellular domain is a determinant of compatibility between connexins. *Journal of Cell Biology* **125**, 879–92.

Wible, B.A., Taglialatela, M., Ficker, E. & Brown, A.M. (1994). Gating of inwardly rectifying K$^+$ channels localized to a single negatively charged residue. *Nature* **371**, 246–9.

Wickman, K.D., Iñiguez-Lluhi, J.A., Davenport, P.A., Taussig, R., Krapivinsky, G.B., Linder, M.E., Gilman, A.G. & Clapham, D.E. (1994). Recombinant G-protein βγ-subunits activate the muscarinic-gated atrial potassium channel. *Nature* **368**, 255–7.

Williams, A.J. (1994). An introduction to the methods available for ion channel reconstitution. In *Microelectrode Techniques: The Plymouth Workshop Handbook*, 2nd edition, ed. D. Ogden, pp. 79–99. Cambridge: Company of Biologists.

Williams, K., Zappia, A.M., Pritchett, D.B., Shen, Y.M. & Molinoff, P.B. (1994). Sensitivity of the *N*-methyl-D-aspartate receptor to polyamines is controlled by NR2 subunits. *Molecular Pharmacology* **45**, 803–9.

Williams, M. & Sills, M.A. (1990). Quantitative analysis of ligand–receptor interactions. In *Comprehensive Medicinal Chemistry*, ed. C. Hansch, P.G. Sammes & J.B. Taylor, vol. 3, pp. 45–80. Oxford: Pergamon Press.

Wistow, G.J., Pisano, M.M. & Chepelinsky, A.B. (1991). Tandem sequence repeats in transmembrane channel proteins. *Trends in Biochemical Sciences* **16**, 170–1.

Wo, Z.G. & Oswald, R.E. (1994). Transmembrane topology of two kainate receptor subunits revealed by N-glycosylation. *Proceedings of the National Academy of Sciences, USA* **91**, 7154–8.

Wo, Z.G. & Oswald, R.E. (1995). Unraveling the modular design of glutamate-gated ion channels. *Trends in Neurosciences* **18**, 161–8.

Woodhull, A. (1973). Ionic blockage of sodium channels in nerve. *Journal of General Physiology* **61**, 687–708.

Woolley, G.A. & Wallace, B.A. (1992). Model ion channels: gramicidin and alamethicin. *Journal of Membrane Biology* **129**, 109–36.

Yamaguchi, K., Nakajima, Y., Nakajima, S. & Stanfield, P.R. (1990). Modulation of inwardly rectifying channels by substance P in cholinergic neurones from rat brain in culture. *Journal of Physiology* **426**, 499–520.

Yang, J., Ellinor, P.T., Sather, W.A., Zhang, J.-F. & Tsien, R.W. (1993). Molecular determinants of Ca^{2+} selectivity and ion permeation in L-type Ca^{2+} channels. *Nature* **366**, 158–61.

Yau, K.-W. (1994). Cyclic nucleotide-gated channels: an expanding new family of ion channels. *Proceedings of the National Academy of Sciences, USA* **91**, 3481–3.

Yau, K.-W. & Baylor, D.A. (1989). Cyclic GMP-activated conductance of retinal photoreceptor cells. *Annual Review of Neuroscience* **12**, 289–327.

Yeager, M. & Gilula, N.B. (1992). Membrane topology and quaternary structure of cardiac gap junction ion channels. *Journal of Molecular Biology* **223**, 929–48.

Yellen, G. (1987). Permeation in potassium channels: implications for channel structure. *Annual Review of Biophysics* **16**, 227–46.

Yellen, G., Jurman, M.E., Abramson, T. & MacKinnon, R. (1991). Mutations affecting internal TEA blockade identify the probable pore-forming region of a K$^+$ channel. *Science* **251**, 939–42.

Yudkin, M. & Offord, R. (1980). *A Guidebook to Biochemistry*, 4th edition. Cambridge: Cambridge University Press.

Zabner, J., Couture, L.A., Gregory, R.J., Graham, S.M., Smith, A.E. & Welsh, M.J. (1993). Adenovirus-mediated gene transfer transiently corrects the chloride transport defect in nasal epithelia of patients with cystic fibrosis. *Cell* **75**, 207–16.

Zagotta, W.N., Hoshi, T. & Aldrich, R.W. (1990). Restoration of inactivation in mutants of *Shaker* potassium channels by a peptide derived from ShB. *Science* **250**, 568–71.

Zagotta, W.N., Hoshi, T., Dittman, J. & Aldrich, R.W. (1994a). *Shaker* potassium channel gating. II: Transitions in the activation pathway. *Journal of General Physiology* **103**, 279–319.

Zagotta, W.N., Hoshi, T. & Aldrich, R.W. (1994b). *Shaker* potassium channel gating. III: Evaluation of kinetic models for activation. *Journal of General Physiology* **103**, 321–62.

Zaman, G.J.R., Flens, M.J., van Leusden, M.R., de Haas, M., Mulder, H.S., Lankelma, J., Pinedo, H.M., Scheper, R.J., Baas, F., Broxterman, H.J. & Borst, P. (1994). The human multidrug resistance-associated protein MRP is a plasma membrane drug-efflux pump. *Proceedings of the National Academy of Sciences, USA* **91**, 8822–6.

Zamponi, G.W., Doyle, D.W. & French, R.J. (1993). Fast lidocaine block of cardiac and skeletal muscle sodium channels: one site with two routes of access. *Biophysical Journal* **65**, 80–90.

Zhang, J.F., Ellinor, P.T., Aldrich, R.W. & Tsien, R.W. (1994). Molecular determinants of voltage-dependent inactivation in calcium channels. *Nature* **372**, 97–100.

Zhang, Z.J., Jurkiewicz, N.K., Folander, K., Lazarides, E., Salata, J.J. & Swanson, R. (1994). K$^+$ currents expressed from the guinea pig cardiac I$_{sK}$ protein are enhanced by activators of protein kinase C. *Proceedings of the National Academy of Sciences, USA* **91**, 1766–70.

Zhou, H., Tate, S.S. & Palmer, L.G. (1994). Primary structure and functional properties of an epithelial K channel. *American Journal of Physiology* **266**, C809–C824.

Zimmerberg, J., Bezanilla, F. & Parsegian, V.A. (1990). Solute inaccessible aqueous volume changes during opening of the potassium channel of the squid giant axon. *Biophysical Journal* **57**, 1049–64.

Zorzato, F., Fujii, J., Otsu, K., Phillips, M., Green, N.M., Lai, F.A., Meissner, G., & MacLennan, D.H. (1990). Molecular cloning of cDNA encoding human and rabbit forms of the Ca^{2+} release channel (ryanodine receptor) of skeletal muscle sarcoplasmic reticulum. *Journal of Biological Chemistry* **265**, 2244–56.

Zukin, R.S. & Bennett, M.V.L. (1995). Alternatively spliced isoforms of the NMDAR1 receptor subunit. *Trends in Neurosciences* **18**, 306–13.

Index

absolute reaction rate theory, 134–7, 264
  applied to gating, 181
acetylcholine, 68
  binding site, in nAChR, 79, 175, 176fig.
  inhibition of heart by, 101, 205
  membrane noise produced by, 40fig., 41, 174
acetylcholine receptor
  muscarinic, 69, 215–17
  nicotinic, see nAChR
aconitine, 240
action potential
  of heart muscle, 93, 203, 213
  of nerve axon, 4, 28, 89, 195
  of skeletal muscle, 34
activation, 195, 198fig.; see also gating
activity of ions, 19
activity coefficient, 19
adenylyl cyclase, 212, 216
aerolysin, 118
affinity, 170, 226
ω-agatoxins, 93, 244
agonists, 244
AKT1, 99
alamethicin, 116–18
allosteric effects
  of blocking agents, 243–4
  in cyclic-nucleotide-gated channels, 200–1
  in gating, 161
  in modulation, 212, 246

  in nAChR, 173, 177
amiloride, 102
amino acid residues, properties of, 62tab., 72
amino acid sequences
  determination of, 59–68
  evolution of, 64
  of glutamate receptors, 87fig.
  of nAChR, 70, 71fig., 86fig.
  of voltage-gated sodium channel, 91
γ-aminobutyric acid receptor see GABA receptor
4-aminopyridine (4-AP), 234
AMPA, 86, 247
AMPA receptors, 86
  pharmacology of, 247–8
anaesthetics
  general, 246, 248
  local, 240–2
anatoxin A, 245
ångström unit (Å), 18
anomalous mole-fraction effect, 132–3, 140, 141
antagonists, 244
antechamber, of channel pore, 231; see also vestibule
apamin, 234
APV, 248
aquaporins, 113
arachidonic acid, 221
assembly of Shaker channels, 97
association constant, 170
asthma, 234, 258

atmosphere, ion, 20
ATP (adenosine triphosphate) as gating agent
  in ATP-sensitive potassium channels, 3, 205
  in CFTR channel, 206–7
ATP as source of cAMP, 212, 216
ATP receptor channel, 89
ATP-binding cassette (ABC) superfamily, 109
ATP-sensitive potassium channels see potassium channels, ATP-sensitive
auxilliary subunits, 93, 192–3, 213

bacteriorhodopsin, 73
ball and chain model, 189–91, 230
barbiturates, 246
batrachotoxin, 183fig., 239
Bay K 8644, 244
Becker's disease see myotonia, generalized
benzodiazepines, 246–7
β-barrel, 74, 78, 114, 115fig., 153
β cell, 3, 234
β-sheet, 73, 115
bicuculline, 246
bilayer, lipid, 2fig.
  artificial, 5, 43–5
binding site models of channels, 134–44
BIR channels, 101, 102tab., 204
BK channels, 97, 98

block, kinetics of, 226–9
Boltzmann constant (k), 11tab., 24
Boltzmann distribution, 181
Boltzmann relation, 182, 183fig.
brevetoxins, 240
bright ideas, value of, 263
α-bungarotoxin, 69, 245
bursts, 166, 227

C terminus, 63, 75
C_9, block by, 229–30
calcium channels, voltage-gated
    binding site model of, 142–4
    blocking agents of, 243–4
    dysfunctional mutants of, 254–5
    inactivation, 220
    modes of, 219, 243–4
    molecular basis of selectivity in,
        154–6
    phosphorylation of, 213–14
    S4 segment of, 186fig.
    selectivity of, 127tab., 142
    structure and diversity of, 93–4
calcium ions
    as gating agents, 201–2
    as intracellular messengers, 25, 93,
        103–5, 213, 220, 221
    as modulators, 219–20
    and surface charges, 222–3
calcium release channels, 103–6, 261
calmodulin, 211, 220, 221
cAMP (cyclic adenosine
    monophosphate), 99, 199, 215–16
cAMP-dependent protein kinase see
    protein kinase A
cAMP-gated channel, olfactory, 99, 199
capacitance, 12
capacity current, 37, 38fig., 180
capsaicin, 236
carbachol, 245
cDNA, 66
cell-attached patch configuration, 48
cell lines, expression in, 81
CFTR (cystic fibrosis conductance
    regulator) channel
    functions of, 257
    gating of, 206–7
    identification of, 67, 108
    mutations affecting, 257–60

structure of, 108–9
cGMP (cyclic guanosine
    monophosphate), 98, 199, 222
cGMP-gated channel, photoreceptor,
    98, 199
    selectivity of, 127
chaos, in gating model, 168
Charcot–Marie–Tooth disease, 261
charge, electric, 9, 10tab.
charge immobilization, 180, 193–4
charybdotoxin, 98, 234–6
chemical bonds, 13–17
chemically gated channels see ligand-
    gated channels
CHIP28, 113
chloride channels
    background, 106–9
    dysfunctional mutants of, 256–60
    neuronal, 128
    neurotransmitter-gated see GABA
        receptor and glycine receptor
chlorpromazine, 77, 245
chord conductance, 30
ciguatoxin, 240
CIR subunit, 101, 102tab.
CLC (ClC) channels, 106–8
    dysfunctional mutants of, 256–7
closed times, distribution of, 55–7,
    163–6
clusters, 166
CMTX, 261
CNG channel see cyclic-nucleotide-
    gated channels
codon, 62
colicins, 118
conantokins, 248
concentration cell, 24
conductance, 10tab., 12
conductive state, 162
conductivity, 12
conformational state, 162
connexins, 110–12, 210, 261
connexons, 110, 210
conotoxins, 238–9, 244, 245, 248
    conantokins, 248
    α-conotoxins, 245
    μ-conotoxins, 239
    ω-conotoxins, 94, 244
constant field model, 122–4, 223fig.

covalent bond, 13, 17tab.
CPP, 248
cromakalim, 234
cryoelectron microscopy see image
    analysis, quantitative
crystal radius of ions, 17–19
curare, 245
current, electric, 10tab., 11
currents, single-channel, 6fig., 28–31
    amplitudes of, 53
    analysis of, 52–7
    time characteristics of, 55–7
curare, 176fig.
cyclic nucleotides see cAMP and cGMP
cyclic-nucleotide-gated channels
    binding sites of, 200
    gating in, 199–201
    molecular basis of selectivity in,
        156–7
    structure of, 98–9
cysteine substitution method, 153–4
cystic fibrosis, 108, 257–60 see also CFTR
cytoskeleton, 209
cytotoxity, 251

dantrolene, 261
Debye length, 20, 222
dendrotoxin, 192, 234, 245tab.
dephosphorylation, 211
desensitization, in nAChR, 173, 245
diabetes, treatment of, 3, 234
diacylglycerol, 105, 211, 213
diazepam, 246
diffusion, 21–2
diffusion models of gating, 167–8
dihydropyridine receptor, 103
dihydropyridines, 243
diltiazem, 243
diphtheria toxin, 118
dipole, electric, 12
dipole–dipole interaction, 15
disease, channel dysfunction in, 251–62
disulphide bridges, 72
    in nAChR, 75
DNA
    complementary DNA (cDNA), 66
    non-coding, 64
    recombinant, 64–8
    structure of, 60–1, 68

DNA library, 66
domoic acid, 247
dose–response curve, 168–9
DRK1 (Kv2.1), 96tab., 153
dwell time, 55
dwell time histogram, 56, 164, 229fig.

eag (ether-à-go-go), 99, 255
EC_{50} and ED_{50}, 169
electric charge, 9, 10tab.
electric organs of fish, 69, 91
electrical distance, $z\delta$, 138, 183, 201, 231–3
electricity, 9–13
electrochemical gradient, 23–6
  in cells, 26–7
  defined 1, 23
electrodiffusion, 21, 23, 121–9
elementary charge (e_0), 10, 11tab., 24
end-plate potential, 34, 35fig.
energy barrier, 121, 134
energy profile, 134, 135fig., 139fig., 141fig., 143fig.
episodic ataxia with myokymia syndrome, 255
epitope protection assay, 76
equilibrium potential, 24, 124–5
exons, 63, 82, 95
expression of channel proteins, 79–81

Faraday constant (F), 10, 11tab.
fatty acids, 209, 220–1
feet, in muscle, 103
first latency, 167
flickering, 174
flickery block, 226
flip/flop segment, 88, 89fig.
fluctuation analysis, 39–43
fluxes, ionic, 123, 131, 136
foot in the door model, 196
fractal models of gating, 167–8

G proteins, 101, 105, 204
  modulation by, 214–19
GABA receptor channel
  anomalous mole-fraction effect in, 133, 140
  binding site model of, 140

molecular basis of selectivity in, 159–60
pharmacology of, 246–7
selectivity of, 128
structure of, 84–5
gap junction channels
  dysfunctional mutants of, 261
  gating of, 209–10
  modulation by calcium ions, 220
  pore size, 147–8
  selectivity of, 128–9
  structure of, 110–12
gas constant (R), 11tab., 24
gating, 2, 7fig., 161
  kinetics of, 161–70, 194–9
  of nAChR channel, 170–8
  of voltage-gated channels, 178–99
gating charges, 179–84
gating currents, 179–80, 195–7
gene, 63
  CFTR, 258–9
  duplication, 83
  dysfunctional, 251–62
  of mammalian potassium channels, 64
  of nAChR, 82
  of *Shaker* channel, 63
gene therapy, in cystic fibrosis, 260
genetic code, 62, 63tab.
gigaseal, 48
GHK (Goldman–Hodgkin–Katz) equations, 123–4
GIRK1 channel, 101, 204, 218
glibenclamide, 234
glutamate receptor channels, 85–8
  calcium permeability of, 160
  molecular basis of selectivity in, 160
  pharmacology of, 247–9
  see also NMDA receptor channels
glycine, as modulator, 219
glycine receptor channel
  block of, 247
  dysfunctional mutants of, 260
  molecular basis of selectivity in, 159–60
  structure of, 84–5
glycosylation site, 72
  in nAChR, 75
  in voltage-gated sodium channel, 92

gramicidin channel
  currents through, 5, 46fig.
  molecular dynamics in, 122, 148–9
  streaming potential in, 151
  structure of, 115–16
  water movement through, 148
grayanotoxins, 240

H5 segment, 90
  consensus sequence in, 154
  C-type inactivation and, 191
  of inward rectifier channels, 100, 152
  of voltage-gated potassium channels, 152–4
hair cells, 209
halothane, 246, 261
headstage, 50
α-helix, 72, 73fig.
hereditary hyperekplexia, 260
HERG, 255–6
heteromultimers in potassium channels, 97
heterotrimeric G protein *see* G protein
hexamers, gap junction, 110
Hill coefficient, 170, 206
Hill–Langmuir equation, 169
hinged lid model, 192–3, 212, 253
histrionicotoxin, 245
Hodgkin–Huxley equations, 4, 38, 195
homology, of sequences, 64, 81
HOPP, 254
5-HT (5-hydroxytryptamine, serotonin)
  see next entry
5-HT-gated channel, 85
hydration, of ions, 20–1, 136
hydrogen bond, 15, 17tab.
  in channel pore, 146, 148
  in proteins, 72, 74
hydrogen ions
  block by, 137–8, 227
  local anaesthetics affected by, 242
  surface charges and, 222–3
hydropathy index, 62tab., 74
hydropathy profile, 74
  of nAChR, 74fig.
  of voltage-gated sodium channel, 91fig.
hydrophobic bond, 16, 17tab.

hydrophobicity profile *see* hydropathy profile

5-hydroxytryptamine *see* 5-HT

hyperkalaemic periodic paralysis (HYPP), 251–3, 255fig.

hypokalaemic periodic paralysis (HOPP), 254–5, 256fig.

I_{Cln} protein, 114

ibotenic acid, 248

image analysis, quantitative, 77, 110, 177, 210

inactivation, 188–94
  ball and chain model, 189–91
  C-type, 191–2
  hinged-lid model of, 192–3, 212
  kinetics of, 164, 196–7
  local anaesthetics and, 242
  mutations affecting, 253, 254–5, 255fig.
  phosphorylation and, 212

influenza virus, 118

inositol trisphosphate receptor *see* IP_3 receptor

inside-out patch configuration, 48

$InsP_3$ *see* IP_3

introns, 63, 82

ion channels
  in β cells, 3
  different types of, 7–8
  discovery of, 3–6
  in muscle cells, 34, 37
  nature of, 1
  in nerve axons, 28, 37
  papers on, 263

ion pairs, in *Shaker* channel, 187

ion–dipole interaction, 15, 136

ionic bond, 14, 17tab.

ionic concentrations in cells, 26

ionotropic receptors, defined, 84

ions
  block by, 133–4, 248
  in crystals, 17–19
  hydration of, 20–1
  nature of, 14
  rates of movement, 23
  size of, 17–19, 21
  in solution, 19–23

IP_3 (inositol 1,4,5-trisphosphate) receptor, 104–6, 219

IRK1 channel, 100, 102tab., 203

IsK protein, 113

interdependence of ion movements, 129–34

jump experiments, 166–7

kainate receptors, 86, 247

kainic acid, 247

KAT1, 99

K_{ATP} channels *see* potassium channels, ATP-sensitive

K(Ca) channels *see* potassium channels, calcium-activated

kinetics of gating, 161–8
  in cyclic-nucleotide-gated channels, 200–1
  in nAChR, 171–4
  in voltage-gated channels, 194–9

Kir subfamily, 101, 102tab.

Kv channel nomenclature, 96

leucine heptad, 188

library, recombinant DNA, 66

lidocaine, 240

ligand-gated channels, 7fig., 8
  kinetics of, 168–70

lipid bilayer, artificial, 5, 43–5

liposomes, 45, 260

long QT syndrome, 256

long-pore effect, 131–2

lophotoxin, 245

M current, 216–17

M_2 protein, influenza virus, 118

M2 segment
  of GABA and glycine receptors, 159
  of glutamate receptors, 88, 160, 248
  of inward rectifier potassium channels, 204fig.
  of nAChR, 77, 78, 157–9

magnesium ions, block by, 203, 205fig.

margainin, 118

Markov process, 162

maxi-K channels, 97, 98

MDR1, 109

mechanosensitive channels, 207–9

mellitin, 118

membrane, plasma, 1

membrane-crossing segments *see* transmembrane segments

messenger RNA (mRNA), 62

metabotropic receptors, 84

Michaelis constant, K_m, 130

Michaelis–Menten relation, 130

microelectrodes, intracellular, 33–5

minK protein, 113–14

MIP, 112

mitochondrial VDAC, 115

modes, 219, 243–4

modulation, 161, 210–23

molecular biological techniques, 59–68, 60tab.

molecular dynamics, 122, 149fig.

mouth, of channel pore, 152, 154, 189fig., 230, 231, 234, 237

mRNA, 62

multiple drug resistance (MDR) protein, 109

multi-state channel, kinetics of, 164–6

mutagenesis, site-directed, 68, 69fig.

mutations, 68

myasthenia gravis, 251

myokymia, 255

myotonia, generalized, 256–7; *see also* potassium-aggravated myotonia

myotonia congenita, 107–8, 187, 256–7

N terminus, 63, 75

nAChR (nicotinic acetylcholine receptor) channel
  acetylcholine binding site, 175, 176fig.
  allosteric nature of, 173, 177
  amino acid sequence of, 70, 71fig.
  block by acetylcholine, 173–4
  blocking agents of, 245
  conformational change in, 172–3, 178
  desensitization of, 173
  evolution of, 81–4
  foetal subunit, 177
  gating of, 170–8
  ion binding site model of, 140
  kinetics of gating, 171–4

molecular basis of selectivity, 157–9
neuronal, 82, 158
phosphorylation of, 214
pore size, 147, 148fig., 151–2
streaming potential in, 151–2
structure of, 68–79, 175–9
subconductance states of, 174
subunits of, 69, 77, 81–4, 176–7, 178fig.
transition time of, 167
nAChR-related receptor family, 84, 86fig.
neosurugatoxin, 245
nereistoxin, 245
Nernst equation, 24, 124
Nernst potential, 24, 26tab., 27
neurotransmitter-gated channels, 7, 84–9
pharmacology of, 244–9
neurotransmitters, modulation by, 214–19
NGK2 (Kv3.1), 96tab., 153
nicorandil, 234
nicotine, 245
nicotinic acetylcholine receptor see nAChR
nifedipine, 243
nitrendipine, 243, 244
nitric oxide, 221–2
NMDA receptor channels, 86
calcium influx through, 221
modulation by nitric oxide, 222
potentiation by glycine, 219, 249
N-methyl-D-aspartate (NMDA), 248
noise, membrane, 46, 54
noise analysis see fluctuation analysis
noxiustoxin, 234, 235tab.
nucleotides, in DNA and RNA, 60–1

Ohm's law, 12, 29
okadaic acid, 217
olfactory transduction, 99
oocytes, *Xenopus*, expression in, 79–80
open channel block, 230, 241
open times, distribution of, 55–7, 163–6
openers of potassium channels, 234
organic cations, use in permeability studies, 144, 145fig., 145tab.

outside-out patch configuration, 48
oxytocin, 102

P region, 90, 237; *see also* H5 segment *and* SS1–SS2 section
P_{2X} ATP receptor channel, 89
$P/4$ method for gating current, 180
P_{400} protein, 105
PAM, 252
papers on channels, 263
parallel-activation models, 197–9
paramyotonia congenita (PC), 193, 252–3
partition coefficient, 123
patch clamp technique, 6fig., 46–52
discovery of, 5
pentameric structure
of nAChR, 69, 77
of related neurotransmitter-gated channels, 85
peptabiols, 118
perforated patch, 49
periodic table, 13
permeability
binding site models of, 134–44
complex theories of, 122
defined, 121
independent electrodiffusion model, 122–9
permeability coefficient, P, 123
permeability sequence for different ions
CFTR channel, 109
pertussis toxin, 217
pH see hydrogen ions
pharmacology of channels, 225–49
phencyclidine, 248
phenylalkylamines, 243
phospholemman, 114
phospholipase C, 105
phosphorylation, modulation by, 210–14
phosphorylation site, 72
in nAChR, 75
in voltage-gated sodium channel, 92
photosynthetic reaction centre, 72
picrotoxin, 246
pinacidil, 234
plasma membrane 1, 2fig.

point-amplitude histogram, 53, 54fig.
polyamines, 203, 249
polymerase chain reaction (PCR), 67
pore, channel, dimensions of, 4
porins, 73, 114–15, 148
potassium channels, ATP-sensitive
in β cell, 3
blocking agents of, 234
conductance of, 129
gating of, 205–6
kinetics of block, 228–9
reversal potential of, 125, 126fig.
sulphonylurea receptor and, 109
potassium channels, calcium activated
gating in, 201–2
as multi-ion pore, 141
phosphorylation of, 217
streaming potential in, 151
structure of, 97–8
potassium channels, inward rectifier
as multi-ion pores, 142
gating in, 202–5
nomenclature of, 102tab.
phosphorylation of, 217
questions about, 264–5
structure and diversity of, 100–2
potassium channels, sarcoplasmic reticulum
selectivity of, 127
streaming potential in, 151
potassium channels, voltage-gated
blocking agents of, 229–32, 234–6, 237fig.
dysfunctional mutants of, 255–6
gating of, 184, 187, 195, 197–8
H5 segment of, 152–4
kinetics of block, 228
long pore effect in, 131–2
molecular basis of selectivity in, 152–4
as multi-ion pores, 140–2
nomenclature of, 96tab.
parallel-activation model of, 197–8
pore lining of, 152
pore size, 146
selectivity of, 127tab., 146
structure and diversity of, 94–8
β-subunits of, 192, 193fig.

potassium-aggravated myotonia (PAM), 252–3, 254fig.
potential difference, 10tab., 11
primary structure, defined, 77
probes, for DNA library, 67
procaine, 240
prolactin fusion protein, 76
pronase, removal of inactivation by, 189
protein kinase, 211
protein kinase A, 206–7, 211–14, 216
protein kinase C, 211, 212
protein phosphatase, 211
proteins, sequence determination of, 59–68
putative, meaning of, 74
pyrethrins, 240

Q/R site, 88, 248
quaternary ammonium ions, block by, 229–33
quaternary structure of nAChR, 77–9
quinoxalinediones, 248
quisqualic acid, 86, 247

R (gas constant), 11tab., 24
RACTK1, 102
rate constant, 163
    in nAChR, 172
rate theory models, 134–7, 181, 264
recombinant DNA technology, 64–8
reconstitution, 44
rectification
    defined, 12
    intrinsic, 203
red tides, 237, 240
resistance, 12
resting potential, 34, 203
restriction enzymes, 64–5
reversal potential, 29, 125
RNA, structure of, 61
RNA editing, 88
rod outer segment channels, 98
ROMK channels, 100, 102tab., 203
ryanodine, 103
ryanodine receptor (RyR), 103–4, 219, 261–2

S4 segment, 90
    of cyclic-nucleotide-gated channels, 156, 201
    role in gating, 185–8, 197
sarcoplasmic reticulum, 103, 261
saturation of conductance, 130
saxitoxin, 237, 238
scorpion toxins, 92, 234, 235tab., 238tab., 240
screening, of DNA library, 66
sea anemone toxins, 240
second messengers, 214–18
secondary structure, defined, 77
selectivity, 2, 7fig., 121
    GHK equations used to describe, 124–9
    molecular basis of, 152–60
selectivity filter, size of, 144–8, 157
serotonin *see* 5-HT
seven-transmembrane segment receptors, 215
Shab, Shaw and *Shal*, 95, 96tab.
Shaker potassium channels,
    gene encoding, 95
    identification of, 67, 95
    sister genes, 95
shellfish poisoning, 237, 240
sialic acid, 222
signal sequence, 75
single channel conductance (γ), 29–31
single channel current (i), 28–31
single-strand conformational polymorphism (SSCP), 253
SK channels, 98
sliding helix model of S4 movement, 186, 187fig.
slo gene, 98, 201
slope conductance, 30
sodium channels, amiloride-sensitive, 102–3
sodium channels, voltage-gated
    auxiliary subunits in, 93, 193
    binding site model of, 138–9
    block by hydrogen ions, 137–9
    blocking agents of, 236–42
    calcium-permeable mutants of, 155
    dysfunctional mutants of, 251–3, 254–5fig.
    gating of, 179–80, 183–9, 195, 197

modulation of, 212–13, 239–40
molecular basis of selectivity in, 154–6
neurotoxin binding sites, 236, 238tab.
phosphorylation of, 92fig., 212–13
pore size, 144–6
selectivity of, 126, 127tab., 144–6
structure of, 91–93
somatostatin, 217
spermine, 203–4, 249
spider toxins, 93, 244, 248
splicing, alternative, 64
    of glutamate receptors, 87–8
    of *Shaker* potassium channels, 95
splicing, of RNA, 64
SS1–SS2 section, 90
SSCP, 253
startle disease, 260
states, kinetic, 162
stochastic nature of channel gating, 5, 162
streaming potentials, 149–52
stretch-activated channels, 207–9
strychnine, 247
subconductance levels, 55, 174, 203, 210
subunits, 59
    auxiliary, 93, 192–3
    of GABA and glycine receptors, 84–5
    of glutamate receptors, 86–7
    of nAChR, 69, 77, 81–4
    role in inactivation, 192–3
    of voltage-gated channels, 90, 93, 192–3
succinylcholine, 245, 261
sulphonylurea drugs, 234
sulphonylurea receptor, 109
synthetic peptides, channel-forming, 118–19

techniques, value of, 264
tertiary structure, defined, 77
tetraethylammonium (TEA), 226, 229
tetrameric structure
    of ryanodine receptor, 103
    of voltage-gated channels, 90fig., 92, 95

of voltage-gated potassium channels, 95

tetrodotoxin
   block by, 237, 238
   discovery of, 4
   uses, 91, 225–6, 238
Thomsen's disease *see* myotonia congenita
toxins
   bacterial, 118
   biological functions of, 225
transfection, 81
transition rate constant, 163
transition time, 167
transmembrane segments, 72
   channels with two, 99–103
   of CLC channel, 107
   of connexins, 111
   identification of, 75
   of nAChR, 74–7
   of ryanodine receptor, 103
   of voltage-gated channels, 90

transcription, 61
translation, 62
triethylnonylammonium *see* C_9
tunnel, in potassium channel pore, 231
two-state channel
   block in, 226–9
   kinetics of, 162–4, 168–70, 181

uK_{ATP}–1 channel, 101, 102tab.
use-dependent block, 241

van der Waals forces, 16, 17tab.
vasopressin, 102
VDAC (voltage-dependent anion channel), 115
vector, 64
verapamil, 243
veratridine, 240
vestibule, of channel pore, 152, 236, 237fig., 241; *see also* antechamber
visual transduction, 99
voltage clamp technique, 35–9

voltage sensor, 7fig., 90–1; *see also* S4 segment
voltage-gated channels
   gating in, 178–99
   pore size of, 144–8
   selectivity of, 126–7
   structure of, 89–99
   see also under calcium channels, potassium channels, *and* sodium channels

wasp toxins, 248
water
   in channel pore, 148–9, 188
   in hydrated ions, 20, 136
   movement in gating, 188, 197
   structure of, 15, 16, 21, 148
water channels, 113
WCH-CD, 113
whole-cell patch recording, 49

$z\delta$ *see* electrical distance